Sustainable Agricultural Engineering Technologies and Applications

Sustainable Agricultural Engineering Technologies and Applications

Editors

Muhammad Sultan
Yuguang Zhou
Walter Den
Uzair Sajjad

MDPI • Basel • Beijing • Wuhan • Barcelona • Belgrade • Manchester • Tokyo • Cluj • Tianjin

Editors
Muhammad Sultan
Department of Agricultural Engineering
Bahauddin Zakariya University
Multan
Pakistan

Yuguang Zhou
Department of Agricultural Engineering
China Agricultural University
Beijing
China

Walter Den
Department of Mathematical, Physical and Engineering Sciences
Texas A&M University
San Antonio, Texas
United States

Uzair Sajjad
Department of Energy and Refrigeration Air-Conditioning Engineering
National Taipei University of Technology
Taipei
Taiwan

Editorial Office
MDPI
St. Alban-Anlage 66
4052 Basel, Switzerland

This is a reprint of articles from the Special Issue published online in the open access journal *Sustainability* (ISSN 2071-1050) (available at: www.mdpi.com/journal/sustainability/special_issues/Sustainable_Agricultural_Technologies_Applications).

For citation purposes, cite each article independently as indicated on the article page online and as indicated below:

LastName, A.A.; LastName, B.B.; LastName, C.C. Article Title. *Journal Name* **Year**, *Volume Number*, Page Range.

ISBN 978-3-0365-5890-5 (Hbk)
ISBN 978-3-0365-5889-9 (PDF)

Contents

About the Editors

Muhammad Sultan

Dr. Muhammad Sultan is an Associate Professor at the Department of Agricultural Engineering, Bahauddin Zakariya University, Multan (Pakistan). He holds a B.Sc. (2008) and M.Sc. (2010) in Agricultural Engineering from the University of Agriculture, Faisalabad (Pakistan), and Ph.D. (2015) in Energy & Environmental Engineering from Kyushu University (Japan). He did postdoctoral research in Energy & Environmental Engineering (2017) at Kyushu University (Japan) and in Mechatronic Systems Engineering (2019) at Simon Fraser University (Canada). He has published more than 300 articles in international journals, conferences, books, and book chapters. He has been a reviewer for more than 100 SCIE journals and holds an editor role for 10 renowned journals with publishers such as SAGE, MDPI, and Frontiers. His research focuses on developing energy-efficient temperature and humidity control systems for agricultural applications including greenhouse, fruit/vegetable storage, livestock, and poultry applications.

Yuguang Zhou

Dr. Zhou Yuguang is an Associate Professor at the College of Engineering, China Agricultural University. He is carrying out investigations on biogas engineering technology and clean combustion of biomass and its emission control, based on the platforms including "National Center for International Research on BioEnergy Science and Technology"(iBEST), Ministry of Science and Technology, and "Key Laboratory of Clean Production and Utilization of Renewable Energy"(CPURE), Ministry of Agriculture and Rural Affairs. He has conducted more than 20 projects that were granted by the National Science Foundation of China, the United Nations Development Programme, Worldbank, etc.

Walter Den

Dr. Walter Den is presently a Professor and Director of Water Resources Science and Technology at Texas A&M University-San Antonio. He has a Ph.D. and an M.S. degree in Environmental Engineering from the University of Southern California, and a Mechanical Engineering degree from the University of California, Santa Barbara. His research interests span across the treatment and reuse of impaired water from industrial, energy extraction, and agricultural operations; microalgal carbon capture and biomass conversation; and water conservation practices for industrial manufacturing and irrigation. He has published in more than 70 peer-reviewed journals and has served as a consultant for industries and water regulating agencies in the U.S., Taiwan, and Singapore.

Uzair Sajjad

Dr. Uzair Sajjad is working as a Research Assistant Professor at the Department of Energy and Refrigeration Air-Conditioning Engineering, National Taipei University of Technology, Taiwan. He completed his Ph.D. from National Chiao Tung University Taiwan and has been awarded two science and technology awards by CTCI Taiwan for his research excellence. He is carrying out investigations on different topics of heat transfer, energy, applications of machine learning and artificial intelligence, surface engineering, heat pump and vacuum freeze-drying, and additive manufacturing. Currently, he has been granted a project by the National Science and Technology Council Taiwan. In these areas, he has published more than 50 SCI articles. Furthermore, he has served as a reviewer and guest editor for many well-reputed journals.

sustainability

MDPI

Article

Evaluating Evaporative Cooling Assisted Solid Desiccant Dehumidification System for Agricultural Storage Application

Ghulam Hussain [1], Muhammad Aleem [1], Muhammad Sultan [1,*], Uzair Sajjad [2], Sobhy M. Ibrahim [3], Redmond R. Shamshiri [4,*], Muhammad Farooq [5], Muhammad Usman Khan [6] and Muhammad Bilal [1]

1 Department of Agricultural Engineering, Faculty of Agricultural Sciences & Technology, Bahauddin Zakariya University, Multan 60800, Pakistan; aaeamri@gmail.com (G.H.); muhammadaleem189@gmail.com (M.A.); bilalranauni@gmail.com (M.B.)

2 Department of Mechanical Engineering, National Yang Ming Chiao Tung University, Hsinchu 300, Taiwan; energyengineer01@gmail.com

3 Department of Biochemistry, College of Science, King Saud University, P.O. Box 2455, Riyadh 11451, Saudi Arabia; syakout@ksu.edu.sa

4 Department of Engineering for Crop Production, Leibniz Institute for Agricultural Engineering and Bioeconomy, 14469 Potsdam, Germany

5 Department of Mechanical Engineering (New Campus-KSK), University of Engineering and Technology, Lahore 54890, Pakistan; engr.farooq@uet.edu.pk

6 Department of Energy Systems Engineering, Faculty of Agricultural Engineering and Technology, University of Agriculture, Faisalabad 38040, Pakistan; usman.khan@uaf.edu.pk

* Correspondence: muhammadsultan@bzu.edu.pk (M.S.); rshamshiri@atb-potsdam.de (R.R.S.); Tel.: +92-333-610-8888 (M.S.)

Abstract: The study aims to investigate Maisotsenko cycle evaporative cooling assisted solid desiccant air-conditioning (M-DAC) system for agricultural storage application. Conventional air-conditioning (AC) systems used for this application are refrigeration-based which are expensive as they consume excessive amount of primary-energy. In this regard, the study developed a lab-scale solid silica gel-based desiccant AC (DAC) system. Thermodynamic performance of the developed system was investigated using various adsorption/dehumidification and desorption/regeneration cycles. The system possesses maximum adsorption potential i.e., 4.83 g/kg-DA at higher regeneration temperature of 72.6 °C and long cycle time i.e., 60 min:60 min. Moreover, the system's energy consumption performance was investigated from viewpoints of maximum latent, sensible, and total heat as well as latent heat ratio (LHR), which were found to be 0.64 kW, 1.16 kW, and 1.80 kW, respectively with maximum LHR of 0.49. Additionally, the study compared standalone DAC (S-DAC), and M-DAC system thermodynamically to investigate the feasibility of these systems from the viewpoints of temperature and relative humidity ranges, cooling potential (Q_p), and coefficient of performance (*COP*). The S-DAC system showed temperature and relative humidity ranging from 39 °C to 48 °C, and 35% to 66%, respectively, with Q_p and *COP* of 17.55 kJ/kg, and 0.37, respectively. Conversely, the M-DAC system showed temperature and relative humidity ranging from 17 °C to 25 °C, and 76% to 93%, respectively, with Q_p and *COP* of 41.80 kJ/kg, and 0.87, respectively. Additionally, the study investigated respiratory heat generation rate (Q_{res}), and heat transfer rate (Q_{rate}) by agricultural products at different temperature gradient (ΔT) and air velocity. The Q_{res} and Q_{rate} by the products were increased with ΔT and air velocity, respectively, thereby generating heat loads in the storage house. Therefore, the study suggests that the M-DAC system could be a potential AC option for agricultural storage application.

Keywords: evaporative cooling; desiccant dehumidification; agricultural storage; air conditioning; system performance

Citation: Hussain, G.; Aleem, M.; Sultan, M.; Sajjad, U.; Ibrahim, S.M.; Shamshiri, R.R.; Farooq, M.; Usman Khan, M.; Bilal, M. Evaluating Evaporative Cooling Assisted Solid Desiccant Dehumidification System for Agricultural Storage Application. *Sustainability* **2022**, *14*, 1479. https://doi.org/10.3390/su14031479

Academic Editor: Emanuele Radicetti

Received: 29 December 2021
Accepted: 24 January 2022
Published: 27 January 2022

1. Introduction

The world's population has grown rapidly over the past few decades and is predicted to move to about 9.8 billion by 2050. Due to this population growth, the demand for food will rise about 70% by 2050 [1]. The agriculture sector plays a significant role in managing global food demand. Being an agriculture-dependent country, Pakistan produces 13.67 million tonnes of fruits and vegetables, but about 35–40% of total production is lost owing to post-harvest losses [2]. The post-harvest losses refer to degradation of quality as well as quantity of the agricultural products. One of the most important reasons for the post-harvest losses is high amount of moisture content available in the agricultural products [3,4]. Moreover, several factors are responsible for the post-harvest losses as shown in Figure 1 [5]. After harvesting, fruits and vegetables perform respiration, transpiration, and ripening processes. These decay processes of agricultural products can be minimized by providing optimum storage conditions i.e., temperature and relative humidity [6]. Conventional storage techniques are equipped with refrigeration-based systems which consume huge amount of primary energy as well as degrading the environment by utilization of harmful refrigerants [7]. Moreover, these systems are incapable of providing the appropriate temperature and relative humidity conditions especially for agricultural products [8]. In this regard, alternative energy-efficient and environment-friendly options are available including evaporative cooling (EC), and desiccant air-conditioning (DAC) systems. The EC options are direct EC, indirect EC, and Maisotsenko cycle EC (MEC) systems i.e., an advanced form of the indirect EC [9]. The MEC system thermodynamically lowers the ambient air temperature by utilizing psychrometric accessible renewable energy [10]. Two thermodynamic operations which involve EC and heat transfer stipulate a cooling effect where ambient air almost approaches the dew point temperature rather than wet bulb temperature [9,11]. The MEC system has been investigated in literature for heating, ventilation, and air-conditioning applications [12–16]. However, the potential of the standalone EC systems is limited in humid areas [16,17]. Therefore, to overcome this limitation, the DAC system is an emerging option because of its potential to provide promising results in humid climates [18–20].

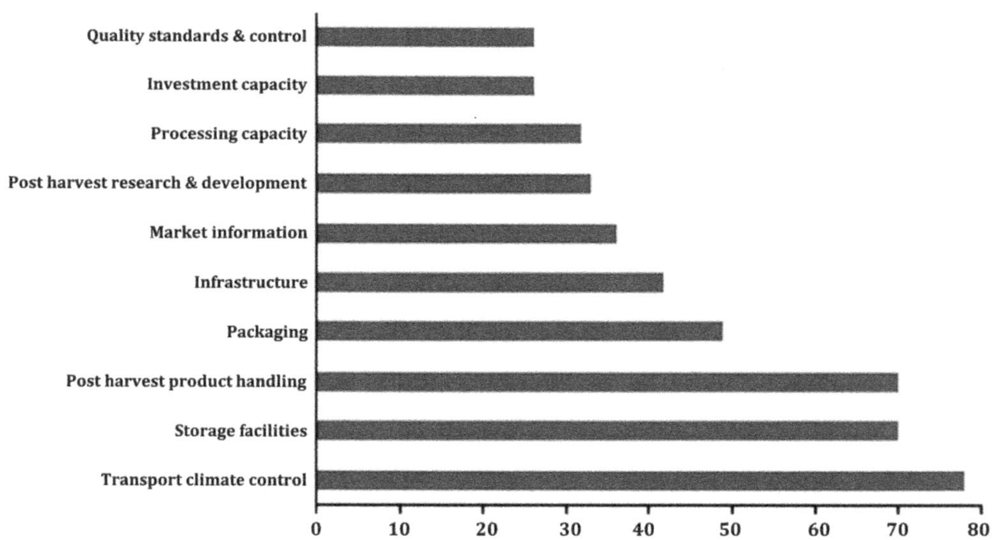

Figure 1. Major categories for causes of post-harvest losses, reproduced from ref. [5].

The main component of the DAC system is desiccant material which adsorb moisture from ambient/process air and thereby dehumidify the air. The adsorption/dehumidification and desorption/humidification processes are mainly dependent on characteristics of adsorbents [21–23]. The DAC systems are energy efficient as they are operated or regenerated with thermal energy that could be available through low-grade waste heat and renewable energy sources. Niu et al. [24] experimentally compared the energy saving potential of the DAC system with conventional refrigeration-based AC system. The study concluded that the DAC system had a maximum energy saving potential of about 58.9% [24]. Jia et al. [25] conducted an experimental study on the DAC system's ability to remove moisture from the air and its energy saving potential as compared with a conventional AC system. The study showed an energy saving potential of about 37.5% as compared with a conventional AC system [25]. Similar studies are reported in the literature which showed good energy saving potential by the DAC systems for various AC applications [26–29]. However, the standalone DAC (S-DAC) system i.e., without the integration of MEC with the DAC system, could not achieve appropriate temperature and relative humidity conditions particularly for agricultural product storage [9,30–32]. In this regard, an MEC system integrated with a DAC (M-DAC) system combines characteristics of both S-DAC and MEC systems, and thereby has potential to achieve sensible load via MEC and latent load through the DAC system [8]. The M-DAC system has been investigated in literature for various agricultural applications which involve greenhouse AC [33,34], livestock AC [35,36], and poultry AC [37–40]. The M-DAC system has shown promising results for these agricultural applications.

In this study, a lab-scale solid silica gel-based DAC system was developed. The performance of the developed system was investigated thermodynamically, using various adsorption/dehumidification and desorption/regeneration cycles. Furthermore, performance of the developed system regarding energy consumption was investigated from the viewpoints of latent heat (Q_L), sensible heat (Q_S), and total heat (Q_T) as well as latent heat ratio (LHR) The study proposed two kinds of AC system involving standalone DAC, and MEC-assisted DAC systems for potential application in agricultural product storage. A thermodynamic analysis was conducted for both systems to explore applicability of these systems considering temperature and relative humidity conditions, cooling potential (Q_p), and coefficient of performance (COP). In addition, the study investigates respiratory heat generation rate (Q_{res}) and heat transfer rate (Q_{rate}) to investigate heat loads by the agricultural products.

Temperature/Humidity Requirements for Agricultural Storage Application

The storage and handling of agricultural products is one of the key issues of the 21st century because of the intricate mechanisms of transpiration, respiration as well as chilling injury. Appropriate temperature and relative humidity conditions are a primary concern to reduce post-harvest losses. The shelf life of agricultural products is a function of storage temperature. The storage temperature effect on the shelf life of fruits and vegetables is represented in Figure 2. A higher storage temperature reduces the shelf life of fruits and vegetables, while a lower storage temperature could increase the shelf life as represented in Figure 2. Moreover, the quality, composition, and texture of the products are significantly influenced by the temperature [41–43]. Figure 3 shows the effect of storage temperature on quality of peach, Boston-type lettuce, and asparagus. As the storage temperature increases the quality of the products decreases.

After harvesting, fruits and vegetables behave like living organisms and perform respiration processes according to Equation (1) available in ref. [46]. When ambient air oxygen combines with the product's conserved sugar/starch, carbon dioxide and water are generated, thereby releasing respiration heat. In order to minimize respiration rate, the agricultural products should be stored at low oxygen level. However, as the temperature rises, the respiration rate increases, thereby reducing the shelf life of the product. The influence of temperature on respiration rate by some fruits and vegetables is shown in Figure 4. The effect of temperature on some fruits is shown in Figure 4a, which shows respiration rate increases with temperature. Similarly, the effect of temperature on vegetables in

Figure 4b shows that the respiration rate increases with temperature. This respiration rate increases heat i.e., respiratory heat in a storage house. The quantity of moisture present in the ambient air, expressed as relative humidity, is the primary driver of water loss from harvested goods. The agricultural products preserve their nutritional quality, flavour, and appearance at high relative humidity. In contrast, shrivelling occurs at low relative humidity ranges because of excessive transpiration [47,48]. In this regard, the optimum temperature and relative humidity conditions for fruits and vegetables are available in refs. [19,49] ranging from −5 °C to 25 °C, and 85% to 95%, respectively. Figure 5 represents the optimum temperature and relative humidity zones for storage of fruits and vegetables with the climatic conditions of Multan.

$$C_6H_{12}O_6 + 6O_2 \rightarrow +6CO_2 + 6H_2O + \textit{respiration heat} \tag{1}$$

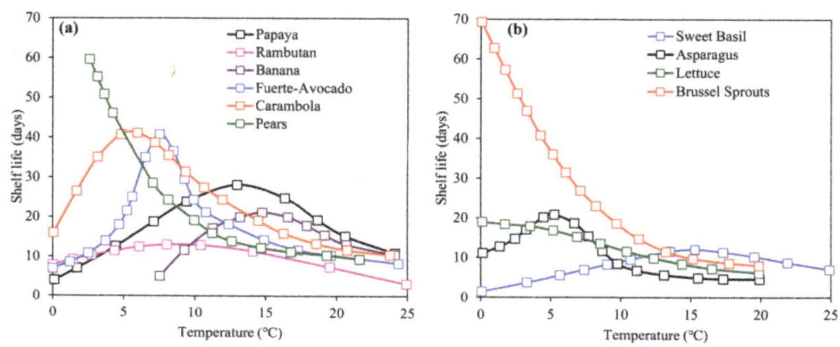

Figure 2. Effect of storage temperature on shelf life of some (**a**) fruits, and (**b**) vegetables, reproduced from ref. [44].

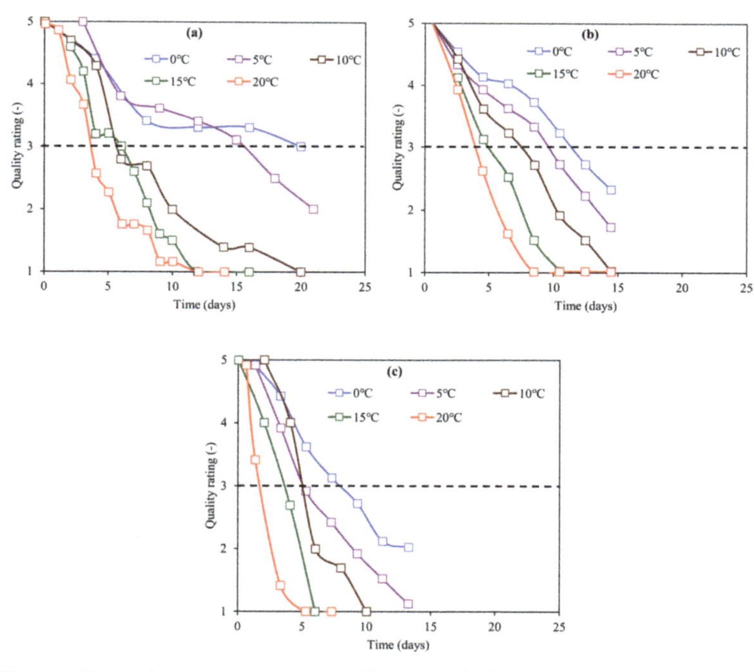

Figure 3. Effect of temperature on the quality (**a**) peach, (**b**) Boston-type lettuce, and (**c**) asparagus. The dotted lines represent acceptable limits of the product, reproduced from ref. [45].

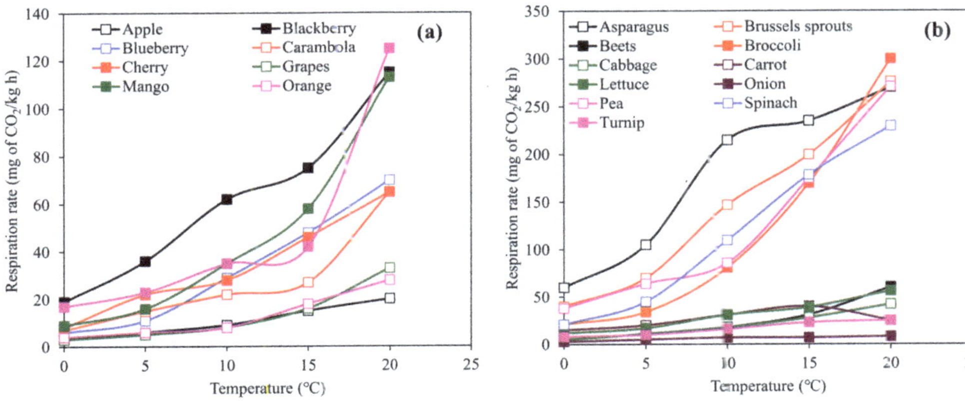

Figure 4. Effect of temperature on respiration rate of some (**a**) fruits, and (**b**) vegetables. data obtained from ref. [50].

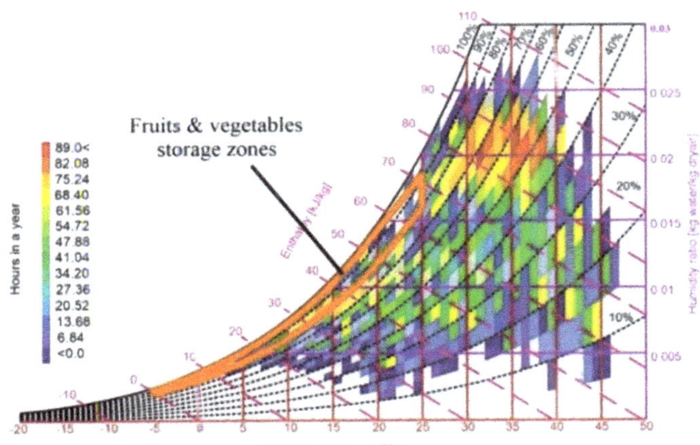

Figure 5. Optimum temperature and humidity zones for storage of fruits and vegetables at climatic conditions of Multan.

2. Proposed Systems

2.1. Standalone Desiccant Air-Conditioning (S-DAC) System

Referring to the Figure 6a, components of the S-DAC system consist of desiccant block, a sensible heat exchanger (SHX), and a heating source. Ambient/process air is entered into the block during the adsorption process and gets dehumidified, which thereby increases the temperature of the air due to the release of the heat of adsorption. After that, the dehumidified air is entered into the SHX which ideally reduces the temperature to ambient air temperature. During the desorption process of the S-DAC system, ambient air was utilized as a return air and directed towards the SHX, which thereby gets heated. Then, this processed heated air enters a heat source which can be operated with thermal energy options like low-grade waste heat as well as renewable energy options such as solar thermal heat and biogas/biomass, etc. In the end, this heated air is directed to a desiccant block which carries away water vapours from the desiccant material to regenerate and hence remove humid air.

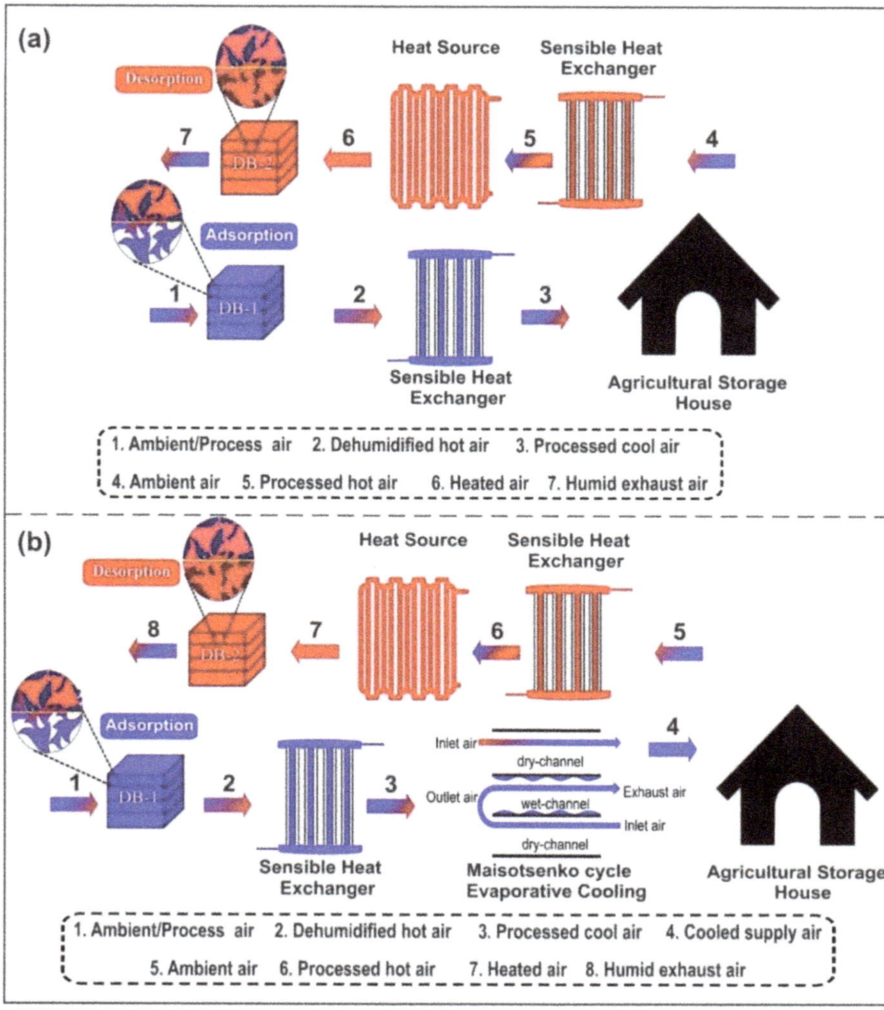

Figure 6. Working scheme of proposed (**a**) standalone DAC (S-DAC), and (**b**) M-cycle integrated DAC (M-DAC) system.

2.2. M-Cycle Integrated Desiccant Air-Conditioning (M-DAC) System Collection

Referring to Figure 6b, components of the M-DAC system consist of desiccant block, SHX, MEC, and a heating source. Ambient air is directed into the block during the adsorption process of the M-DAC system and gets dehumidified, thereby increasing the temperature due to the release of the heat of adsorption. After that, the dehumidified air is directed towards the SHX which decreases the process air temperature to nearly that of the ambient air without change in the humidity ratio. This cooled process air is further directed towards the MEC which lowers the process air temperature sensibly which could be supplied to the agriculture storage house. During the desorption process of the M-DAC system, ambient air, as a return air, is employed and enters the SHX, which thereby gets heated. Then, this processed heated air is directed towards the heat source which can be operated with thermal energy options like low-grade waste heat and renewable energy options such as solar thermal heat and biogas/biomass, etc. In the end, this heated air is directed towards desiccant block which carries away water vapours from the block to regenerate it and hence remove the humid air.

3. Materials and Methods

3.1. Experimental Setup

The solid silica gel-based DAC system was developed at laboratory scale The experimental setup was mainly consisted of solid silica gel, sieves, heat exchanger (i.e., fan heater), temperature, relative humidity, pressure, and velocity (anemometer) measuring sensors. Figure 7a shows the photographic view of the developed experimental setup. The sieves are fabricated by using polyacrylic plastic material frame and mesh to support the material and easy crossing of air. A total of 18 sieves were used to develop the desiccant unit and each sieve has a dimension of 220 mm × 145 mm × 3 mm and carrying ~68 g of silica gel. These sieves are stacked over each other to form a desiccant unit and then insulated on the sides. The important parameters for desiccant unit and silica gel used in this study is shown in Table 1. The fan was used to continuously throw the process air at velocity of about 3.5 m/s, resulting in a mass flow rate ($\dot{m}_a$) of about 0.14 kg/s. The temperature, relative humidity, and pressure sensors were placed at the inlet and outlet sides of the desiccant unit to measure the experimental conditions. The measurement range and absolute accuracy for temperature, relative humidity, pressure, and velocity sensors is about −40 °C to 80 °C; ±2 °C, 0% to 100%; ±5%, 75 kPa to 110 kPa; ±0.3 kPa, and 0 3 m/s to 30 m/s; ±1.5 m/s, respectively.

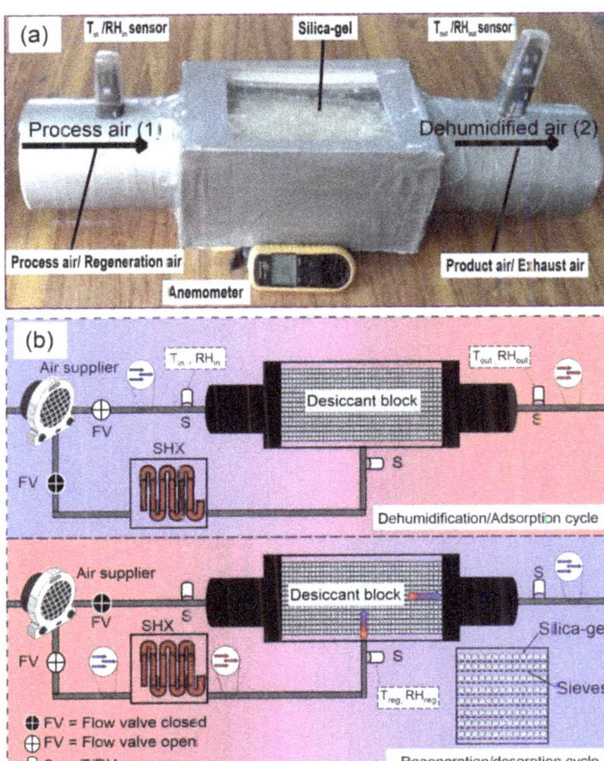

Figure 7. (**a**) Photographic view of the developed experimental system, and (**b**) working scheme of the developed experimental system.

Table 1. Important parameters of desiccant unit and silica gel used in this study.

Parameters	Symbol	Value	Unit
	Parameters related to desiccant unit		
Width	W	0.145	m
Length	L	0.220	m
Height	H	0.045	m
Total silica gel		1.22	kg
	Parameters related to silica gel		
Molar mass	M	60.08	g/kg
Bulk density	ρ_b	919.5	kg/m^3
Specific heat capacity	C_p	0.85	kJ/kg·K
Specific surface area	SSA	750–850	m^2/g
Pore size	φ	2–2.6	nm
Heat of adsorption	Δh	2800	kJ/kg
Thermal conduction	k	0.175	W/m·K

The experimental process is divided into two stages: desorption/regeneration and adsorption/dehumidification. Figure 7b shows a schematic representation of the developed experimental system. During the regeneration process, hot air (achieved by heat exchanger) is supplied to the desiccant unit to primarily reactivate the adsorbent. After that, the dehumidification process commences, and process air is directed to the desiccant unit and moisture adsorbs on the material's surface because of its hygroscopic nature, thereby increasing the temperature of process air due to the release of heat of adsorption [51]. In this study, four desorption as well as adsorption cycles were conducted at different regeneration temperatures. The regeneration temperature was set at 68 °C, 69.2 °C, 70.5 °C, and 72.6 °C for cycles 1, 2, 3 and 4, respectively. Figure 8 shows the performance of the experimental system at different regeneration temperatures by conducting various desorption/adsorption cycles. The system has shown maximum dehumidification potential at higher regeneration temperature of 72.6 °C. In this regard, the study focuses on the desorption and adsorption cycle-4 to thermodynamically explore the performance of the system.

Ideally, the adsorption process follows isenthalpic behaviour, which means that the heat of water vapor condensation following the vapor stage to liquid stage is the same as the heat of adsorption. However, experimental studies have shown that the heat of adsorption is higher than the ideal adsorption heat because of the isosteric heat of adsorption [51,52]. The difference between an ideal isenthalpic line and the actual dehumidification line is mainly due to the type of adsorption mechanism and interaction between adsorbent pairs. Figure 9 represents the psychrometric behaviour of the ideal and experimental dehumidification line and cyclic behaviour of proposed DAC systems. The points 1–8 were calculated by considering the experimental inlet and outlet conditions.

3.2. Data Reduction

The experimental data of the developed DAC based on various desorption and adsorption cycles were obtained to investigate the performance thermodynamically. The inlet and outlet air conditions of both desorption and adsorption cycles were measured during the experiments. After that, Equation (2) was used to calculate the output air condition of the SHX [19,53]. The modelled Equation (3) of the MEC was used to calculate the performance of the MEC system [36,54]. Ambient air as a return air was used to increase the air temperature to some extent by Equation (4). Maximum desorption potential (ΔX_{des}) and adsorption (ΔX_{ads}) potential were calculated from Equations (5) and (6) [31].

$$T_{3,DB} = T_{2,DB} - \varepsilon_{SHX}(T_{2,DB} - T_{1,DB}) \tag{2}$$

$$T_{4,DB} = 6.70 + 0.2630(T_{3,DB}) + 0.5298(X_3) \tag{3}$$

$$T_{6,DB} = T_{5,DB} + \varepsilon_{SHX}(T_{2,DB} - T_{5,DB}) \tag{4}$$

$$\Delta X_{des} = X_{in,\ des} - X_{out,des} \tag{5}$$

$$\Delta X_{ads} = X_{in,\ ads} - X_{out,ads} \tag{6}$$

where, the subscripts 1–6, and DB, are process air conditions represented in Figure 6, and dry-bulb, respectively. T is air temperature (°C), ε_{SHX} is the effectiveness of sensible heat exchanger defined by the American Society of Heating, Refrigerating, and Air-Conditioning Engineers (ASHRAE) [46], X is humidity ratio (g/kg-DA), ΔX_{des} is desorption potential (g/kg-DA), ΔX_{ads} is adsorption potential (g/kg-DA), $X_{in,\ des}$ is inlet humidity ratio during desorption process (g/kg-DA), $X_{out,des}$ is outlet humidity ratio after desorption process (g/kg-DA), ΔX_{ads} is adsorption potential (g/kg-DA), $X_{in,\ ads}$ is inlet humidity ratio during adsorption process (g/kg-DA), and $X_{out,\ ads}$ is outlet humidity ratio after adsorption process (g/kg-DA), respectively.

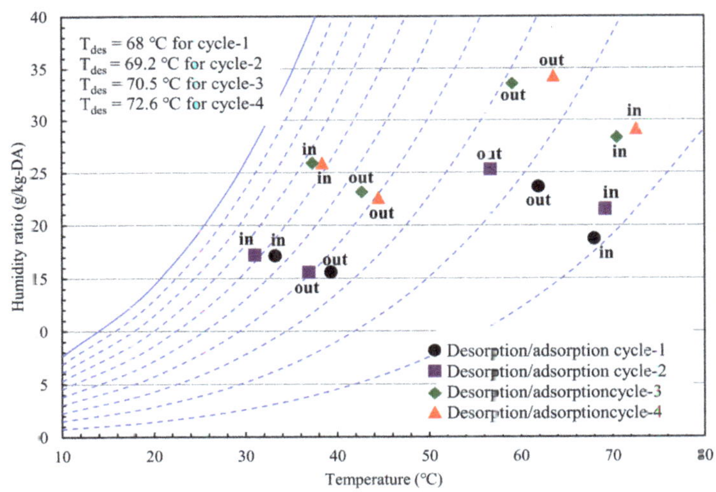

Figure 8. Performance of the experimental system at different regeneration temperatures by conducting various desorption/adsorption cycles.

The performance of the developed desiccant unit regarding energy consumption can be evaluated by latent heat ratio (LHR) [55]. The LHR is the ratio of latent heat consumed to the total energy consumed by the system which is given in Equation (7). The total energy (Q_T) consumed by the system was calculated from Equations (8)–(11) based on sensible heat (Q_S), latent heat (Q_L), and electricity consumed by the fan (Q_F) to supply the process air [55]. However, Equations (12)–(14) were used to calculate cooling potential (Q_p) by the proposed S-DAC and M-DAC system, respectively. Similarly, Equations (13) and (14) were used to calculate the coefficient of performance (COP) by the proposed S-DAC, and M-DAC system, respectively. The air enthalpy was calculated by Equation (15) [46].

$$LHR = \left(\frac{Q_L}{Q_T}\right) \tag{7}$$

$$Q_T = Q_S + Q_L + Q_F \tag{8}$$

$$Q_S = \dot{m}_a \, C_p \, (T_{des,in} - T_{des,out}) \tag{9}$$

$$Q_L = \dot{m}_a \, \Delta X_{ads} \, \gamma = \dot{m}_a \, (h_{ads,in} - h_{ads,out}) \tag{10}$$

$$Q_F = \frac{\dot{V}_{air} dP}{\varepsilon_F} \tag{11}$$

$$Q_{p,\, S-DAC} = h_1 - h_3 \tag{12}$$

$$Q_{p,\, M-DAC} = h_1 - h_4 \tag{13}$$

$$COP_{\,S-DAC} = \frac{\dot{m}_a}{\dot{m}_a}\left(\frac{Cooling\ potential}{Heat\ input}\right) = \frac{\dot{m}_a}{\dot{m}_a}\left(\frac{h_1 - h_3}{h_6 - h_5}\right) \tag{14}$$

$$COP_{\,M-DAC} = \frac{\dot{m}_a}{\dot{m}_a}\left(\frac{Cooling\ potential}{Heat\ input}\right) = \frac{\dot{m}_a}{\dot{m}_a}\left(\frac{h_1 - h_3}{h_7 - h_6}\right) \tag{15}$$

$$h = 1.006 T_{DB} + X(2501 + 1.86 T_{DB}) \tag{16}$$

where, LHR is latent heat ratio (-), Q_L is latent heat consumption (kW), Q_T is total heat consumption (kW), Q_S is sensible heat consumption (kW), Q_F is electricity consumed by fan (kW), $\dot{m}_a$ is mass flow rate (kg/s), C_p is specific heat capacity of air (kJ/kg K), $T_{des,in}$ is the inlet temperature of desorption process (K), $T_{des,out}$ is the outlet temperature of the desorption process (K), ΔX_{ads} is adsorption potential (g/kg-DA), γ is latent heat of vaporization with respect to temperature (kJ/kg), $h_{ads,in}$ is inlet enthalpy of adsorption process (kJ/kg), $h_{ads,out}$ is outlet enthalpy of adsorption process (kJ/kg), $\dot{V}_{air}$ is volume flow rate (m^3/s), dP is pressure drop (kPa), and ε_F is efficiency of the fan taken as 70%. The subscripts 1, 3, 4, 5, 6, 7 *S-DAC*, and *M-DAC*, and *DB* refer to air states as shown in Figure 6, standalone desiccant air-conditioning, M-cycle integrated desiccant air-conditioning system, and dry bulb, respectively. Q_p is cooling potential (kJ/kg), h is air enthalpy (kJ/kg), and COP is coefficient of performance (-), respectively.

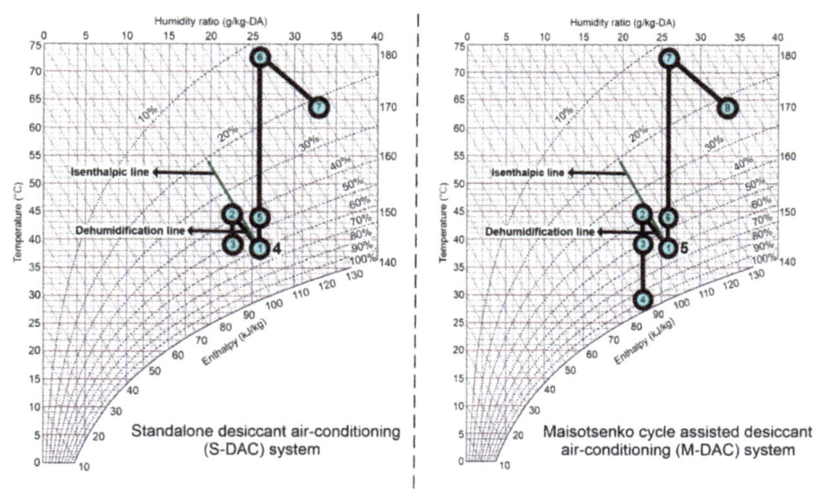

Figure 9. Ideal and experimental dehumidification lines and cyclic behaviour of the proposed DAC systems on Psychrometric/Mollier diagram.

3.3. Uncertainty Analysis

In the experimentation process, there are always errors in calculating/measuring variables due to the inaccuracy of instruments, design limitations, ambient conditions, observations, human errors, and other associated important factors. After experiments, an uncertainty analysis was performed to determine the level of uncertainty in estimating the variables (temperature, relative humidity). Estimating the experimental uncertainty gives confidence in the calculated results. The method is named as the root of the sum of square methods and Equation (17) is used to measure the uncertainty [56–58].

$$\sigma_R = \sqrt{\left(\frac{\partial R}{\partial N_1}\alpha_1\right)^2 + \left(\frac{\partial R}{\partial N_2}\alpha_2\right)^2 + \cdots + \left(\frac{\partial R}{\partial N_n}\alpha_N\right)^2} \quad (17)$$

where, σ_R is the total uncertainty (%) and α_1–α_N represent the uncertainty in independent variables, N_1–N_n represent the independent variables R is given a function of the independent variable. Table 2 shows the uncertainty analysis of various parameters used in this study.

Table 2. Uncertainty analysis of various parameters used in this study.

Parameter	Symbol	Value
Temperature	T	±2.6%
Relative humidity	RH	±5.1%
Pressure	P	±1.6%
Velocity	V	±3.4%
Humidity ratio	X	±1.6%
Enthalpy	h	±2.9%
Cooling potential	Q_P	±5.1%
Coefficient of performance	COP	±2.7%
Sensible heat	Q_S	±2.2%
Latent heat	Q_L	±5.4%
Latent heat ratio	LHR	±5.37%

3.4. Heat Load Calculations for Agricultural Storage Application

The storage of agricultural products offers considerable challenges as they generate heat during storage by consuming oxygen and emitting carbon dioxide, water vapours and heat transfer between product and ambient air. The respiratory process varies with temperature as well as type of the product. Becker et al. [59] developed a relationship of measuring respiratory heat generation rate as a function of the temperature of the agricultural product, given by Equation (18). Furthermore, heat is removed from the product due to heat transfer between product and ambient air which was calculated by Equation (19) [60].

$$Q_{res} = \frac{10.7f}{3600}\left(\frac{9T}{5} - 32\right)^g \quad (18)$$

$$Q_{rate} = h_s \times A_s \times \Delta T = h_s \times A_s \times (T_s - T_a) \quad (19)$$

where, Q_{res} is respiratory heat generation rate (W/kg), T is temperature of the product (°C), f and g are respiratory coefficients depending upon the agricultural product. Table 3 shows the respiratory coefficients of some selected fruits and vegetables. Q_{rate} is heat transfer rate from the product (W), h_s is surface heat transfer coefficient (W/m^2 °C), A_s is surface area covered by a product (m^2), ΔT is temperature difference between product surface and ambient air (°C), T_s is surface temperature of product (°C), and T_a ambient or surrounding air temperature (°C). The h_s vary with velocity of ambient air, product orientation and geometry. Several studies have investigated the h_s against ambient air velocity, and the A_s for various agricultural products presented in Table 4. In this principle, behavior of the Q_{rate} from the agricultural products has been explored by considering literature data of h_s against air velocity, and A_s.

Table 3. Respiratory coefficients of some fruits and vegetables, reproduced from ref. [59].

Product	Respiration Coefficients	
	f	g
	Fruits	
Apple	5.687×10^{-4}	2.5977
Grape	7.056×10^{-5}	3.033
Orange	2.8050×10^{-4}	2.6840
Pear	6.3614×10^{-5}	3.2037
Tomato	2.0074×10^{-4}	2.8350
	Vegetables	
Cabbage	6.0803×10^{-4}	2.6183
Carrot	0.050018	1.7926
Onion	3.668×10^{-4}	2.538
Potato	0.01709	1.769

Table 4. Surface heat transfer coefficient with air velocity, and surface area for some fruits and vegetables.

Product	Air Velocity (m/s)	Surface Heat Transfer Coeff., h_s (W/m^2 °C)	Surface Area, A_s (m^2)	Reference
		Fruits		
Apple	0	11.1	0.0116	[60,61]
	0.39	17		
	0.91	27.3		
	2	45.3		
	5.1	53.4		
Grape	1	30.7	0.0008	[60,62,63]
	1.25	33.8		
	1.5	37.8		
	1.75	40.7		
	2	42.3		
Orange	0.11	66.4	0.01622	[60,64,65]
	0.33	69.3		
Pear	1	12.6	0.007598	[60,62,66]
	1.25	14.2		
	1.5	15.8		
	1.75	16.1		
	2	19.5		
Tomato	1	10.9	0.008	[60,62,67]
	1.25	13.1		
	1.5	13.6		
	1.75	14.9		
	2	17.3		
		Vegetables		
Cabbage	0.431	53.8033	0.000415	[68,69]
Carrot	0.0158	547	0.007758	[70,71]
Onion	0.431	54.6125	0.001404	[61,72]
Potato	0.431	39.0599	0.000134	[69,73]

4. Results and Discussion

The performance of the desiccant unit was examined thermodynamically for various desorption and adsorption cycles. Figure 10 shows experimental desorption and adsorption cycles profiles of temperature for the developed DAC system at different desorption temperatures (T_{des}). The switching time of 1:1 (equal time for both stages) was considered for the desorption and adsorption cycles. These cycles were conducted to explore optimum switching time, and desorption/regeneration temperature for higher coefficient of performance and maximum adsorption potential. The temperature cycles represent that 20 min:20 min, 30 min:30 min, 45 min:45 min, and 60 min:60 min was set for the desorption and adsorption cycle 1, 2, 3 and 4, respectively. The outlet temperature (after adsorption process) ranges from 39 °C to 62 °C (cycle-1), 37 °C to 57 °C (cycle-2), 43 °C to 59 °C (cycle-3), and 44 °C to 63 °C (cycle-4), respectively. The process air temperature increases due to release of heat of adsorption. Figure 11 shows experimental desorption and adsorption cycles profiles of relative humidity for the DAC system. The outlet relative humidity (after adsorption process) ranges from 17% to 35% (cycle-1), 22% to 40% (cycle-2), 23% to 43% (cycle-3), and 20% to 38% (cycle-4), respectively.

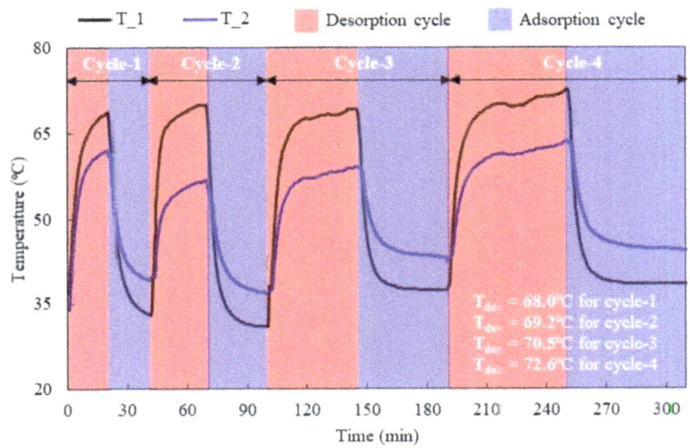

Figure 10. Experimental desorption/regeneration and adsorption/dehumidification cycle profiles of temperature for the DAC system at different desorption temperature (T_{des}).

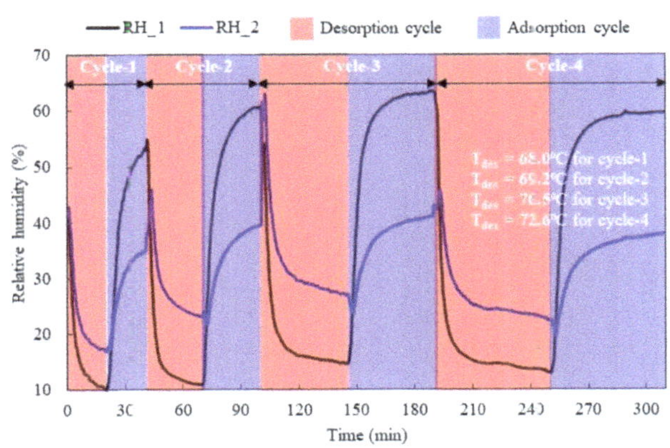

Figure 11. Experimental desorption/regeneration and adsorption/dehumidification cycle profiles of relative humidity for the DAC system at different desorption temperature (T_{des})

Figure 12 shows experimental desorption and adsorption cycle profiles of humidity ratio for the developed DAC system. The adsorption cycle-4 represents maximum adsorption potential due to higher desorption temperature. Figure 13 shows desorption and adsorption potential (ΔX) in different experimental cycles In the desorption process, the amount of water vapours desorbed increases rapidly, and then reduces when saturation pressure of the process air and desiccant material surface approach equilibrium. Consequently, when the adsorption process starts, the amount of water vapours adsorbed is quite high, but gradually decreases and becomes steady due to saturation condition of the desiccant material. In Figure 13, the 60 min adsorption cycle-4 represents the maximum adsorption potential of 4.8 g/kg-DA via desiccant unit at desorption temperature of 72.6 °C, thereby showing the significance of switching time for the system and desorption temperature, which are the most important parameters for the feasibility of the DAC system [19]. Moreover, Aleem et al. [53] investigated the performance of laboratory scale silica gel-based DAC system thermodynamically. According to the results, maximum adsorption potential and coefficient of performance was observed at higher desorption temperature of 70 °C and 60 min:60 min cycle time. In this regard, the adsorption cycle-4 has been explored for performance evaluation of the developed system.

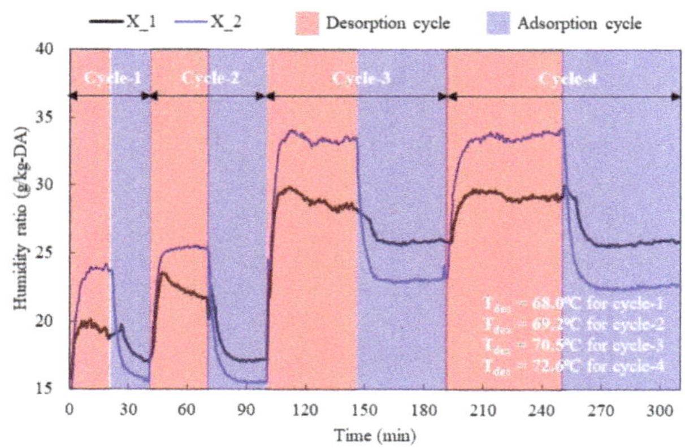

Figure 12. Experimental desorption/regeneration and adsorption/dehumidification cycle profiles of humidity ratio for the DAC system at different desorption temperature (T_{des}).

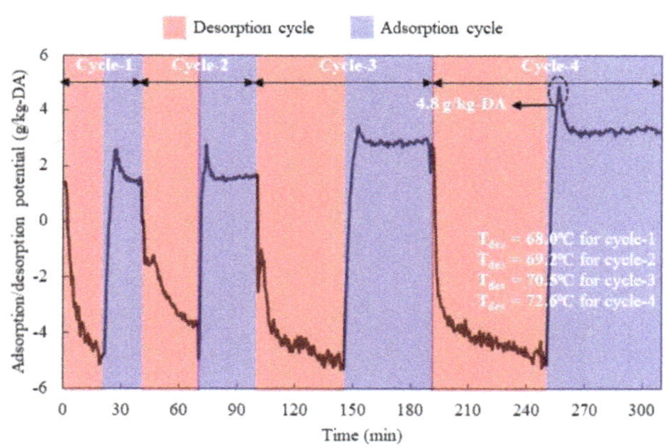

Figure 13. Adsorption and desorption potential in various experimental adsorption and desorption cycles of the DAC system at different desorption temperatures (T_{des}).

The performance of the developed desiccant unit regarding energy consumption was explored from viewpoints of latent heat (Q_L), sensible heat (Q_S), total heat (QT) consumed by the system, and latent heat ratio (LHR) as shown in Figure 14. The Q_L consumed by the system during the adsorption process varied between 0.02 kWh and 0.64 kWh. The Q_S consumed by the system during desorption process varied between 0.20 kW and 1.16 kW. The Q_S was higher due to heat provided for desorption of water vapours from the desiccant material to regenerate it. The energy consumed to reach the high desorption temperature of 72.6 °C directly influenced the total energy consumed by the system. In this regard, the Q_T consumed by the system varied between 0.15 kW and 1.80 kW during both adsorption and desorption processes. However, the Q_T consumption could be reduced by maintaining a lower desorption temperature while, according to the experimental results, a higher desorption temperature was necessary for maximum adsorption potential. The electricity consumed by the fan was found to be 6.38 W. The LHR by the developed system varied between 0.02 and 0.49. The proposed air-conditioning (AC) systems are energy-efficient as compared with conventional refrigeration-based AC systems as discussed in Section 1.

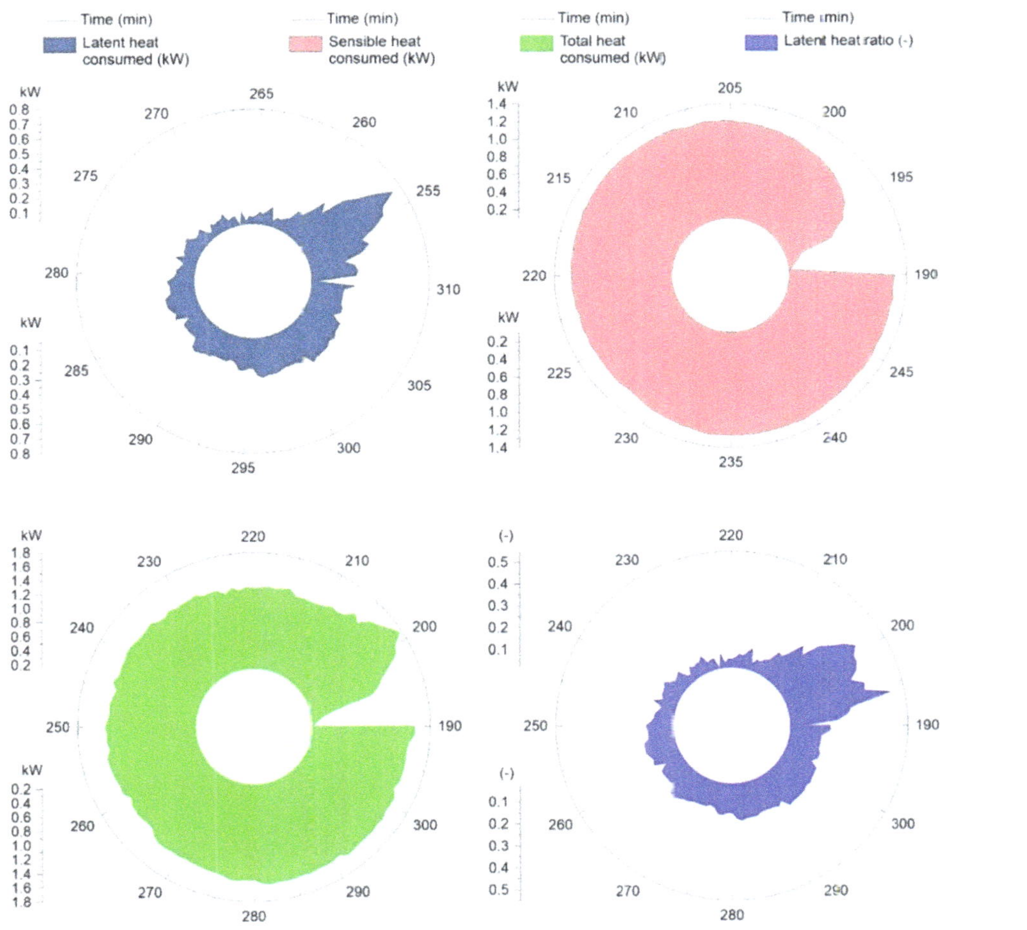

Figure 14. Latent heat (Q_L), sensible heat (Q_S), and total heat (Q_T) consumed by the experimental system, and latent heat ratio (LHR) achieved by the system during desorption and adsorption cycle-4.

Figure 15 shows temperature and relative humidity profiles for the adsorption cycle-4 using S-DAC, and M-DAC system. The temperature and relative humidity of the S-DAC system ranges between 39 °C and 48 °C, and 35% and 66%, respectively. Therefore, the S-DAC system was unable to achieve the appropriate temperature and relative humidity conditions for storage of the agricultural products. Similarly, the temperature and relative humidity of the M-DAC system ranges from 17 °C to 25 °C, and 76% to 98%, respectively, and could thereby be a potential option for the storage of agricultural products. The temperature and relative humidity profiles of the S-DAC, and M-DAC system for adsorption cycle-1, cycle-2, and cycle-3 is available in Appendix A: Figures A1–A3, respectively.

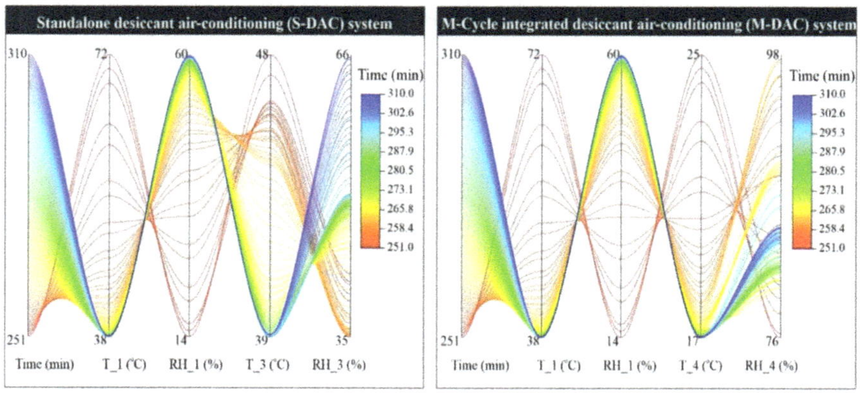

Figure 15. Profiles of temperature and relative humidity by the proposed S-DAC, and M-DAC system for adsorption cycle-4.

Figure 16 shows the cooling potential (Q_p) of the S-DAC, and M-DAC systems for adsorption cycle-4. The Q_p of the S-DAC, and M-DAC systems varied between 3.30 kJ/kg and 17.50 kJ/kg, and 29.40 kJ/kg to 41.80 kJ/kg, respectively. Figure 17 shows COP of the S-DAC and M-DAC systems for adsorption cycle-4. The COP for S-DAC and M-DAC systems varied between 0.058 and 0.368, and 0.513 and 0.868, respectively. Figure 18 shows psychrometric representation for performance of the proposed DAC systems. In Figure 18, the S-DAC system could not achieve appropriate temperature and relative humidity conditions for the storage of agricultural products. In contrast, the M-DAC system has achieved appropriate temperature and relative humidity conditions for the storage of agricultural products. Therefore, the M-DAC system could be utilized for the potential application of agricultural products storage.

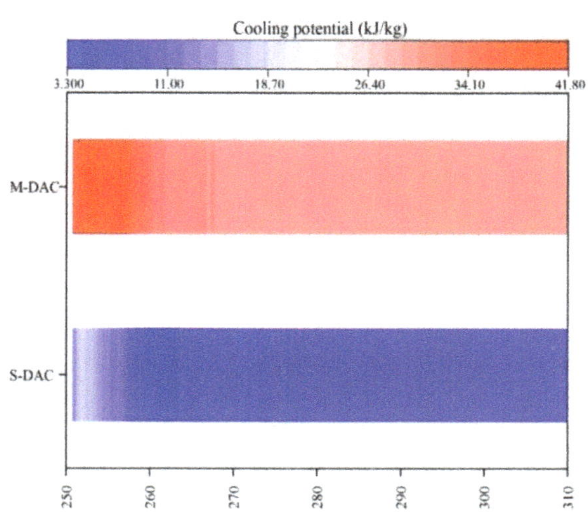

Figure 16. Profiles of cooling potential (Q_p) by the proposed S-DAC, and M-DAC systems for adsorption cycle-4.

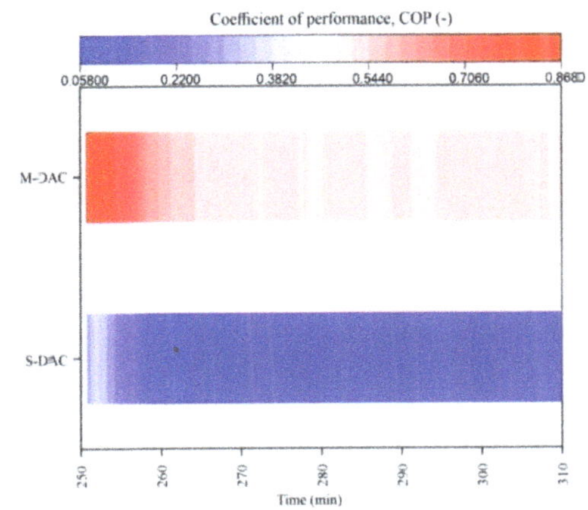

Figure 17. Profiles of COP by the proposed S-DAC, and M-DAC systems for adsorption cycle-4.

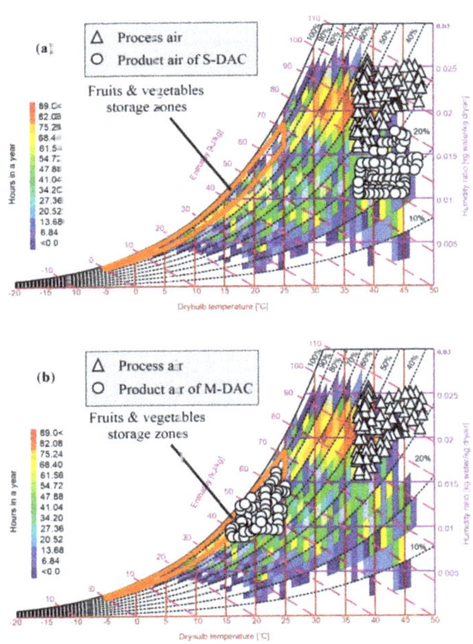

Figure 18. Psychrometric representation of process air and product air by the proposed (**a**) S-DAC, and (**b**) M-DAC systems.

The fruits and vegetables respire as living organism by emitting carbon dioxide to surrounding environment, thereby causing respiratory heat generation (Q_{res}). The Q_{res} vary with temperature and type of the product. Figure 19 shows the effect of temperature on the Q_{res} by some fruits and vegetables. The Q_{res} increases with temperature, hence during storage respiration process should be kept as low as possible to avoid heat generation from the product. In addition, heat is removed from the product due to the Q_{rate} between product and ambient or surrounding air i.e., the difference between surface temperature of product and surrounding air temperature (ΔT), and velocity of air.

Figure 20 shows the effect of the ΔT and air velocity on the Q_{rate} by some selected fruits (apple, grape, orange, pear, and tomato). The Q_{rate} by fruits increases with increase in air velocity and ΔT. In case of apple the Q_{rate} at 0.91 m/s, with ΔT of 5 °C was 1.58 W, but at same velocity with ΔT of 10 °C, 15 °C, and 20 °C, the Q_{rate} was 3.16 W, 4.75 W, and 6.33 W, respectively. A similarly increasing trend of the Q_{rate} at different air velocity and ΔT was observed by other selected fruits. Likewise, Figure 21 shows the effect of ΔT on the Q_{rate} by some selected vegetables (cabbage, carrot, onion, and potato). The Q_{rate} by vegetables increases with ΔT. In case of cabbage, Q_{rate} at 0.431 m/s with ΔT of 5 °C was 0.116 W, but at same velocity with ΔT of 10 °C, 15 °C, and 20 °C the Q_{rate} was 0.22 W, 0.33 W, and 0.45 W, respectively. A comparable trend was observed for other selected vegetables. Furthermore, higher air velocities and higher relative humidity of surrounding air caused wilting of fruits and vegetables. The appropriate air velocity is varied with size of storage house. The recommended air velocity is about 0.003 m/s for storage house having area of 10 m^2, and relative humidity of about 90% to 95% [50].

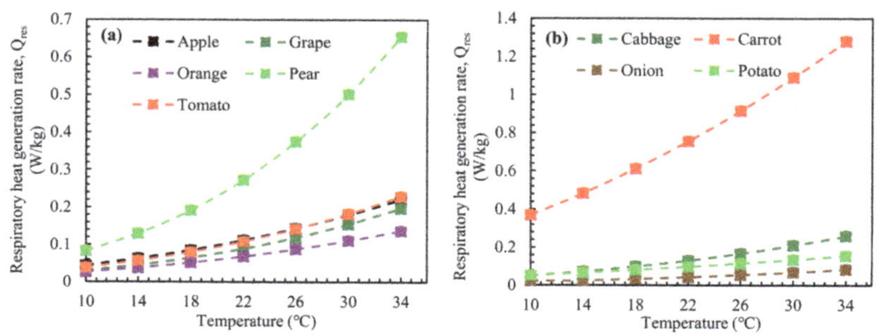

Figure 19. Effect of temperature on respiration rate (Q_{res}) by some (**a**) fruits, and (**b**) vegetables.

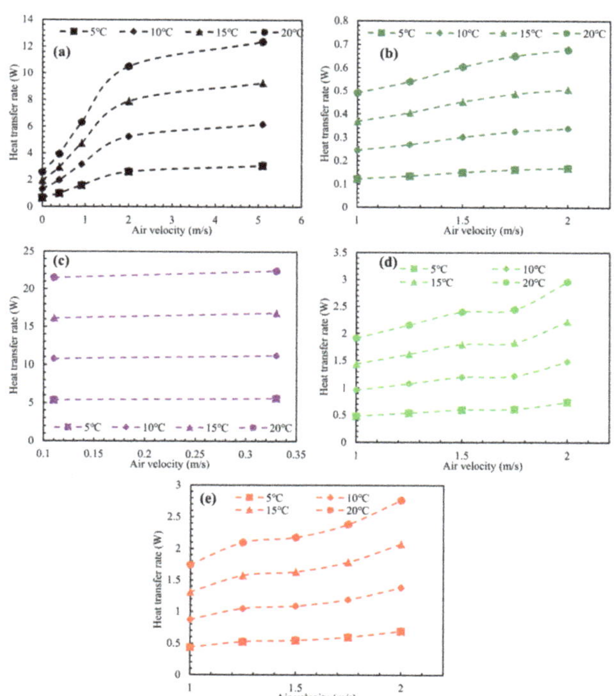

Figure 20. Effect of temperature gradient (ΔT) and air velocity on the heat transfer rate (Q_{rate}) by (**a**) apple, (**b**) grape, (**c**) orange, (**d**) pear, and (**e**) tomato.

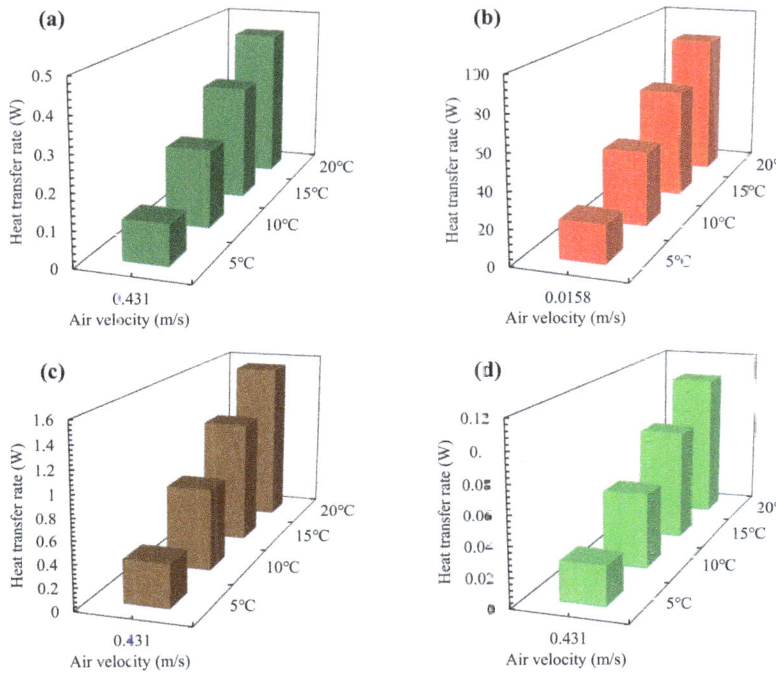

Figure 21. Effect of temperature gradient (ΔT) on the heat transfer rate (Q_{rate}) by (**a**) cabbage, (**b**) carrot, (**c**) onion, and (**d**) potato.

5. Conclusions

The present study aims to evaluate the Maisotsenko cycle evaporative cooling assisted solid desiccant air-conditioning (M-DAC) system for agricultural storage application. The agricultural products required appropriate temperature and relative humidity conditions for their efficient storage to avoid post-harvest losses. In this regard, the air-conditioning (AC) system is principally required to optimize the temperature and relative humidity conditions, which could potentially reduce post-harvest losses. Therefore, in this study, a lab-scale solid silica gel-based desiccant AC (DAC) system was developed. The performance of the developed system was investigated thermodynamically, using various adsorption/dehumidification and desorption/regeneration cycles. The system possesses better performance in terms of adsorption potential (Q_p) i.e., 4.88 g/kg-DA at higher desorption temperature i.e., 72.6 °C and comparatively long cycle time i.e., 60 min:60 min. Furthermore, performance of the developed system regarding energy consumption was investigated from viewpoints of latent heat (Q_L), sensible heat (Q_S), and total heat (Q_T) as well as latent heat ratio (*LHR*). The maximum consumption of Q_L, Q_S, and Q_T, by the system was found to be 0.64 kW, 1.16 kW, and 1.80 kW, respectively by achieving the maximum LHR of 0.49.

Furthermore, the study proposed two kinds of AC options involving standalone DAC (S-DAC), and M-DAC systems to investigate applicability of these systems for agricultural storage application. The proposed systems were compared thermodynamically from perspectives of temperature as well as relative humidity ranges, cooling potential (Q_p), and coefficient of performance (COP). The S-DAC system showed temperature and relative humidity ranging from 39 °C to 48 °C, and 35% to 66%, respectively, with Q_p and COP of 17.55 kJ/kg, and 0.37, respectively, which were inappropriate conditions for agricultural storage. Conversely, the M-DAC system showed temperature and relative humidity ranging from 17 °C to 25 °C, and 76% to 98%, respectively, with Q_p and COP of 41.80 kJ/kg, and

0.87, respectively which lies somehow between appropriate conditions for the agricultural products storage application. Additionally, the study investigated respiratory heat generation rate (Q_{res}), and heat transfer rate (Q_{rate}) by some fruits and vegetables at different temperature gradient (ΔT) and air velocity. The Q_{res} and Q_{rate} by the fruits and vegetables were increased with temperature, and ΔT as well as air velocity, respectively. These heat loads reduced shelf life and quality of agricultural products. Therefore, the M-DAC system could be a potential AC option for agricultural storage application.

Author Contributions: Conceptualization, G.H.; methodology, G.H.; software, G.H. and M.A.; validation, M.S., U.S. and M.F.; formal analysis, G.H. and M.A.; investigation, G.H., M.A., M.S. and R.R.S.; resources, M.S.; data curation, G.H., M.A., U.S., R.R.S., M.U.K. and M.B.; writing—original draft preparation, G.H.; writing—review and editing, M.A., M.S., S.M.I., R.R.S., M.F., M.U.K. and M.B.; visualization, M.A., S.M.I., R.R.S., M.F. and M.U.K.; supervision, M.S.; project administration, M.S.; funding acquisition, S.M.I. and R.R.S. All authors have read and agreed to the published version of the manuscript.

Funding: This work was supported by Researchers Supporting Project number (RSP-2021/100), King Saud University, Riyadh, Saudi Arabia.

Institutional Review Board Statement: Not applicable.

Informed Consent Statement: Not applicable.

Data Availability Statement: The data is contained within the article.

Acknowledgments: This work was supported by Researchers Supporting Project number (RSP-2021/100), King Saud University, Riyadh, Saudi Arabia. All this work is part of the Ph.D. research of Ghulam Hussain (1st author). This research work was carried in the Department of Agricultural Engineering, Bahauddin Zakariya University, Multan, Pakistan. The authors acknowledge the financial support by the Open Access Publication Fund of the Leibniz Association, Germany.

Conflicts of Interest: The authors declare no conflict of interest.

Nomenclature

AC	air-conditioning
A_s	surface area of product (m^2)
COP	coefficient of performance
C_p	specific heat capacity of air (kJ/kg K)
DAC	desiccant air-conditioning
dP	pressure drop (kPa)
EC	evaporative cooling
f	respiratory coefficient (-)
g	respiratory coefficient (-)
h	air enthalpy (kJ/kg)
h_s	surface heat transfer coefficient (W/m^2 °C)
LHR	latent heat ratio (LHR)
M-DAC	Maisotsenko-Cycle integrated desiccant air-conditioning
MEC	Maisotsenko-Cycle evaporative cooling
$\dot{m}_a$	mass flow rate (kg/s)
N_1–N_n	independent variables
p	cooling potential (kJ kg)
Q_T	total heat consumption (kW)
Q_L	latent heat consumption (kW)
Q_S	sensible heat consumption (kW)
Q_F	electricity consumption by fan (kW)
Q_p	cooling potential (kJ/kg)
Q_{res}	respiratory heat generation rate (W/kg)
Q_{rate}	heat transfer rate (W)
S-DAC	standalone desiccant air-conditioning

SHX	sensible heat exchanger
T_a	ambient air temperature (°C)
T_s	surface temperature of product (°C)
$\dot{V}_{air}$	volume flow rate (m^3/s)
X	humidity ratio (g/kg-DA)
ΔX_{ads}	adsorption/dehumidification potential (g/kg-DA)
ΔX_{des}	desorption/regeneration potential (g/kg-DA)
ΔT	temperature gradient (°C)
α_1–α_N	uncertainty in independent variables
α_R	total uncertainty (%)
ε_F	efficiency of fan (%)
ε_{SHX}	effectiveness of sensible heat exchanger (-)
γ	latent heat of vaporization (kJ/kg)
Subscripts	
ads	adsorption
DB	dry bulb
des	desorption
in	inlet condition
out	outlet condition

Appendix A

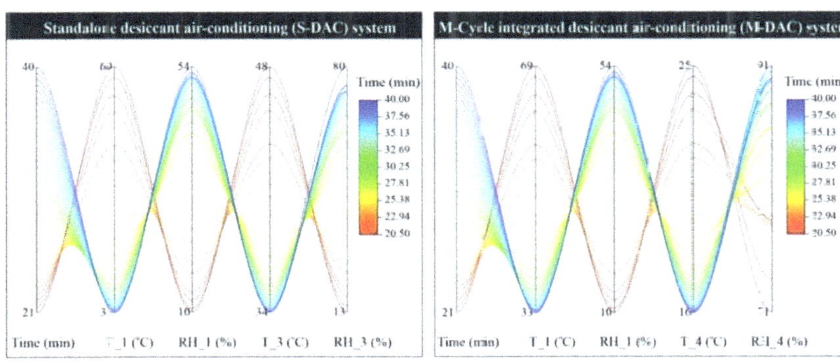

Figure A1. Profiles of temperature and relative humidity of the S-DAC, and M-DAC system for adsorption/dehumidification cycle-1.

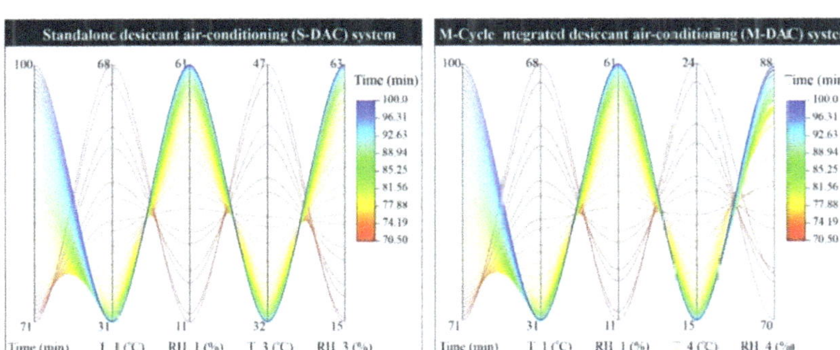

Figure A2. Profiles of temperature and relative humidity of the S-DAC, and M-DAC system for adsorption/dehumidification cycle-2.

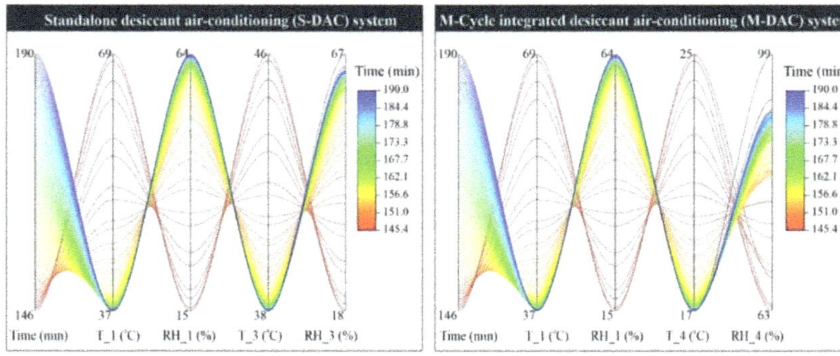

Figure A3. Profiles of temperature and relative humidity of the S-DAC, and M-DAC system for adsorption/dehumidification cycle-3.

References

1. Noya, I.; González-García, S.; Bacenetti, J.; Fiala, M.; Moreira, M.T. Environmental Impacts of the Cultivation-Phase Associated with Agricultural Crops for Feed Production. *J. Clean. Prod.* **2018**, *172*, 3721–3733. [CrossRef]
2. Abbas, M.; Maqsood, W.; Ijaz, M.; Hashim, M.Y. Post-Harvest Losses in Fruits and Vegetables in Pakistan. Available online: https://agrihunt.com/articles/pak-agri-outlook/post-harvest-losses-in-fruits-and-vegetables-in-pakistan/ (accessed on 9 June 2021).
3. Ishaq, M.; Sultan, M.; Aleem, M.; Bilal, M.; Ullah, H.S.; Mahmood, M.H.; Miyazaki, T. Desiccant Dehumidification System for Storage of Fruits and Vegetables. In *Energy-Efficient Systems for Agricultural Applications*; Sultan, M.M.T., Ed.; Springer: Cham, Switzerland, 2022; pp. 65–83. [CrossRef]
4. Ishaq, M.; Aleem, M.; Ashraf, H.; Hafiz, S.U.; Sultan, M. Study on Desiccant Dehumidification System Using Experiments and Steady-State Model. In Proceedings of the International Exchange and Innovation Conference on Engineering & Sciences (IEICES), Fukoka, Japan, 22—23 October 2020; Volume 6, pp. 92–97.
5. FAO; CIHEAM. Mediterra 2016: Zero Waste in the Mediterranean. In *Natural Resources, Food and Knowledge*; International Centre of Advanced Mediterranean Agronomic Studies (CHIEAM); FAO; Presses de Sciences PO: Paris, France, 2016.
6. Brasil, I.M.; Siddiqui, M.W. *Chapter 1—Postharvest Quality of Fruits and Vegetables: An Overview*; Siddiqui, M.W., Ed.; Academic Press: Amsterdam, The Netherlands, 2018; pp. 1–40.
7. Sultan, M.; El-Sharkawy, I.I.; Miyazaki, T.; Saha, B.B.; Koyama, S. An Overview of Solid Desiccant Dehumidification and Air Conditioning Systems. *Renew. Sustain. Energy Rev.* **2015**, *46*, 16–29. [CrossRef]
8. Mahmood, M.H.; Sultan, M.; Miyazaki, T. Significance of Temperature and Humidity Control for Agricultural Products Storage: Overview of Conventional and Advanced Options. *Int. J. Food Eng.* **2019**, *15*, 20190063. [CrossRef]
9. Mahmood, M.H.; Sultan, M.; Miyazaki, T.; Koyama, S.; Maisotsenko, V.S. Overview of the Maisotsenko Cycle—A Way towards Dew Point Evaporative Cooling. *Renew. Sustain. Energy Rev.* **2016**, *66*, 537–555. [CrossRef]
10. Pandelidis, D.; Anisimov, S. Numerical Analysis of the Heat and Mass Transfer Processes in Selected M-Cycle Heat Exchangers for the Dew Point Evaporative Cooling. *Energy Convers. Manag.* **2015**, *90*, 62–83. [CrossRef]
11. Asfahan, H.M.; Sajjad, U.; Sultan, M.; Hussain, I.; Hamid, K.; Ali, M.; Wang, C.-C.; Shamshiri, R.R.; Khan, M.U. Artificial Intelligence for the Prediction of the Thermal Performance of Evaporative Cooling Systems. *Energies* **2021**, *14*, 3946. [CrossRef]
12. Pandelidis, D.; Anisimov, S.; Worek, W.M. Performance Study of the Maisotsenko Cycle Heat Exchangers in Different Air-Conditioning Applications. *Int. J. Heat Mass Transf.* **2015**, *81*, 207–221. [CrossRef]
13. Caliskan, H.; Hepbasli, A.; Dincer, I.; Maisotsenko, V. Thermodynamic Performance Assessment of a Novel Air Cooling Cycle: Maisotsenko Cycle. *Int. J. Refrig.* **2011**, *34*, 980–990. [CrossRef]
14. Zhan, C.; Duan, Z.; Zhao, X.; Smith, S.; Jin, H.; Riffat, S. Comparative Study of the Performance of the M-Cycle Counter-Flow and Cross-Flow Heat Exchangers for Indirect Evaporative Cooling—Paving the Path toward Sustainable Cooling of Buildings. *Energy* **2011**, *36*, 6790–6805. [CrossRef]
15. Chua, K.J.; Chou, S.K.; Yang, W.M.; Yan, J. Achieving Better Energy-Efficient Air Conditioning—A Review of Technologies and Strategies. *Appl. Energy* **2013**, *104*, 87–104. [CrossRef]
16. Lal Basediya, A.; Samuel, D.V.K.; Beera, V. Evaporative Cooling System for Storage of Fruits and Vegetables—A Review. *J. Food Sci. Technol.* **2013**, *50*, 429–442. [CrossRef] [PubMed]
17. Sultan, M.; Miyazaki, T. Energy-Efficient Air-Conditioning Systems for Nonhuman Applications. In *Refrigeration*; Ekren, O., Ed.; InTech: London, UK, 2017; pp. 97–117.
18. Sultan, M.; Miyazaki, T.; Koyama, S.; Khan, Z.M. Performance Evaluation of Hydrophilic Organic Polymer Sorbents for Desiccant Air-Conditioning Applications. *Adsorpt. Sci. Technol.* **2018**, *36*, 311–326. [CrossRef]

19. Kashif, M.; Sultan, M.; Khan, Z.M. Alternative Air-Conditioning Options for Developing Countries. *Eur. J. Eng. Res. Sci.* **2017**, *2*, 76. [CrossRef]
20. Sultan, M.; Miyazaki, T.; Koyama, S. Optimization of Adsorption Isotherm Types for Desiccant Air-Conditioning Applications. *Renew. Energy* **2018**, *121*, 441–450. [CrossRef]
21. Xue, B.; Tahara, K.; Nakashima, K.; Noda, A.; Oktariani, E.; Wijayanta, A.T.; Nakaso, K.; Fukai, J. Numerical Simulation for Steam Generation Process in a Novel Zeolite–Water Adsorption Heat Pump. *J. Chem. Eng. Japan* **2012**, *45*, 408–416. [CrossRef]
22. Xue, B.; Ye, S.; Zhang, L.; Wei, X.; Nakaso, K.; Fukai, J. High-Temperature Steam Generation from Low-Grade Waste Heat from an Adsorptive Heat Transformer with Composite Zeolite-13X/$CaCl_2$. *Energy Convers. Manag.* **2019**, *186*, 93–102. [CrossRef]
23. Oktariani, E.; Tahara, K.; Nakashima, K.; Noda, A.; Xue, B.; Wijayanta, A.T.; Nakaso, K.; Fukai, J. Experimental Investigation on the Adsorption Process for Steam Generation Using a Zeolite–Water System. *J. Chem. Eng. Japan* **2012**, *45*, 355–362. [CrossRef]
24. Niu, X.; Xiao, F.; Ge, G. Performance Analysis of Liquid Desiccant Based Air-Conditioning System under Variable Fresh Air Ratios. *Energy Build.* **2010**, *42*, 2457–2464. [CrossRef]
25. Jia, C.X.; Dai, Y.J.; Wu, J.Y.; Wang, R.Z. Analysis on a Hybrid Desiccant Air-Conditioning System. *Appl. Therm. Eng.* **2006**, *26*, 2393–2400. [CrossRef]
26. Bassuoni, M.M. Experimental Performance Study of a Proposed Desiccant Based Air Conditioning System. *J. Adv. Res.* **2014**, *5*, 87–95. [CrossRef]
27. Qi, R.; Lu, L.; Huang, Y. Parameter Analysis and Optimization of the Energy and Economic Performance of Solar-Assisted Liquid Desiccant Cooling System under Different Climate Conditions. *Energy Convers. Manag.* **2015**, *106*, 1387–1395. [CrossRef]
28. Qi, R.; Lu, L. Energy Consumption and Optimization of Internally Cooled/Heated Liquid Desiccant Air-Conditioning System: A Case Study in Hong Kong. *Energy* **2014**, *73*, 801–808. [CrossRef]
29. Luo, Y.; Shao, S.; Xu, H.; Tian, C.; Yang, H. Experimental and Theoretical Research of a Fin-Tube Type Internally-Cooled Liquid Desiccant Dehumidifier. *Appl. Energy* **2014**, *133*, 127–134. [CrossRef]
30. Mahmood, M.H.; Sultan, M.; Miyazaki, T. Solid Desiccant Dehumidification-Based Air-Conditioning System for Agricultural Storage Application: Theory and Experiments. *Proc. Inst. Mech. Eng. Part A J. Power Energy* **2020**, *234*, 534–547. [CrossRef]
31. Mahmood, M.H.; Sultan, M.; Miyazaki, T. Experimental Evaluation of Desiccant Dehumidification and Air-Conditioning System for Energy-Efficient Storage of Dried Fruits. *Build. Serv. Eng. Res. Technol.* **2020**, *41*, 454–465. [CrossRef]
32. Hussain, G.; Sultan, M.; Aleem, M. Study on Solid Desiccant Dehumidification Cooling System for the Storage of Fruits and Vegetables. In Proceedings of the International Exchange and Innovation Conference on Engineering & Sciences (IEICES), Fukoka, Japan, 21 October 2021; Volume 7, pp. 295–301.
33. Ashraf, H.; Sultan, M.; Shamshiri, R.R.; Abbas, F.; Farooq, M.; Sajjad, U.; Md-Tahir, H.; Mahmood, M.H.; Ahmad, F.; Taseer, Y.R.; et al. Dynamic Evaluation of Desiccant Dehumidification Evaporative Cooling Options for Greenhouse Air-Conditioning Application in Multan (Pakistan). *Energies* **2021**, *14*, 1097. [CrossRef]
34. Sultan, M.; Miyazaki, T.; Saha, B.B.; Koyama, S. Steady-State Investigation of Water Vapor Adsorption for Thermally Driven Adsorption Based Greenhouse Air-Conditioning System. *Renew. Energy* **2016**, *86*, 785–795. [CrossRef]
35. Niaz, H.; Sultan, M.; Khan, A.A.; Miyazaki, T.; Feng, Y.; Khan, Z.M.; Niaz, Y.; Mahmood, M.H.; Ali, I.; Imran, M.A. Study on Evaporative Cooling Assisted Desiccant Air Conditioning System for Livestock Application in Pakistan. *Fresenius Environ. Bull.* **2019**, *28*, 8623–8633.
36. Kashif, M.; Niaz, H.; Sultan, M.; Miyazaki, T.; Feng, Y.; Usman, M.; Shahzad, M.W.; Niaz, Y.; Waqas, M.M.; Ali, I. Study on Desiccant and Evaporative Cooling Systems for Livestock Thermal Comfort: Theory and Experiments. *Energies* **2020**, *13*, 2675. [CrossRef]
37. Aleem, M.; Sultan, M.; Mahmood, M.H.; Miyazaki, T. *Desiccant Dehumidification Cooling System for Poultry Houses in Multan (Pakistan) BT—Energy-Efficient Systems for Agricultural Applications*; Sultan, M., Miyazaki, T., Eds.; Springer International Publishing: Cham, Switzerland, 2022; pp. 19–42. [CrossRef]
38. Aleem, M.; Hussain, G.; Sultan, M.; Miyazaki, T. Lab-Scale Study of Desiccant Dehumidification System for Poultry Shed Air-Conditioning in Pakistan. In Proceedings of the International Exchange and Innovation Conference on Engineering & Sciences (IEICES), Fukoka, Japan, 22—23 October 2020; Volume 6, pp. 85–91.
39. Aleem, M.; Sultan, M. Towards the Development of Energy-Efficient Poultry Air Conditioning System. Available online: https://www.researchgate.net/publication/354138144_Towards_the_development_of_energy-efficient_poultry_air_conditioning_system (accessed on 9 June 2021).
40. Sultan, M.; Aleem, M.; Miyazaki, T. Energy-Efficient Humidity Pump System for Poultry Houses. In *Synergy Development in Renewables Assisted Multi-Carrier Systems*; Amidpour, M., Ebadollahi, M., Jabari, F., Kolahi, M.R., Ghaebi, H., Eds.; Springer: Cham, Switzerland, 2022. [CrossRef]
41. Xu, C.-J.; Guo, D.-P.; Yuan, J.; Yuan, G.-F.; Wang, Q.-M. Changes in Glucoraphanin Content and Quinone Reductase Activity in Broccoli (*Brassica oleracea* var. *italica*) Florets during Cooling and Controlled Atmosphere Storage. *Postharvest Biol. Technol.* **2006**, *42*, 176–184. [CrossRef]
42. Nampan, K.; Techavuthiporn, C.; Kanlayanarat, S. Hydrocooling Improves Quality and Storage Life of "Rong-Rein" Rambutan (*Nephellium lappaceum* L.) Fruit. *Acta Hortic.* **2006**, *712*, 763–769. [CrossRef]
43. Nunes, M.C.N.; Émond, J.P.; Brecht, J.K. Predicting Shelf Life and Quality of Raspberries under Different Storage Temperatures. *Acta Hortic.* **2003**, *628*, 599–606. [CrossRef]

44. Paull, R.E. Effect of Temperature and Relative Humidity on Fresh Commodity Quality. *Postharvest Biol. Technol.* **1999**, *15*, 263–277. [CrossRef]
45. Nunes, M.C.D.N. Impact of Environmental Conditions on Fruit and Vegetable Quality. *Stewart Postharvest Rev.* **2008**, *4*, 1–14. [CrossRef]
46. ASHRAE. *Handbook of Fundamentals*; American Society of Heating, Refrigerating and Air-Conditioning Engineers, Inc.: Atlanta, GA, USA, 2017.
47. Arah, I.K.; Amaglo, H.; Kumah, E.K.; Ofori, H. Preharvest and Postharvest Factors Affecting the Quality and Shelf Life of Harvested Tomatoes: A Mini Review. *Int. J. Agron.* **2015**, *2015*. [CrossRef]
48. Mahajan, P.V.; Rodrigues, F.A.S.; Motel, A.; Leonhard, A. Development of a Moisture Absorber for Packaging of Fresh Mushrooms (*Agaricus bisporous*). *Postharvest Biol. Technol.* **2008**, *48*, 408–414. [CrossRef]
49. ASHRAE. *Handbook of Heating, Ventilation, and Air-Conditioning Application*; American Society of Heating, Refrigerating and Air-Conditioning Engineers, Inc.: Atlanta, GA, USA, 2019.
50. Gross, K.C.; Wang, C.Y.; Mikal, S. *The Commercial Storage of Fruits, Vegetables, and Florist and Nursery Stocks*, 66th ed.; Kenneth, C.G., Wang, C.Y., Mikal, S., Eds.; United States Department of Agriculture, Agricultural Research Service: Washington, DC, USA, 2016.
51. Sultan, M.; El-Sharkawy, I.I.; Miyazaki, T.; Saha, B.B.; Koyama, S.; Maruyama, T.; Maeda, S.; Nakamura, T. Water Vapor Sorption Kinetics of Polymer Based Sorbents: Theory and Experiments. *Appl. Therm. Eng.* **2016**, *106*, 192–202. [CrossRef]
52. Sultan, M.; El-Sharkawy, I.I.; Miyazaki, T.; Saha, B.B.; Koyama, S.; Maruyama, T.; Maeda, S.; Nakamura, T. Insights of Water Vapor Sorption onto Polymer Based Sorbents. *Adsorption* **2015**, *21*, 205–215. [CrossRef]
53. Aleem, M.; Hussain, G.; Sultan, M.; Miyazaki, T.; Mahmood, M.H.; Sabir, M.I.; Nasir, A.; Shabir, F.; Khan, Z.M. Experimental Investigation of Desiccant Dehumidification Cooling System for Climatic Conditions of Multan (Pakistan). *Energies* **2020**, *13*, 5530. [CrossRef]
54. Sultan, M.; Miyazaki, T.; Niaz, H.; Shabir, F.; Ashraf, S.; Khan, Z.M.; Mahmood, M.H.; Raza, H.M.U. Thermodynamic Assessment of Solar Chimney Based Air-Conditioning System for Agricultural and Livestock Applications. In Proceedings of the 4th International Conference on Energy, Environment and Sustainable Development 2016 (EESD-2016), Jamshoro, Pakistan, 1–3 November 2016; Volume 2016, pp. 1–9.
55. Islam, M.R.; Alan, S.W.L.; Chua, K.J. Studying the Heat and Mass Transfer Process of Liquid Desiccant for Dehumidification and Cooling. *Appl. Energy* **2018**, *221*, 334–347. [CrossRef]
56. Esen, H.; Inalli, M.; Esen, M. Technoeconomic Appraisal of a Ground Source Heat Pump System for a Heating Season in Eastern Turkey. *Energy Convers. Manag.* **2006**, *47*, 1281–1297. [CrossRef]
57. Holman, J.P. *Experimental Methods for Engineers*; McGraw-Hill: New York, NY, USA, 2012.
58. Hepbasli, A.; Akdemir, O. Energy and Exergy Analysis of a Ground Source (Geothermal) Heat Pump System. *Energy Convers. Manag.* **2004**, *45*, 737–753. [CrossRef]
59. Becker, B.R.; Fricke, B.A. Simulation of Moisture Loss and Heat Loads in Refrigerated Storage of Fruits and Vegetables. *New Dev. Refrig. Food Saf. Qual.* **1996**, *6*, 210–221.
60. ASHRAE, *ASHRAE Handbook—Refrigeration*; American Society of Heating, Refrigerating, and Air-Conditioning Engineers, Inc.: Atlanta, GA, USA, 2014.
61. Rafiee, S.; Keyhani, A.; Tabatabaeefar, A. Some Physical Properties of Apple Cv. 'Golab'. Available online: https://cigrjournal.org/index.php/Ejounral/article/view/1124/1167 (accessed on 9 June 2021).
62. Dincer, I. Development of New Effective Nusselt-Reynolds Correlations for Air-Cooling of Spherical and Cylindrical Products. *Int. J. Heat Mass Transf.* **1994**, *37*, 2781–2787. [CrossRef]
63. Zhang, Y.; Keller, M. Grape Berry Transpiration Is Determined by Vapor Pressure Deficit, Cuticular Conductance, and Berry Size. *Am. J. Enol. Vitic.* **2015**, *66*, 454–462. [CrossRef]
64. Topuz, A.; Topakci, M.; Canakci, M.; Akinci, I.; Ozdemir, F. Physical and Nutritional Properties of Four Orange Varieties. *J. Food Eng.* **2005**, *66*, 519–523. [CrossRef]
65. Bennett, A.H.; Soule, J.; Yost, G.E. Temperature Response of Florida Citrus to Forced-Air Precooling. *ASHRAE J.* **1966**, *4*, 48–54.
66. Babic, L.; Matic-Kekic, S.; Dedovic, N.; Babic, M.; Pavkov, I. Surface Area and Volume Modeling of the Williams Pear (*Pyrus communis*). *Int. J. Food Prop.* **2012**, *15*, 880–890. [CrossRef]
67. Beuchat, L.R.; Farber, J.N.; Garrett, E.H.; Harris, L.J.; Parish, M.E.; Suslow, T.V.; Busta, F.F. Standardization of a Method to Determine the Efficacy of Sanitizers in Inactivating Human Pathogenic Microorganisms on Raw Fruits and Vegetables. *Compr. Rev. Food Sci. Food Saf.* **2003**, *2*, 174–178. [CrossRef]
68. Phungamngoen, C.; Rittisak, S. Surface Characteristics of Leafy Vegetables and Their Effects on Salmonella Attachment. *E3S Web Conf.* **2020**, *141*, 1–5. [CrossRef]
69. Parashar, V.; Trivedi, S.K.; Haleem, A. Heat Transfer Parameterization towards Enhancing Shelf Life of Vegetables in Low Cost Cold Chain with FACCC. *Int. J. Innov. Technol. Explor. Eng.* **2020**, *9*, 822–829. [CrossRef]
70. Jahanbakhshi, A.; Abbaspour-Gilandeh, Y.; Gundoshmian, T.M. Determination of Physical and Mechanical Properties of Carrot in Order to Reduce Waste during Harvesting and Post-Harvesting. *Food Sci. Nutr.* **2018**, *6*, 1898–1903. [CrossRef] [PubMed]
71. Chang, S.Y.; Toledo, R.T. Simultaneous Determination of Thermal Diffusivity and Heat Transfer Coefficient during Sterilization of Carrot Dices in a Packed Bed. *J. Food Sci.* **1990**, *55*, 199–205. [CrossRef]

72. Kaveri, G.; Thirupathi, V. Studies on Geometrical and Physical Properties of CO_4 Onion Bulb (*Allium cepa* Lvar. Aggregatum Don.). *Int. J. Rec. Sci. Res.* **2015**, *6*, 2897–2902.
73. Gamea, G.R.; Abd El-Maksoud, M.A.; Abd El-Gawad, A.M. Physical Characteristics and Chemical Properties of Potato Tubers Under Different Storage Systems. *Misr J. Agric. Eng.* **2009**, *26*, 385–408. [CrossRef]

sustainability

Article

On the Classification of a Greenhouse Environment for a Rose Crop Based on AI-Based Surrogate Models

Showkat Ahmad Bhat [1,*], Nen-Fu Huang [1,*], Imtiyaz Hussain [2], Farzana Bibi [3], Uzair Sajjad [4,*], Muhammad Sultan [5], Abdullah Saad Alsubaie [6] and Khaled H. Mahmoud [6]

1 Institute of Communication Engineering, National Tsing Hua University, Hsinchu 300044, Taiwan
2 Department of Power Mechanical Engineering, National Tsing Hua University, No. 101, Section 2, Guangfu Road, East District, Hsinchu 300044, Taiwan; imtiyazkou@yahoo.com
3 Institute of Computer Science and Information Technology, The Women University, Multan 154-8533, Pakistan; farzanabatool1997@gmail.com
4 Department of Mechanical Engineering, National Yang Ming Chiao Tung University, 1001 University Road, Hsinchu 300093, Taiwan
5 Department of Agricultural Engineering, Bahauddin Zakariya University, Bosan Road, Multan 60800, Pakistan; muhammadsultan@bzu.edu.pk
6 Department of Physics, College of Khurma University College, Taif University, P.O. Box 11099, Taif 21944, Saudi Arabia; asubaie@tu.edu.sa (A.S.A.); k.hussein@tu.edu.sa (K.H.M.)
* Correspondence: showkatbhat1994@gmail.com (S.A.B.); nfhuang@cs.nthu.edu.tw (N.-F.H.); energyengineer01@gmail.com (U.S.)

Abstract: A precise microclimate control for dynamic climate changes in greenhouses allows the industry and researchers to develop a simple, robust, reliable, and intelligent model. Accordingly, the objective of this investigation was to develop a method that can accurately define the most suitable environment in the greenhouse for an optimal yield of roses. Herein, an optimal and highly accurate BO-DNN surrogate model was developed (based on 300 experimental data points) for a quick and reliable classification of the rose yield environment considering some of the most influential variables including soil humidity, temperature and humidity of air, CO_2 concentration, and light intensity (lux) into its architecture. Initially, two BO techniques (GP and GBRT) are used for the tuning process of the hyper-parameters (such as learning rate, batch size, number of dense nodes, number of dense neurons, number of input nodes, activation function, etc.). After that, an optimal and simple combination of the hyper-parameters was selected to develop a DNN algorithm based on 300 data points, which was further used to classify the rose yield environment (the rose yield environments were classified into four classes such as soil without water, correct environment, too hot, and very cold environments). The very high accuracy of the proposed surrogate model (0.98) originated from the introduction of the most vital soil and meteorological parameters as the inputs of the model. The proposed method can help in identifying intelligent greenhouse environments for efficient crop yields.

Keywords: greenhouse; microclimate; Bayesian optimization; deep neural network; roses yield; Gaussian process; gradient boosting

Citation: Bhat, S.A.; Huang, N.-F.; Hussain, I.; Bibi, F.; Sajjad, U.; Sultan, M.; Alsubaie, A.S.; Mahmoud, K.H. On the Classification of a Greenhouse Environment for a Rose Crop Based on AI-Based Surrogate Models. *Sustainability* **2021**, *13*, 12166. https://doi.org/10.3390/su132112166

Academic Editor: Dalia Štreimikienė

Received: 3 October 2021
Accepted: 29 October 2021
Published: 4 November 2021

1. Introduction and Motivation

1.1. Introduction

Climate change throughout the globe is affecting agricultural production due to increasing temperatures, fluctuating precipitation patterns, and rising corban dioxide concentrations in the atmosphere. In these changing environmental conditions, greenhouse crop cultivation is preferred compared with open field growing. The cultivation of crops in the greenhouse prolongs the agricultural growing season, protect yields against weather variations, offers a reliable growing ecosystem, and thus maximizes productivity. Thus, it is essential to adopt precision agriculture techniques in order to maintain the ideal environmental parameters such as humidity, carbon dioxide, and temperature along with soil moisture and nutrients in accordance with the crop growth cycle [1,2]. Exposure to uneven environmental factors produces stress, disease, or even a fall in the crops, resulting in substantial financial losses to growers [3]. Greenhouse weather control mechanisms need to consider multivariate and non-linear systems with variables greatly dependent on the external environment and the design of the greenhouse [4,5], even though the greenhouse cannot be independently controlled. Thus, developing a precise climate model in a greenhouse is an essential approach to control these dynamic climate changes and attain proficient climate management.

Greenhouse environment models can be developed on either the physical laws driving ecological cycles, or the interpretation of data obtained from such processes. With the development of high-performance computational systems, several analytical models [6–8] have been developed. Yet, this methodology may produce inconsistent outcomes when applied to true environmental conditions due to the complexity of these models and the frequent need for calculation and the approximation of unmeasurable parameters, for example, water vapor pressure, biological factors, rate of photosynthesis, soil heat flux density, and other factors [9]. On the contrary, due to the advancement of existing computational strategies, deep leaning prediction models based on big data [10] are being progressively applied to several fields. ANN models are incredible predicting tools [11,12] because of their capabilities to model systems without making assumptions [13] and to evaluate nonlinear systems. The most significant benefits of deep learning models over several classes of nonlinear models is that ANN models can approximate a vast group of functions with a high level of precision [14]. This approach delivers swift and reliable results for precision agriculture applications, namely, the climate estimation of greenhouses [15], the growth of plants, and the detection of stress compared to existing physical models [6,7].

For the generation and collection of data, smart greenhouses are equipped with IoTs, wireless sensor networks (WSNs), and actuators [16]. Sensors sense the atmosphere in the greenhouse and measure temperature, light intensity, humidity, CO_2 levels, pressure, etc. If any irregularity is detected in the environmental conditions of the greenhouse, the ANN-based central control station directs actuators to execute required actions such as watering the crops, increasing or decreasing the light intensity, opening and closing windows, etc.

Besides, an appropriate comprehension of the variations of different parameters in the greenhouse climate related to the requirements of the particular crop at various development phases needs more consideration. As rose plants are susceptible to large variations in temperature, light, and humidity, the cultivation of greenhouse roses in geographical areas with environmental conditions that are not satisfactorily near the base prerequisites will encompass added risks and costs of production [17]. Exclusively relying upon the parameter measurement data from sensors is insufficient to obtain solid harvests in the greenhouse. Having a profound learning model for forecasting the future air parameters will assist in keeping up with the climate [18]. For instance, having the predicted values of temperature, CO_2, and humidity assist in maintaining the flower size and a high yield, and can prevent the growth of pests that harm the rose plants. Additionally, predicting greenhouse climate changes will help in the event of sensor breakdown and will reduce the energy utilization in the greenhouse [3].

1.2. Aims and Motivation

Roses are amongst some of the most highly marketed flowers globally and have ruled the flower market since the 1990s owing to their year-long availability and the ever-increasing demand in beauty products and from the decoration industry. Roses are a functional food product similar to barley and other crops [19]. Natural environmental conditions are not always optimum to achieve the growing demand of crop requirements [20,21]. Extreme weather conditions such as exposure to direct sun, hail, biotic, and abiotic stresses can critically damage the product quality and yield [22]. Therefore, greenhouses are increasingly being used, since they can adjust the interior environmental parameters through artificial lights, aeration, and heating and ventilation systems [23]. Thus, crop growing cycles can be designed based on market demands. The environmental parameters required for the appropriate growth of roses are relative humidity, CO_2 concentration, soil humidity, air temperature, light intensity, and the electrical conductivity of soil (see Figure 1).

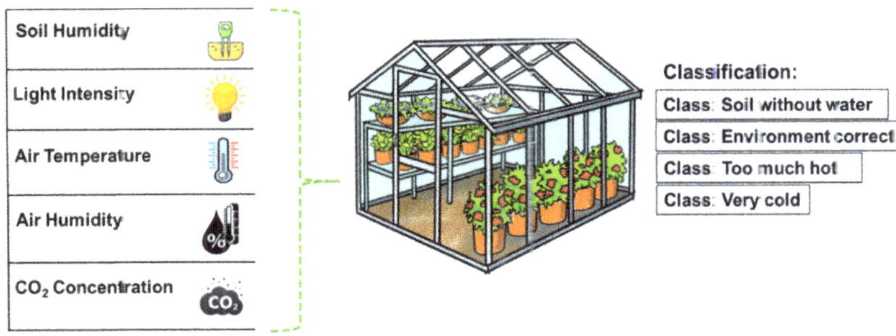

Figure 1. Parameters affecting the greenhouse rose yield environment.

There are several analytical models for the interpretation of the data collected from wireless sensor networks or IoTs, but these models may produce inconsistent outcomes when applied to true environmental conditions due to the high complexity of these models and the frequent need for calculation and the approximation of unmeasurable parameters. Based on the aforementioned discussion, it is extremely crucial to develop a method particularly for AI-based methods [24] that can accurately define the most suitable environment in the greenhouses for rose yield production. This is because the AI-based methods have gained lot of success in agriculture during recent years in relation to crop yield production, detection, precision agriculture, and so on [25–29].

To the best of the authors' knowledge, only a single study is available in the literature regarding the use of AI in the rose's greenhouse environment. This study presents the ANN and ANFIS methods to forecast the risk level for pests in the rose greenhouse [30]. Other than this, no study is available in the literature on this subject. The present study is the first of its kind in classifying the greenhouse environment for rose crops based on AI-based surrogate models. The proposed models are deep neural networks based on the optimal set of hyper-parameters defined by the Bayesian optimization scheme. AI-based surrogate models can be a reliable, simple, and robust solution. For instance, Bayesian optimization (BO) techniques such as the Gaussian process (GP) and Gradient boosting (GBRT) can be employed to provide optimal hyper-parameters to be integrated with deep neural networks (DNN). In line with this, the objective of this study was to develop an optimal and highly accurate BC-DNN surrogate model (based on 300 experimental data points) for a quick and reliable classification of the rose yield environment considering some of the most influential variables including soil humidity, the temperature and humidity of air, CO_2 concentration, and light intensity (lux) into its architecture. The rose yield environments (outputs) are classified into four classes such as soil without water, correct environment, too hot, and very cold environments. Initially, two BO techniques (GP and GBRT) were used for the tuning process of the hyper-parameters (such as learning rate, batch size, number of dense nodes, number of dense neurons, number of input nodes, activation function, etc.). The most accurate set of hyper-parameters was selected to build the DNN model based on 300 data points, which was further used to classify the rose yield environment. The very high accuracy of the proposed surrogate model originates from the introduction of the most vital soil and meteorological parameters as the inputs of the model.

2. Materials and Methods

2.1. Data Collection

A total of 300 experimental data points from various sensors regarding soil humidity, light intensity, temperature, air humidity, and CO_2 concentration for 04 different classes of greenhouse rose yield environments were taken from the open literature [31]. The data were acquired by an autonomous robot integrating the sensors including soil humidity, light intensity, temperature, air humidity, and CO_2 concentration. Table 1 shows that a wide range of experimental data have been included in this study to discuss greenhouse rose yield environments.

Table 1. Investigated parameters and their data range.

Parameter	Data Range
Soil humidity (kPa)	124–821
Light intensity (lux)	0–54612.5
Temperature (°C)	15.9–40.2
Air humidity (%)	39.2–96.9
CO_2 concentration (ppm)	34–243
Environment	Class 0, 1, 2, and 3

2.2. Data Visualization

The experimental data have been visualized in terms of heat maps, correlation charts, pairs, and violin plots. The heat map and correlation chart represent the relationship between input and output features while the data distribution has been visualized by pairs, violins, and distplot. In addition, the data density for each class has been shown. A heat map showing the correlation between the input and output variables Figure 2. The dependency of the various input variables on the output parameters can be visualized by using a correlation chart as provided Figure 3. The data distribution of the input and output parameters including the soil humidity, air temperature and humidity, CO_2 concentration, lux (light intensity), and class (output) is represented by a pair plot (see Figure 4). A clearer picture of the experimental data distribution of various input features with respect to the only output parameter, class, is highlighted in the violin plots (see Figure 5).

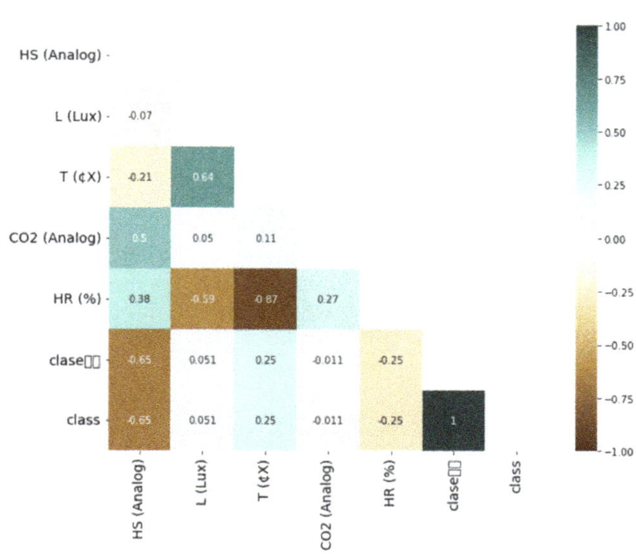

Figure 2. Heat map showing the correlation between the input and output variables.

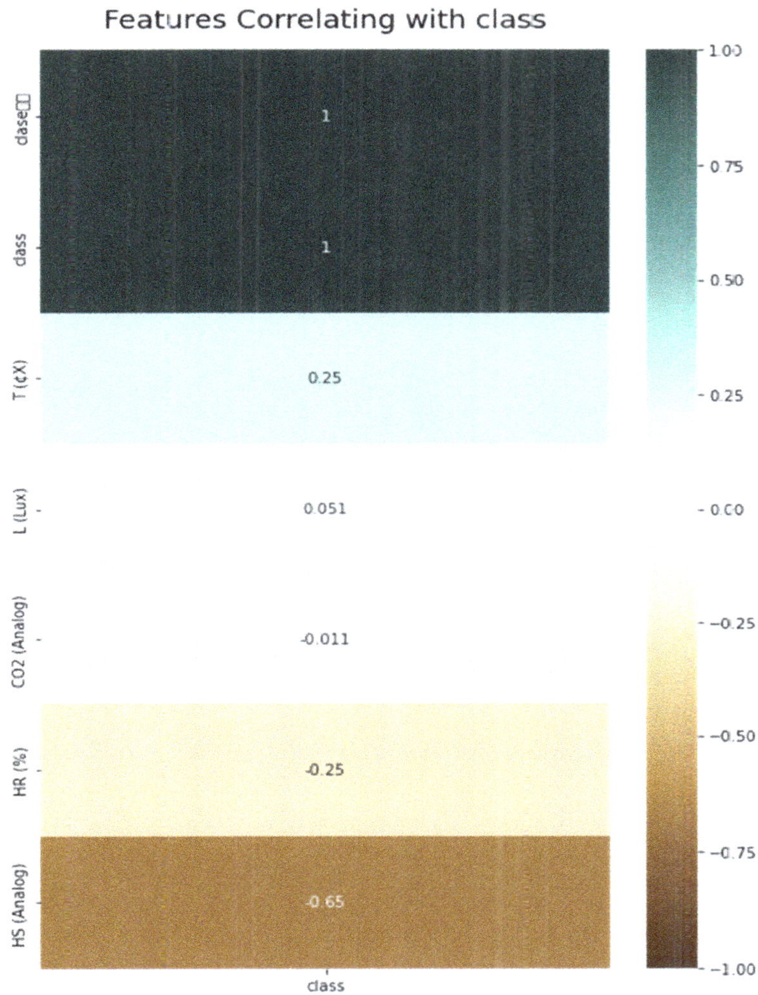

Figure 3. Dependency of the input variables on the output parameter.

Herein, the experimental data were distributed into 04 different classes (namely class 0, class 1, class 2, and class 3). The total number of data points for each class is illustrated in Figure 6.

The density of each input parameter's acquired data is presented by distplot (see Figure 7). The distplot illustrates the data distribution of each parameter in terms of density distribution.

2.3. Bayesian Optimization Integrated with a Deep Neural Network Algorithm

Algorithms of two different Bayesian optimization schemes, namely, Gaussian process regression (GPR) and Gradient boosting regression trees (GBRT) integrated with the deep neural network are illustrated in Figure 8.

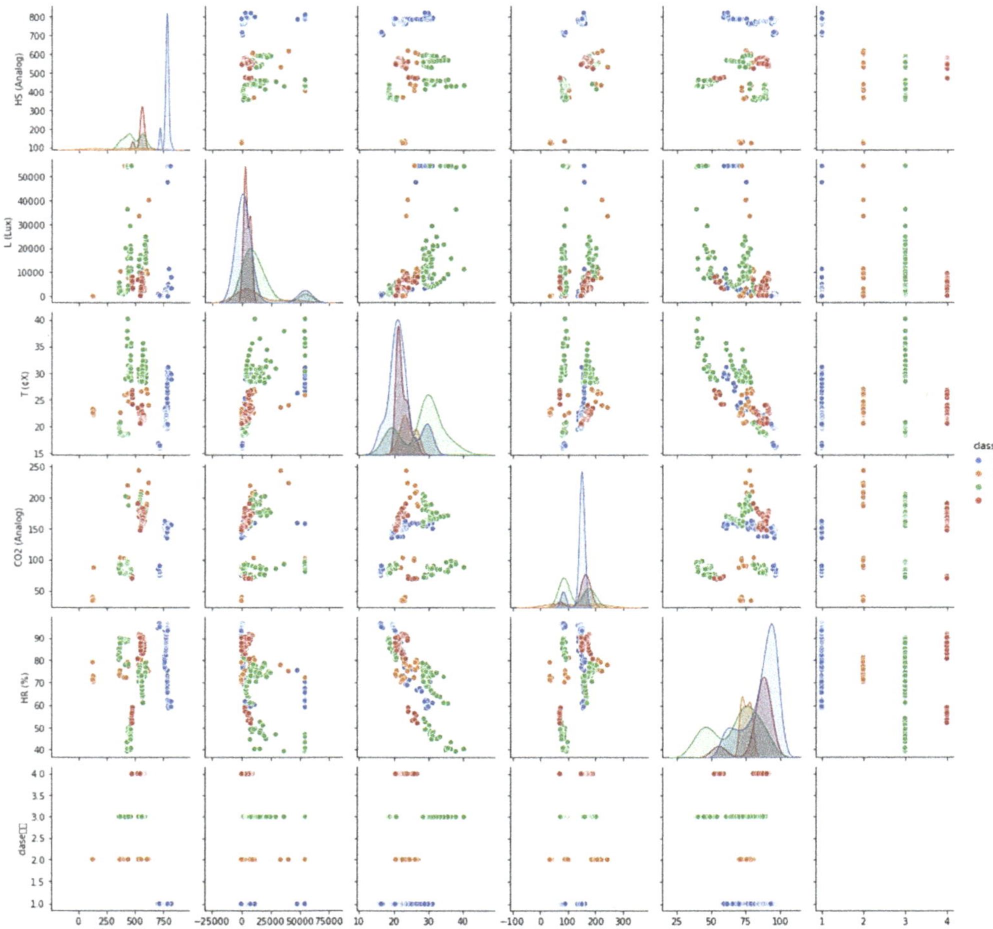

Figure 4. Representation of the data distribution via a pair plot.

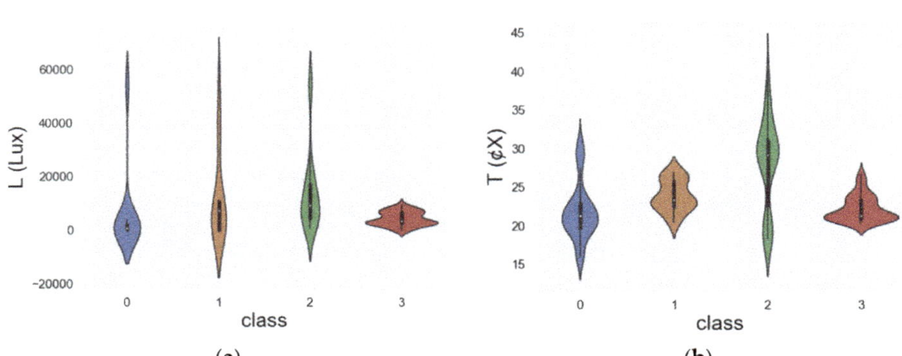

Figure 5. *Cont.*

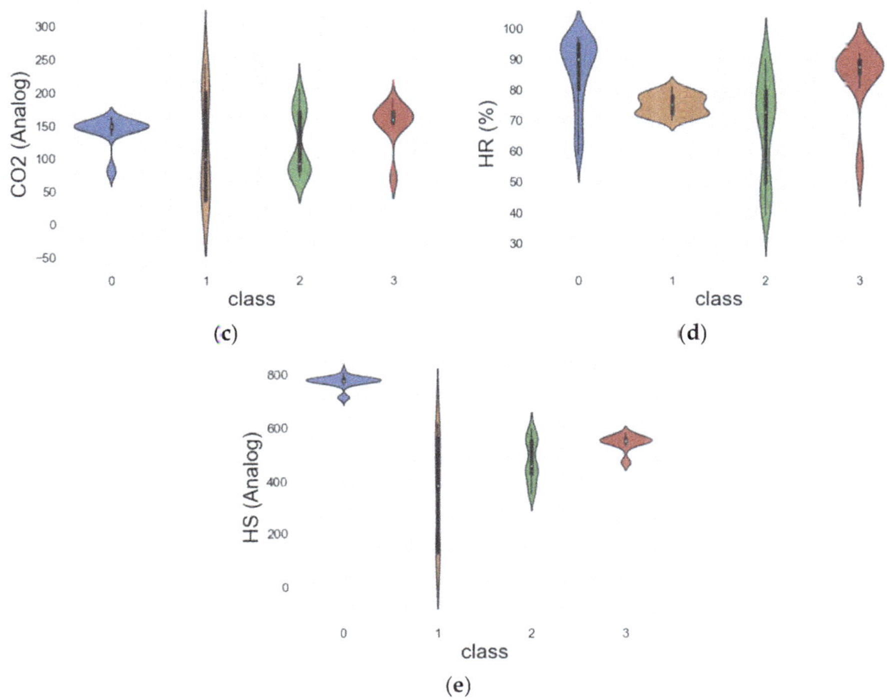

Figure 5. Violin plots for the experimental data distribution of (**a**) light, (**b**) temperature, (**c**) CO_2 concentration, (**d**) air humidity, and (**e**) soil humidity with respect to the class.

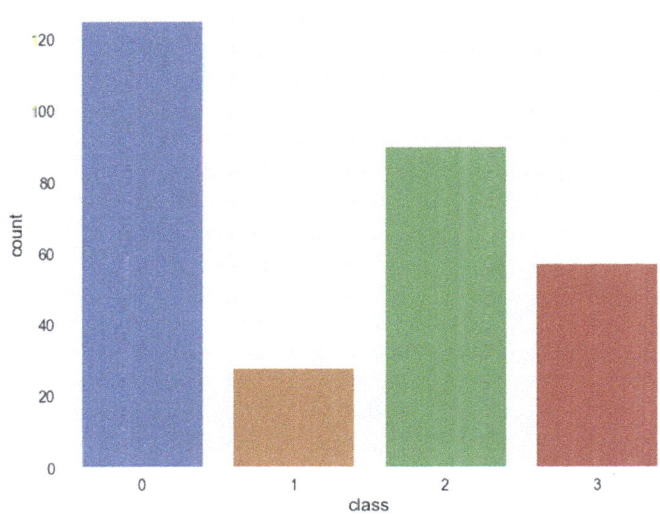

Figure 6. The experimental data distribution for each class.

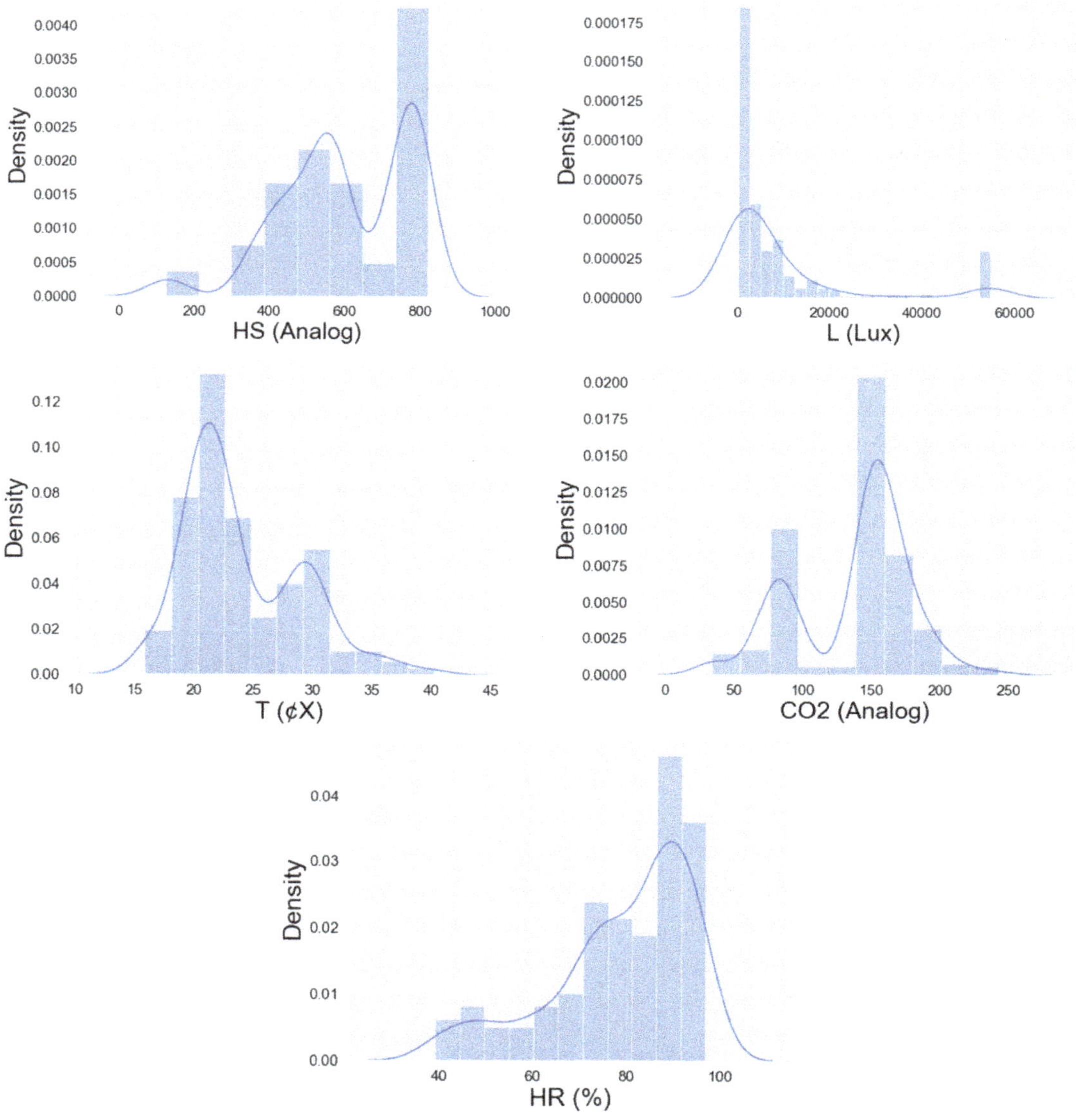

Figure 7. Illustration of the data distribution of each parameter in terms of density distribution.

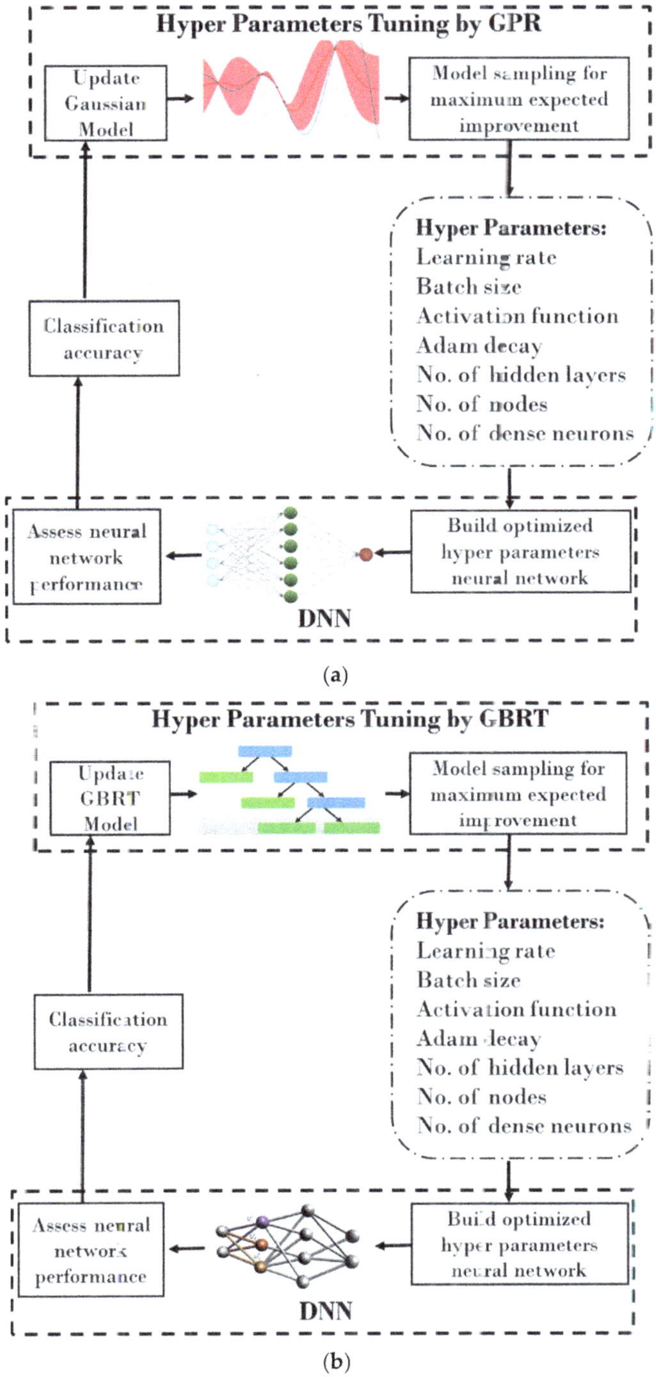

Figure 8. Bayesian optimization integrated in the deep neural network (**a**) Gaussian process regression and (**b**) Gradient boosting regression trees.

3. Results and Discussion

In this section, the range of the considered hyper-parameters is first provided followed by the tuning processes of two different Bayesian optimization schemes (GP and GBRT). Furthermore, the way that the maximum convergence was achieved is explained. In addition, the optimal combination of the hyper parameters is chosen. The chosen optimal hyper-parameters are then employed to develop a deep neural network model, which is then used to classify the greenhouse environments for rose yields. Moreover, the classification accuracy of the developed model in terms of a confusion matrix and an accuracy table is presented. The details of the input features and their impact on the model's classification accuracy is evaluated in the sensitivity analysis section. Other than that, individual impact of each input variable on the model's classification accuracy is evaluated. More discussions are presented in the subsequent sections.

3.1. Optimization of the Hyper-Parameters

The considered hyper-parameters were tuned by using two different Bayesian optimization schemes (GP and GBRT). The selected hyper-parameters include the learning rate, Adam decay, input nodes, dense layers, dense nodes, batch size, and activation function. The range of all the investigated hyper-parameters for the tuning process is given in Table 2.

Table 2. Range of hyper parameters.

Hyper Parameter	Investigated Range
Learning rate	0.0001–0.1
Adam decay	0.000001–0.01
Input nodes	1–5
Dense layers	1–10
Dense nodes	1–500
Batch size	1–100
Activation function	Softmax, Sigmoid, ReLU, tanh

The range of the considered hyper-parameters along with the hyper-parameter tuning process by the GBRT and GPR algorithms is depicted in Figures 9 and 10, respectively. It is worth mentioning that the blue and orange regions represent the strong and weak dependence of the variable, respectively, while the asterisk sign points towards the optimal point. For further analysis, the GPR algorithm was considered. Detailed information on the finally selected architecture of the optimal model (GPR) is tabulated in Table 3. From the Figure 10, it can be observed that the 'tanh' activation function provided optimal results compared to the Softmax, sigmoid, and ReLU. A comparison between the suitability of these activation functions is provided in Figure 11.

Table 3. Selected architecture of the optimal model.

Optimization Method	Gaussian Process
Learning rate	0.000416
No. of hidden layers	10
No. of neurons in input layer	5
No. of neurons in each hidden layer	265
Activation function	tanh
Batch size	36
Adam decay	0.007963
No. of neurons in output layer	4
No. of iterations	80

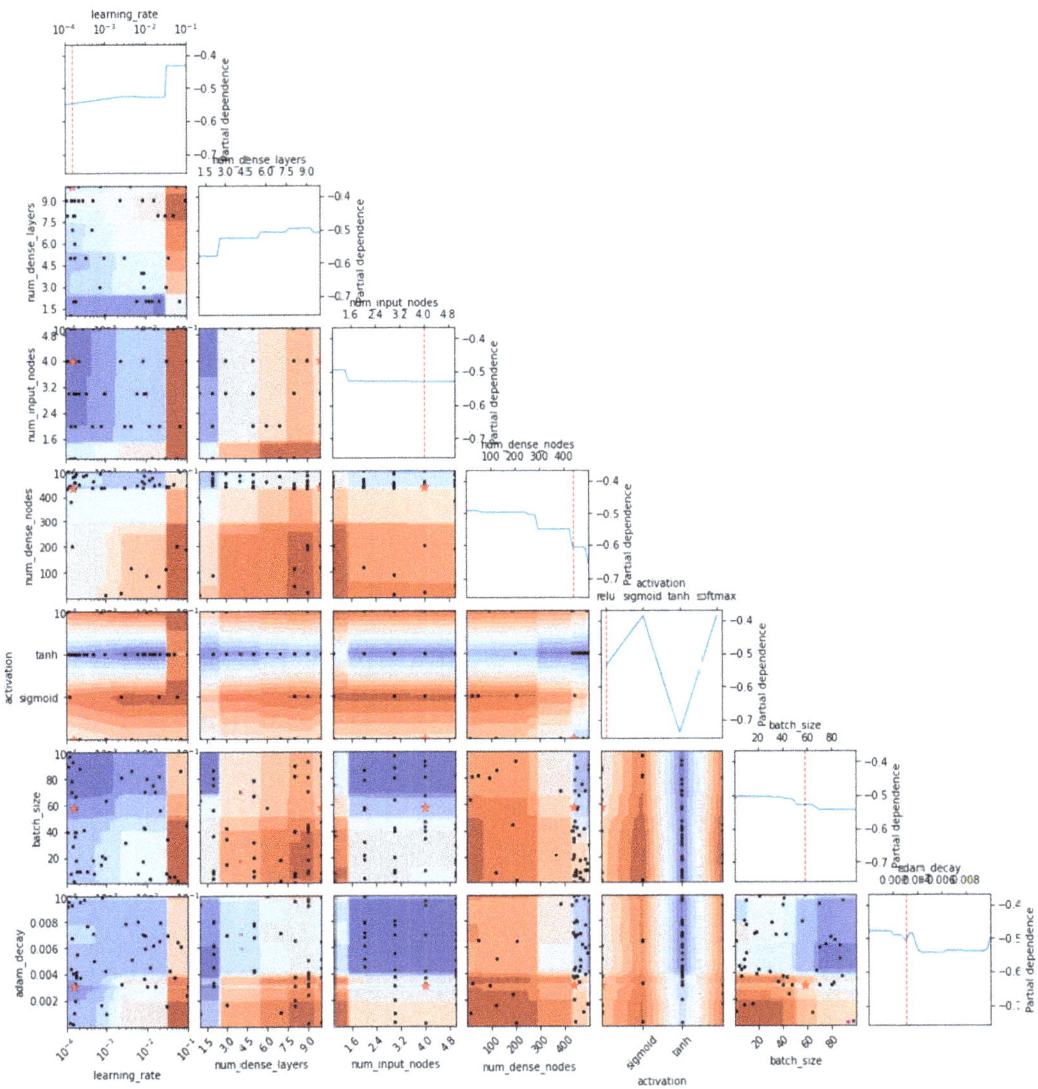

Figure 9. The tuning process of the hyper-parameters in GBRT.

Convergence plots for both optimization schemes such as GP and GBRT provide a clear picture of the way the error was minimized. The initial convergence was reached very fast because the number of input parameters and the amount of training data affected the convergence rate, and in this study the model was evaluated for 300 experimental data points containing five input parameters. For instance, in Figure 12. it can be clearly noticed that the convergence error for both the GP and GBRT algorithms was minimized at the fourth call.

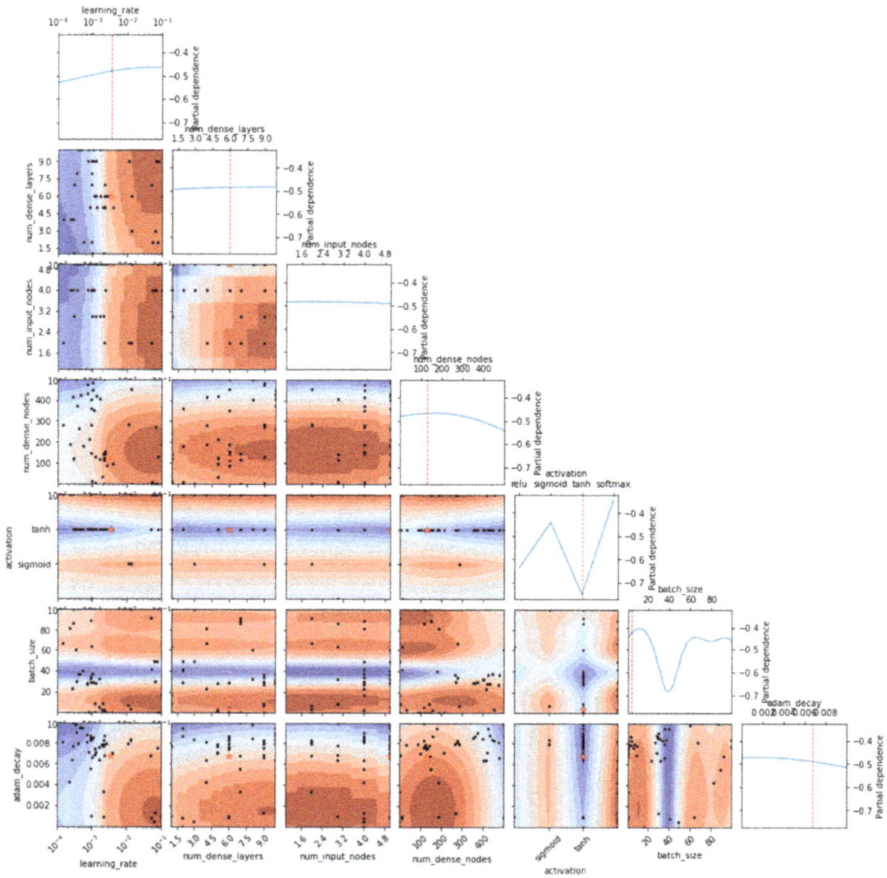

Figure 10. The tuning process of the hyper-parameters in GPR.

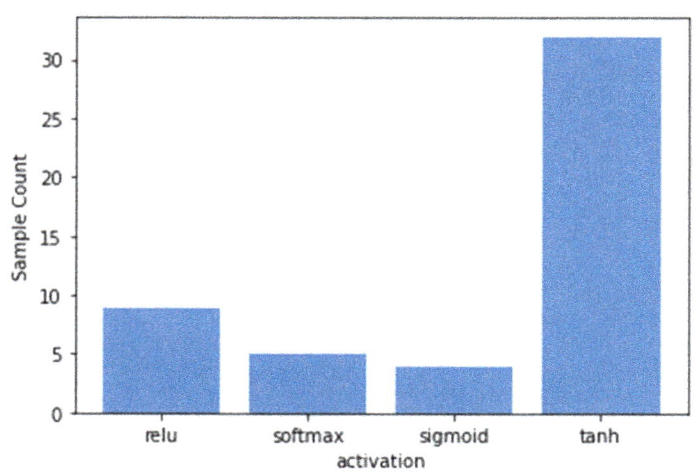

Figure 11. A performance comparison of the selected activation functions.

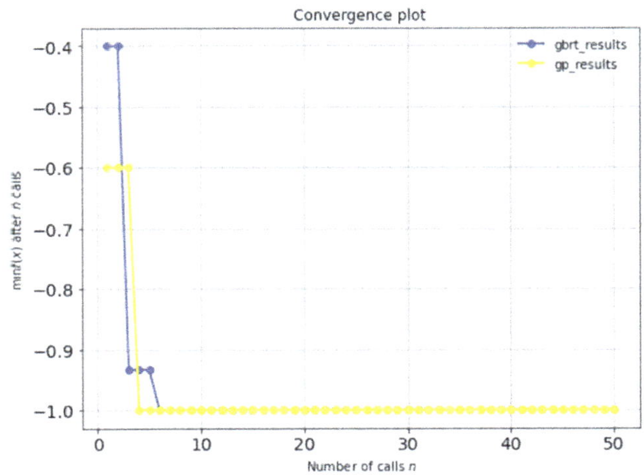

Figure 12. Convergence plots for the GPR and GBRT algorithms.

3.2. Training and Developing the Deep Neural Network

The experimental data were distributed into training (80%) and testing (20%) datasets. The training and validation losses of the developed model are depicted in Figure 13. The total number of iterations was kept up to 80. Apparently, both of the losses were minimized until the 35th iteration, so the training process was stopped. This shows that the training process was computationally economical and quick.

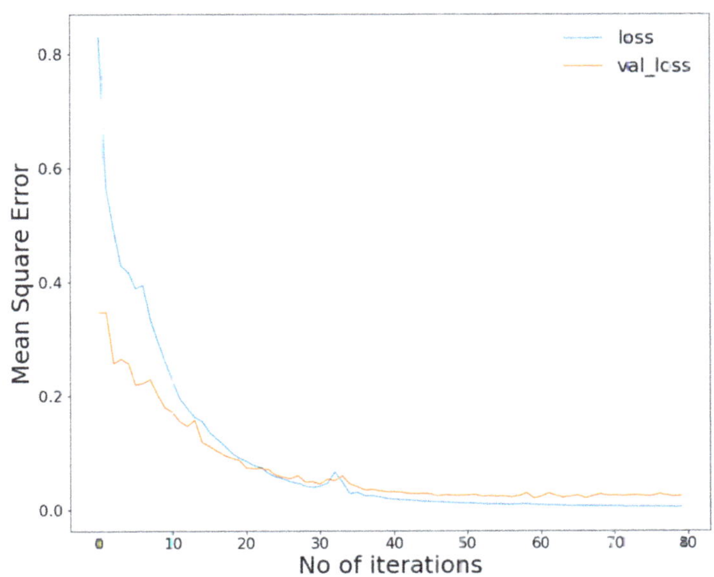

Figure 13. Training and validation losses during training.

Figure 14 illustrates the classification performance of the developed model for each class of the environment. Apparently, the developed model was able to accurately classify 59 out of 60 environments for various classes. This explains how well the model performs for different greenhouse environments within the tested range. The classification accuracy of the selected surrogate model is presented in Table 4.

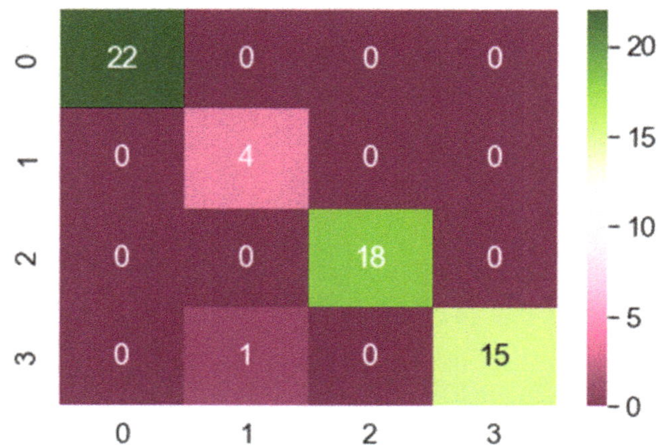

Figure 14. Classification performance of the developed model.

Table 4. Classification accuracy description of the selected surrogate model.

	Precision	Recall	F1-Score
0	1.00	1.00	1.00
1	0.80	1.00	0.89
2	1.00	1.00	1.00
3	1.00	0.94	0.97
Accuracy			0.98
Macro Avg.	0.95	0.98	0.96
Weighted Avg.	0.99	0.98	0.98

The performance of the developed model is highlighted in terms of precision, recall, and F1-score. Precision and recall are the fraction of the relevant instances among the retrieved instances and the fraction of the relevant instances that were retrieved. Both precision and recall are therefore based on relevance. The values of precision and recall from Table 4 show that the proposed model had a high classification efficiency for the rose's greenhouse environment. The F1-score from Table 4 also indicates the perfect precision and recall of the optimal surrogate model. In addition, the overall accuracy of the model along with the macro and weighted averages are described as well. The final model could perform the classification task with an overall accuracy of 0.98.

3.3. Sensitivity Analysis

The individual impact of each input variable on the model's output (i.e., classification of the greenhouse rose yield environments) is portrayed using the SHAP library. More particularly, the ways in which the various input features such as soil humidity, temperature, air humidity, light intensity, and CO_2 concentration affected the model's classification accuracy are shown in Figure 15. It can be clearly seen that the sensitivity of the different features was not the same for various classes. However, some of the factors were sensitive for all the classes. For example, the most influential factor for each class was the soil humidity followed by the temperature. Regarding class 1 (correct environment), the feature with the most impact was air humidity followed by soil humidity, temperature, light intensity, and CO_2 concentration.

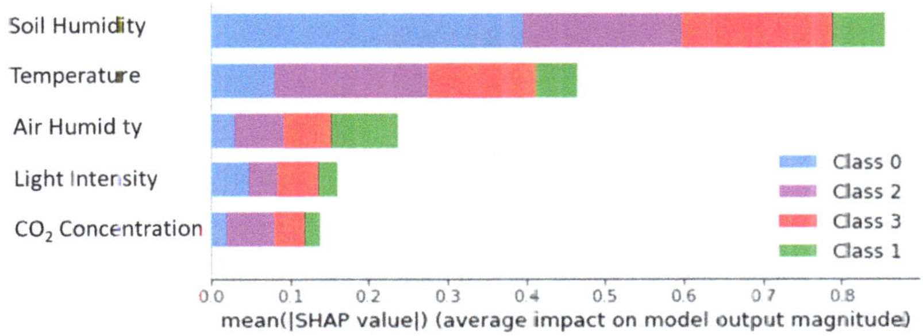

Figure 15. Individual impact of input variables on the model's output.

A tabulated performance comparison of the various developed models is illustrated in Table 5. In the original model, all five input variables were considered for classification while in the rest of the models, each single variable was dropped and rest of the four variables were used to classify the rose yield environment. It is obvious that there was no noticeable impact of dropping a single (any of the variable at a time) variable on the classification accuracy of the models. All of the developed models were able to perform the classification task with an overall accuracy of 0.98.

Table 5. Performance comparison of the various developed models.

Model No.	Input Features	Confusion Heat Map	Overall Accuracy
1	soil humidity light intensity temperature air humidity CO_2 concentration	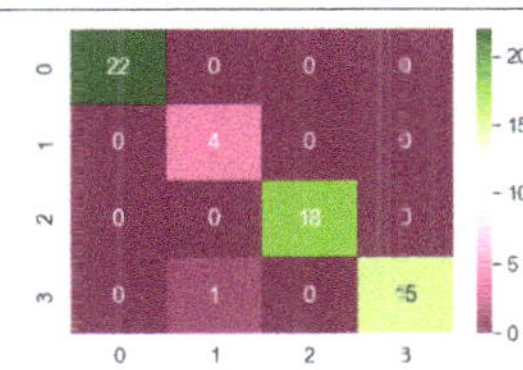	0.98
2	light intensity temperature air humidity CO_2 concentration	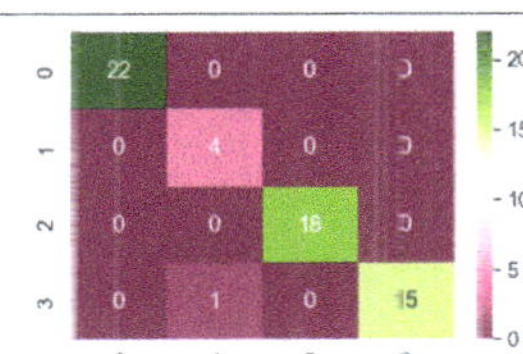	0.98
3	soil humidity temperature air humidity CO_2 concentration	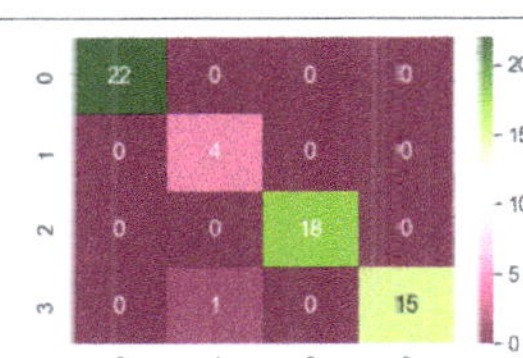	0.98
4	soil humidity light intensity air humidity CO_2 concentration		0.98

Table 5. *Cont.*

Model No.	Input Features	Confusion Heat Map	Overall Accuracy
5	soil humidity light intensity temperature CO_2 concentration	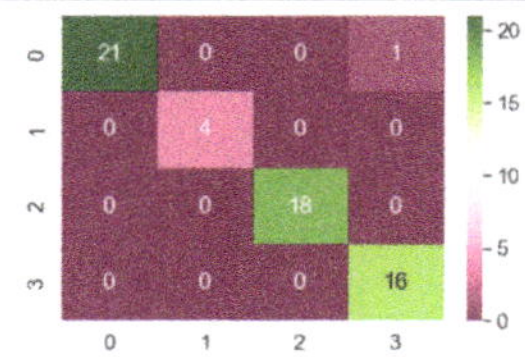	0.98
6	soil humidity light intensity temperature air humidity	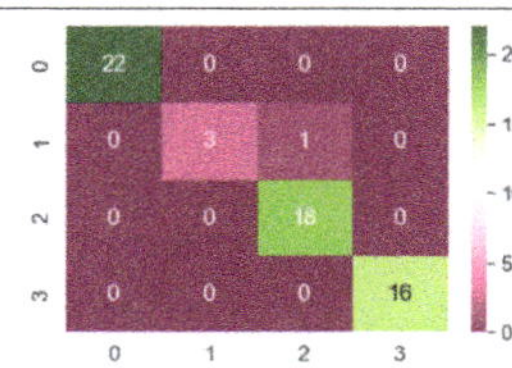	0.98

4. Conclusions

In the current study, surrogate models were developed that can accurately define the most suitable environment in greenhouses for rose yield production. In this regard, Bayesian optimization (BO) techniques such as the Gaussian process (GP) and Gradient boosting (GBRT) were employed to provide optimal hyper-parameters to be integrated with deep neural networks (DNN).

- The optimal set of hyper-parameters includes the learning rate (0.000416), the number of hidden layers (10), the number of neurons in each hidden layer (265), the activation function (tanh), batch size (36), Adam decay (0.007963), and number of iterations (80).
- An optimal and highly accurate BO-DNN surrogate model (based on 300 experimental data points) was developed for a quick and reliable classification of the rose yield environment considering the most influential variables including soil humidity, temperature and humidity of air, CO_2 concentration, and light intensity (lux) into its architecture.
- The proposed surrogate models can accurately classify the rose yield environments (classified into four classes such as soil without water, correct environment, too hot, and very cold environments).
- The developed model can classify different roses yield environments with an overall accuracy of 0.98. The very high accuracy of the proposed surrogate models originates from the inclusion of the most influential parameters as the inputs of the model.
- This study provides an easy, quick, reliable, and intelligent method to identify and perform corrective measures to improve the quality of the roses. With the proposed method, greenhouse environments can be evaluated and selected for an efficient crop yield of roses and other vegetables and fruits.

Author Contributions: Conceptualization, S.A.B. and N.-F.H.; methodology, S.A.B., I.H., F.B., U.S.; software, S.A.B., I.H., F.B.; validation, A.S.A., K.H.M.; formal analysis, S.A.B., M.S.; investigation, S.A.B., U.S., N.-F.H.; resources, U.S., S.A.B.; data curation, S.A.B., N.-F.H., I.H., U.S.; writing—original draft preparation, S.A.B., U.S., N.-F.H., F.B.; writing—review and editing, S.A.B., U.S., A.S.A., M.S., K.H.M., F.B.; visualization, S.A.B., U.S., I.H.; supervision, N.-F.H.; project administration, N.-F.H., M.S., U.S.; funding acquisition, N.-F.H. All authors have read and agreed to the published version of the manuscript.

Funding: This research was funded by the Ministry of Science and Technology (MOST) of Taiwan Grant Number: MOST109-2923-E-007-006. The APC was funded by N.-F.H.

Institutional Review Board Statement: Not applicable.

Informed Consent Statement: Not applicable.

Data Availability Statement: Data are contained within the article.

Acknowledgments: The authors would like to acknowledge the financial support of Taif University Researchers Supporting Project number (TURSP-2020/189), Taif University, Taif, Saudi Arabia. The authors are also indebted to the financial support from the Ministry of Science and Technology, Taiwan under contracts MOST109-2923-E-007-006.

Conflicts of Interest: The authors declare no conflict of interest.

References

1. Shamshiri, R.; Ismail, W.I.W. A review of greenhouse climate control and automation systems in tropical regions. *J. Agric. Sci. Appl.* **2013**, *2*, 176–183. [CrossRef]
2. Shamshiri, R.R.; Hameed, I.A.; Balasundram, S.K.; Ahmad, D.; Weltzien, C.; Yamin, M. Fundamental research on unmanned aerial vehicles to support precision agriculture in oil palm plantations. In *Agricultural Robots—Fundamentals and Application*; Intech Open: London, UK, 2018; pp. 91–116.
3. Jung, D.-H.; Kim, H.S.; Jhin, C.; Kim, H.-J.; Park, S.H. Time-serial analysis of deep neural network models for prediction of climatic conditions inside a greenhouse. *Comput. Electron. Agric.* **2020**, *173*, 105402. [CrossRef]
4. El Ghoumari, M.; Tantau, H.-J.; Serrano, J. Non-linear constrained MPC: Real-time implementation of greenhouse air temperature control. *Comput. Electron. Agric.* **2005**, *49*, 345–356. [CrossRef]
5. Seginer, I.; McClendon, R. Methods for optimal control of the greenhouse environment. *Trans. ASAE* **1992**, *4*, 1299–1307. [CrossRef]
6. Reddy, M.N.; Rao, N. *GIS Based Decision Support Systems in Agriculture*; National Academy of Agricultural Research Management: Rajendranagar, India, 1995; pp. 1–11.
7. Kaiwartya, O.; Abdullah, A.H.; Cao, Y.; Raw, R.S.; Kumar, S.; Lobiyal, D.K.; Isnin, I.F.; Liu, X.; Shah, R.R. T-MQM: Testbed-based multi-metric quality measurement of sensor deployment for precision agriculture—A case study. *IEEE Sens. J.* **2016**, *16*, 8649–8664. [CrossRef]
8. Sarmah, K.; Deka, C.; Sharma, U.; Sarma, R. Role of GIS based technologies in sustainable agriculture resource planning & management using spatial decision support approach. *Int. J. Innov. Res. Eng. Manag.* **2018**, *5*, 30–34.
9. Pawlowski, A.; Sánchez-Molina, J.; Guzmán, J.; Rodríguez, F.; Dormido, S. Evaluation of event-based irrigation system control scheme for tomato crops in greenhouses. *Agric. Water Manag.* **2017**, *183*, 16–25. [CrossRef]
10. Bhat, S.A.; Huang, N.-F. Big Data and AI Revolution in Precision Agriculture: Survey and Challenges. *IEEE Access* **2021**, *9*, 110209–110222. [CrossRef]
11. Sajjad, U.; Hussain, I.; Hamid, K.; Bhat, S.A.; Ali, H.M.; Wang, C.-C. A deep learning method for estimating the boiling heat transfer coefficient of porous surfaces. *J. Therm. Anal. Calorim.* **2021**, *145*, 1911–1913. [CrossRef]
12. Hamid, K.; Sajjad, U.; Yang, K.S.; Wu, S.-K.; Wang, C.-C. Assessment of an energy efficient closed loop heat pump dryer for high moisture contents materials: An experimental investigation and AI based modelling. *Energy* **2022**, *238*, 121819. [CrossRef]
13. Asfahan, H.M.; Sajjad, U.; Sultan, M.; Hussain, I.; Hamid, K.; Ali, M.; Wang, C.-C.; Shamshiri, R.R.; Khan, M.U. Artificial intelligence for the prediction of the thermal performance of evaporative cooling systems. *Energies* **2021**, *14*, 3946. [CrossRef]
14. Sajjad, U.; Hussain, I.; Wang, C.-C. A high-fidelity approach to correlate the nucleate pool boiling data of roughened surfaces. *Int. J. Multiph. Flow* **2021**, *142*, 103719. [CrossRef]
15. Nasrollahi, H.; Ahmadi, F.; Ebadollahi, M.; Nobar, S.N.; Amidpour, M. The greenhouse technology in different climate conditions: A comprehensive energy-saving analysis. *Sustain. Energy Technol. Assess.* **2021**, *47*, 101455.
16. Shamshiri, R.R.; Hameed, I.A.; Thorp, K.R.; Balasundram, S.K.; Shafian, S.; Fatemieh, M.; Sultan, M.; Mahns, B.; Samiei, S. *Greenhouse Automation Using Wireless Sensors and IoT Instruments Integrated with Artificial Intelligence*; InTech Open: London, UK, 2021.
17. Bheemanahalli, R.; Gajanayake, B.; Lokhande, S.; Singh, K.; Seepaul, R.; Collins, F.; Reddy, K.R. Physiological and pollen-based screening of shrub roses for hot and drought environments. *Sci. Hortic.* **2021**, *282*, 110062. [CrossRef]
18. van Klompenburg, T.; Kassahun, A.; Catal, C. Crop yield prediction using machine learning: A systematic literature review. *Comput. Electron. Agric.* **2020**, *177*, 105709. [CrossRef]
19. Zeng, Y.; Pu, X.; Du, J.; Yang, X.; Li, X.; Mandal, M.; Nabi, S.; Yang, T.; Yang, J. Molecular mechanism of functional ingredients in barley to combat human chronic diseases. *Oxid. Med. Cell. Longev.* **2020**, *2020*, 3836172. [CrossRef]
20. Erazo, M.; Rivas, D.; Pérez, M.; Galarza, O.; Bautista, V.; Huerta, M.; Rojo, J.L. Design and implementation of a wireless sensor network for rose greenhouses monitoring. In Proceedings of the IEEE 2015 6th International Conference on Automation, Robotics and Applications (ICARA), Queenstown, New Zealand, 17–19 February 2015; pp. 256–261.
21. Plaut, Z.; Zieslin, N. Productivity of greenhouse roses following changes in soil moisture and soil air regimes. *Sci. Hortic.* **1974**, *2*, 137–143. [CrossRef]

22. Shamshiri, R.R.; Jones, J.W.; Thorp, K.R.; Ahmad, D.; Man, H.C.; Taheri, S. Review of optimum temperature, humidity, and vapour pressure deficit for microclimate evaluation and control in greenhouse cultivation of tomato: A review. *Int. Agrophys.* **2018**, *32*, 287–302. [CrossRef]
23. Shamshiri, R.R.; Kalantari, F.; Ting, K.; Thorp, K.R.; Hameed, I.A.; Weltzien, C.; Ahmad, D.; Shad, Z.M. Advances in greenhouse automation and controlled environment agriculture: A transition to plant factories and urban agriculture. *Int. J. Agric. Biol. Eng.* **2018**, *11*, 1–22. [CrossRef]
24. Zhu, N.; Liu, X.; Liu, Z.; Hu, K.; Wang, Y.; Tan, J.; Huang, M.; Zhu, Q.; Ji, X.; Jiang, Y. Deep learning for smart agriculture: Concepts, tools, applications, and opportunities. *Int. J. Agric. Biol. Eng.* **2018**, *11*, 32–44. [CrossRef]
25. Zheng, Y.-Y.; Kong, J.-L.; Jin, X.-B.; Wang, X.-Y.; Su, T.-L.; Zuo, M. CropDeep: The crop vision dataset for deep-learning-based classification and detection in precision agriculture. *Sensors* **2019**, *19*, 1058. [CrossRef] [PubMed]
26. Elavarasan, D.; Vincent, P.D. Crop yield prediction using deep reinforcement learning model for sustainable agrarian applications. *IEEE Access* **2020**, *8*, 86886–86901. [CrossRef]
27. Mohan, P.; Patil, K.K. Deep learning based weighted SOM to forecast weather and crop prediction for agriculture application. *Int. J. Intell. Eng. Syst.* **2018**, *11*, 167–176. [CrossRef]
28. Altikat, S. Prediction of CO_2 emission from greenhouse to atmosphere with artificial neural networks and deep learning neural networks. *Int. J. Environ. Sci. Technol.* **2021**, *18*, 3169–3178. [CrossRef]
29. Mekonnen, Y.; Namuduri, S.; Burton, L.; Sarwat, A.; Bhansali, S. Machine learning techniques in wireless sensor network based precision agriculture. *J. Electrochem. Soc.* **2019**, *167*, 037522. [CrossRef]
30. Tay, A.; Lafont, F.; Balmat, J.-F. Forecasting pest risk level in roses greenhouse: Adaptive neuro-fuzzy inference system vs artificial neural networks. *Inf. Process. Agric.* **2020**, *8*, 368–397. [CrossRef]
31. Wilmer Champutiz, P.R.-M.; Fuentes, E.; Peluffo, D. Roses Greenhouse Cultivation Database Repository (RosesGreenhDB). In *IEEE Dataport*; IEEE: New York, NY, USA, 2019.

sustainability

MDPI

Article

Conceptualization of Bioreactor Landfill Approach for Sustainable Waste Management in Karachi, Pakistan

Ihsanullah Sohoo [1,2,*], Marco Ritzkowski [1], Muhammad Sultan [3], Muhammad Farooq [4] and Kerstin Kuchta [1]

1 Circular Resource Engineering and Management (CREM), Institute of Environmental Technology and Energy Economics, Hamburg University of Technology, Blohmstr. 15, 21079 Hamburg, Germany; m.ritzkowski@tuhh.de (M.R.); kuchta@tuhh.de (K.K.)

2 Department of Energy and Environment Engineering, Dawood University of Engineering and Technology, Karachi 74800, Pakistan

3 Department of Agricultural Engineering, Bahauddin Zakariya University, Multan 60800, Pakistan; muhammadsultan@bzu.edu.pk

4 Department of Mechanical Engineering (New Campus-KSK), University of Engineering and Technology, Lahore 54890, Pakistan; engr.farooq@uet.edu.pk

* Correspondence: sohoo.ihsanullah@tuhh.de

Abstract: Finding a sustainable approach for municipal solid waste (MSW) management is becoming paramount. However, as with many urban areas in developing countries, the approach applied to MSW management in Karachi is neither environmentally sustainable nor suitable for public health. Due to adoption of an inefficient waste management system, society is paying intangible costs such as damage to public health and environment quality. In order to minimize the environmental impacts and health issues associated with waste management practices, a sustainable waste management and disposal strategy is required. The aim of this paper is to present a concept for the development of new bioreactor landfills for sustainable waste management in Karachi. Furthermore, this paper contributes to estimation of methane (CH_4) emissions from waste disposal sites by employing the First Order Decay (FOD) Tier 2 model of the Intergovernmental Panel on Climate Change (IPCC) and determining of the biodegradation rate constant (k) value. The design and operational concept of bioreactor landfills is formulated for the study area, including estimation of land requirement, methane production, power generation, and liquid required for recirculation, along with a preliminary sketch of the proposed bioreactor landfill. This study will be helpful for stockholders, policy makers, and researchers in planning, development, and further research for establishment of bioreactor landfill facilities, particularly in the study area as well as more generally in regions with a similar climate and MSW composition.

Keywords: municipal solid waste; sanitary landfill; open dumps; waste to energy; climate change

Citation: Sohoo, I.; Ritzkowski, M.; Sultan, M.; Farooq, M.; Kuchta, K. Conceptualization of Bioreactor Landfill Approach for Sustainable Waste Management in Karachi, Pakistan. *Sustainability* **2022**, *14*, 3364. https://doi.org/10.3390/su14063364

Academic Editor: Ming-Lang Tseng

Received: 7 February 2022
Accepted: 11 March 2022
Published: 13 March 2022

1. Introduction

In order to control environmental impacts and maintain better public health, municipal solid waste (MSW) must be managed in a sustainable way [1]. However, sustainable management of huge amounts of MSW is a challenge, especially in developing countries, due to lack of financial and technical resources, increasing population, economic development, and rapid urbanization [2]. According to the study [3], the financial costs to the public of negligence are five to ten times higher than the economic costs of efficient management of the waste The costs to be paid by society if waste is not managed effectively is a 'cost of negligence' which includes public health costs, the cost of environmental deterioration because of uncollected wastes, uncontrolled dumping, open burning, and inefficient resource recovery, productivity loss, flood damage, loss of business and tourism, and long-term cleanup costs [3].

Worldwide, about 2.01 billion metric tonnes of MSW are generated yearly, and this amount is expected to increase more than two- or even three-fold in lower-income countries by 2050 due to significant economic development and rising populations [4,5]. One study [3] estimated that two billion members of the global population lack access to regular waste collection services, and about three billion people have no access to controlled waste landfilling facilities. Despite many years of rising public awareness, the problem of uncollected waste disposal continues to exist in developing countries [6]. The rate of waste collection strongly depends on the income of the citizens in a country. In high-income countries, the collection rate is close to 100%; however, in lower–middle income and low-income countries the waste collection rate is about 51% and 39%, respectively [5].

However, governments are presently proceeding towards sustainable methods of waste disposal after realizing the environmental risks and economical costs of open waste dumping [7]. In this regard, economic conditions, specific legislation, and the geographical location of a country has a significant influence on the adoption of certain waste disposal approaches [8,9]. Generally, effective MSW management practices involve source separation, door-to-door collection, transportation, storage, separation of organic and inorganic waste (plastics, glass and metals) at the storage point, material recycling, biological treatment (anaerobic digestion and composting) of biodegradable wastes, thermal treatment (incineration) with energy recovery, and final disposal of residual waste residues at landfills [10].

Over the years, approaches to MSW disposal on land have evolved from uncontrolled open dumping to engineered landfill systems [11]. Land disposal of MSW accounted for more than 1.5 billion tonnes of the total 2.01 billion tonnes of waste generated globally in 2016 [5]. The total number of waste disposal sites in operation globally is about 300,000–500,000 [12]. In the recent past, uncontrolled dumping was the main approach to waste disposal used worldwide [13]. However, open dumping remains in practice as the main solid waste disposal method for more than half of the global population [14,15].

According to studies [16,17], the MSW generation rate in Karachi is 15,600 tonnes/day, with 53–60% of this the organic fraction. Typically, organic waste is neglected after sorting of recyclables from waste mixture. It is neither collected by scavengers, nor do the municipal authorities utilise it through compositing, anaerobic digestion, or other treatment [17]. Neither municipal authorities nor private companies are willing to separate organic waste for biological treatment due to the lack of vision and policy to utilize it for energy generation and the absence of a market for compost products [18]. Hence, this mismanagement of waste results in the loss of both a valuable energy-containing resource and leads to environmental and public health issues.

One study [19] estimated that the amount of MSW annually disposed of at dumpsites (2.2 million tonnes) has the potential to emit about 3.9 million tonnes of carbon dioxide equivalent ($MtCO_2$-eq.) emissions. In order to minimize the environmental impacts and health issues associated with open waste disposal in Karachi, a sustainable waste management and disposal facility is required. This study intends to present a concept for the development of bioreactor landfills and sustainable waste management in the city.

This paper contributes to estimation of methane (CH_4) emissions from waste disposal sites, determination of the degradation rate constant (k) value under the prevailing climatic conditions in Karachi and formulation of a design and operational concept for a bioreactor landfill. Additionally, estimations concerning land requirements, methane production, power generation, and liquid required for recirculation in order to maintain the required waste degradation rate in bioreactor landfill conditions are reported in this paper.

2. MSW Landfilling Approaches

2.1. Open Dumps

The open dump method is an elementary level of solid waste disposal, and is identified with the uncontrolled deposition of waste with only limited or without any control measures [20]. Overall, 33% of waste is openly disposed of at dumpsites globally, and in lower income countries (where dumpsites are the leading waste disposal facilities) more than 90% of waste is openly disposed of [5,21]. In Pakistan, 70% of waste generated ends up in dumpsites [17].

The operation of open dumps poses serious threats to the environment and human health [22]. The environmental and public health damage caused by open disposal of waste includes ground and surface water contamination through the generation of leachate, contamination of soil by solid waste or leachate, air pollution due to gaseous emissions, provision of breeding grounds to disease vectors such as mosquitos, flies, and rodents, odour problems, and uncontrolled methane emissions [23,24]. Furthermore, open burning of MSW, commonly practiced in developing countries, leads to the release of harmful contaminants including fine particulates ($PM_{2.5}$), and damages the air quality in urban areas [25].

2.2. Anaerobic Landfills

Anaerobic sanitary landfills are known as well-designed waste disposal facilities which do not require any processes to influence waste degradation [26,27]. However, control measures to minimize environmental and public health effects are incorporated at the site, including a bottom liner and surface top cover as well as leachate and gas treatment (heat/power generation or flaring) facilities [26,27].

The sanitary landfill approach is the most popular waste treatment method due to its high volume handling capacity, low investment, and minimal technical requirements [28]. It has been reported [29] that the biodegradation processes of the organic fraction of municipal solid waste are slower under anaerobic conditions than under aerobic conditions in a landfill. Investigation results from old landfills in Germany and other European countries showed noticeable emission potential from landfills operated under anaerobic conditions, and it is estimated that gaseous emissions can last at least for thirty years, and that leachate emissions can last for many decades or even centuries depending on site-specific conditions [30].

2.3. Semi-Aerobic Landfills

The semi-aerobic is the oldest approach regarding landfill aeration; this method was developed in the early 1970s in Japan and is known as the "Fukuoka method" [9,31]. The semi-aerobic landfill process is driven by a natural air ventilation mechanism which provides a speedy waste stabilization solution through the availability of oxygen in the waste mass without demanding high resources and technology [31]. The semi-aerobic landfill system can be a suitable method for meeting the sustainability requirements cost-effectively and with low technical input, especially in developing countries which are lacking in sustainable waste disposal due to funding issues and technical limitations [32].

A semi-aerobic landfill system consists of a horizontally-installed perforated pipe network with an adequate slope at the bottom of landfill for leachate collection, with perforated pipes erected vertically at intersections and at the end of each branch for air ventilation [9,31]. Furthermore, in a semi aerobic landfill system, air flows through the pipe network by means of a natural advection process due to temperature differences between the landfill body and the ambient environment [9,31]. The temperature difference is a result of exothermic biodegradation of the organic fraction of the waste mass; the release of this heat can raise the temperature in the waste body by 50–70 °C [31].

This temperature difference leads to density differences in the gas inside the landfill, creating a buoyance force which allows the gas to flow up through the waste mass and vent out the vertical gas extraction pipes, developing negative pressure as a result that allows more air to be drawn inside the landfill body through the leachate collection pipes [31,33]. In an aerobic environment, organic matter degrades more effectively than in anaerobic conditions; thus, air circulation through the waste mass results in enhanced waste stabilization and improved emission quality and quantity [31].

A study on full-scale aeration in semi-aerobic landfills by [34] has shown that the relationship between airflow rate and ambient temperature is negatively proportional, as in winter a large flow rate was noticed, while no flow of air was observed in summer. In a semi-aerobic landfill system, anaerobic conditions prevail inside the waste mass due to insufficient air distribution, which promotes methane formation. However, the CO_2 and CH_4 emission ratio of a semi-aerobic landfill (4:1) is much lower than an anaerobic landfill system (1:1) [31].

2.4. Aerated Landfills

In situ aeration is a quite new technology for intensified removal of biodegradable organic material left in old landfills [35]. For aeration of landfills, two approaches are applied; one is forced aeration, while in the second air is supplied in natural conditions. Forced aeration is realized by injection of air into the landfilled waste mass through means of different types of blowers [36]. The major objective of the aerobic in situ aeration is to stabilize and change the emission behavior of organic matter deposited in the landfill [37].

Aerobic degradation processes in landfills enable the significantly faster decomposition of organics (e.g., hydrocarbons) compared with anaerobic processes, resulting in increased carbon discharge in the gas phase and decreased leachate concentration [38,39].

A study by [35] reported that when landfill gas production is decreased to such a level that energy generation is not economically feasible and even flaring of extracted gas is not practical, there will be up to 10–20% residual gas production potential remaining of the total production potential. Moreover, it may take decades to stabilize the remaining organic material in the anaerobic environment; by providing aerobic conditions, the residual organic matter can be degraded in a limited time (<10 years under a conducive environment) [35].

The in situ aeration approach goes beyond the concept of injecting air into the landfill, including a well design and spacing options for the suitable volume and pressure of air, air distribution, temperature, and moisture control as well as pollution discharge in the leachate and gas phases [9]. The major objective of aerobic in situ aeration is to oxidize and change the emission behavior of organic material deposited in landfill, and in the end to significantly reduce the emission potential in a more appropriate way [37].

Aerobic degradation processes in landfills enable the significantly faster decomposition of organics (e.g., hydrocarbon) compared with anaerobic processes; as result, carbon discharge in the gas phase increases and leachate concentration decreases [38,39]. In all, nitrogen elimination is the most significant advantage that can be obtained from aeration technology [40,41]. Several authors [9,42] mention that the aeration of waste material in the landfill body is an essential and unavoidable pretreatment step in the landfill mining process to prevent uncontrolled gaseous emissions from waste during excavation activity. Presently, various approaches and concepts are applied in the aeration of landfills, such as semi-aerobic landfills, high pressure aeration, low pressure aeration (including active aeration with and without off-gas extraction), passive aeration via air venting, and energy self-sufficient landfills [9].

2.5. Bioreactor Landfills

A bioreactor landfill is an engineered and modern shape of a conventional anaerobic/aerobic landfill where moisturization of the waste takes place by injecting water (fresh or wastewater) and recirculating the leachate to optimize waste degradation processes [43–45]. The recirculation of leachate facilitates cycling of microbes and nutrients into the waste mass and maintains an optimal moisture content in the landfilled waste [46]. The cycling of microbes and nutrients is intended to enhance microbial processes for transformation and stabilisation of easily and moderately degradable organic waste fractions, within the timeframe of 5–10 years for bioreactor process execution [47].

Various studies [48–51] have reported the positive effects of moisturization of the waste and leachate recycling during landfill operation, which includes speedy waste biodegradation and stabilization, increasing LFG (methane) production, rapid settlement, reduced leachate quantity, and leachate treatment cost savings. Furthermore, bioreactor landfills and their variations represent a sustainable alternative approach to conventional sanitary (dry tomb) landfills [52]. However, bioreactors can have drawbacks, e.g., odours and physical instability of the waste material due to increased moisture [53].

Moreover, establishment of infrastructure for leachate recirculation and/or aeration may cause increased capital and operational costs [53]. Studies have suggested that the high upfront costs involved in operation and construction of bioreactor landfills can be balanced by future economic benefits, including an increase in the active life of the landfill (waste disposal period), more efficient use of airspace [54], lower minimum leachate treatment/disposal costs, delay in the need to construct a new cell and cap, savings in the post-closure care period thanks to less need for monitoring and lower financial guarantee obligations, and higher efficiency in landfill gas collection, resulting in larger revenues generated from production [55].

According to [53], the bioreactor approach can be applied when the waste to be deposited possesses a high quantity of biodegradable organics. Bioreactor landfills can be designed as anaerobic, aerobic, semi-aerobic, and hybrid landfills [36,56]. The basic differences between these designs of bioreactor landfills are linked with their operations, layouts, and arrangements for leachate recirculation, landfill gas collection, and (optional) air injection system [45]. Bioreactor landfills are mostly operated under anaerobic conditions [57,58]. In a hybrid bioreactor landfill, a series of aerobic and anaerobic conditions are observed [53,59]. The aeration of the bioreactor landfill is realized through injection of air/oxygen to establish an environment for aerobic biodegradation of the landfilled waste in order to control methane emissions and accelerate waste stabilization [60].

However, hindrances in oxygen distribution in the waste mass due to high moisture content and leachate recirculation have been reported by various research studies [61–63]. Moreover, other studies [64,65] have stated that degradation of waste is significantly influenced by the rate of oxygen distribution. The pros and cons associated with the different waste disposal approaches discussed in the above sections are summarized in Table 1.

Table 1. Summary of pros and cons of different landfill approaches.

Landfilling Approach	Pros	Cons	Reference
Open disposal	No or low cost is involved in the short-term. Income source for waste scavengers.	Long-term environmental costs such as uncontrolled emissions of toxic gases due to open decomposition of waste, ground water contamination, and soil contamination due to toxic and concentrated leachate release. Public health problems.	[66]
Anaerobic landfills	LFG with high methane concentration can be used as an energy source. Relatively low cost is involved in the short term.	High COD, BOD_5 and VFA concentrations in leachate. High level of ammonia in leachate. Formation of hydrogen sulphide (H_2S) gas from the decomposition of gypsum wall board in waste. Long duration in waste stabilization. Long term LFG (methane) emissions.	[59,67]
Semi-aerobic landfills	Promotes waste and leachate stabilization Reduced biological stabilization time of landfilled waste. In situ leachate treatment. Low-cost system.	Careful management and operation needed for optimal performance	[33,59,68]
Aerobic landfills	Speedy waste stabilization. No or low methane production with reduced GHG emissions. Low or no residual methane emissions. In situ leachate treatment. Moisture removal by air stripping. Nitrogen removal. Better waste settlement.	High energy demand.	[35,59,68]

3. Methods and Data

3.1. Estimation of Methane Emissions from Waste Disposal Sites in Karachi

The estimations of methane emissions from waste disposal sites in Karachi provided here are based on the LFG production model by Tabasaran and Rettenberger, (1987) [69] as given in Equation (1). This model is considered a simple method for prognosis of methane from waste disposal sites, and depicts the anaerobic degradation of degradable organic carbon (*DOC*) as in the first-order decay (FOD) Tier 2 model of the IPCC [70,71]. This model is used by various studies to estimate landfill gas production rates, such as [37,70]:

$$G_t = 1.868C_{org}(0.014T + 0.28)\left(1 - e^{-kt}\right) \quad (1)$$

where G_t is the LFG production during a specific time, t (m^3/tonne fresh waste); C_{org} is total organic carbon in waste (kg/tonne); T is the temperature (35 °C); k is the degradation rate constant, ($k = \ln 2/T_{0.5}$); and t is the landfill operation time (years).

The C_{org} was determined by considering the degradable organic content (*DOC*) according to the organic fraction of MSW in Karachi (as reported by [72–74]), and is provided in Table 2. The degradable organic content (*DOC*) of MSW used in this study was determined using Equation (2), as per the Intergovernmental Panel on Climate Change (IPCC), 2001 [75]:

$$DOC = (0.4 \times A) + (0.2 \times B) + (0.15 \times C) + (0.43 \times D) + (0.24 \times E) + (0.24 \times F) \quad (2)$$

where *A*, *B*, *C*, *D*, *E*, and *F* represent the fractions of paper, green waste, food waste, wood, textile, and nappies, respectively, present in MSW generated in Karachi, as shown in Table 3.

Table 2. Composition of MSW generation in Karachi.

Waste Component	FW	GW	Paper	Glass	Metal	Plastic	Fines	Nappies	Textile	TP	Wood
Fraction in sample [% *w*/*w*]	26.10	17.04	7.97	5.6	1.1	8	3.7	9.8	5.57	10	3.11

Table 3. Determination of *DOC* in the synthetic waste sample using IPCC default values.

Waste Components	%	*DOC* Default Value	*DOC* %
Paper (A)	7.97	0.4	3.2
Green waste (B)	17.04	0.2	3.4
Food (C)	26.10	0.15	3.9
Wood (D)	3.11	0.43	1.3
Nappies (E)	9.8	0.24	2.4
Textile (F)	5.57	0.24	1.3
Total			15.5

For selection of the degradation rate constant (k) value of waste disposed at dumpsites in Karachi, three different k values were analysed. In the first, an average of half-lives of easily (four years), moderately (nine years), and hardly (twenty years) degradable wastes were considered. The k value determined in this approach was 0.095/year.

In the second, the default k value 0.05 suggested by IPCC 2000 [76] was applied, and the third k value for conventional landfills reported in the literature [43,77], 0.04, was used to model the landfill gas emissions. The data utilized for the estimation of landfill gas emissions from waste disposal sites in Karachi are provided in Table 4.

Table 4. Data used for estimation of methane emissions from waste disposal sites in Karachi.

Data	Unit	Value	Reference
MSW generation	[tonnes/day]	15,600	[17]
MSW landfilled	[%]	70	[17]
MSW landfilled-FM	[tonnes/day]	10,920	
MSW landfilled-FM	[million-tonnes/year]	4	
Density of methane	[kg/m^3]	0.66	[78]
Methane fraction in LFG	[%] average	50	
Global warming potential of methane (over 100 years horizon)	[CO_2-eq]	25	[79,80]
Total DOC in the waste	kg/tonne FM	155	
Default k value for waste disposal sites		0.05	[76]

3.2. Estimation of Land Requirement for Bioreactor Landfill

The landfill requirements for bioreactor development were estimated using Equation (3), as reported by previous studies [81,82]:

$$Total\ required\ disposal\ area = \left[\left(\frac{Waste\ quantity\ (t)}{waste\ density\ (\frac{t}{m3})}\right)/landfill\ height\ (m)\right] \quad (3)$$

3.3. Estimation of Power Generation from Bioreactor Landfill

The electric power generation from recovered methane during anaerobic operation of a bioreactor landfill was estimated using Equation (4), as reported by [78,83]:

$$P_e = \Psi \times f_{methane} \times \rho \times \omega \times \frac{1\ \text{kWh}}{3.6\ Mj} \times \eta_e. \quad (4)$$

where $P_{el.}$ is the electrical power generated (kWh), Ψ is landfill gas collection rate (m^3/h), $f_{metahne}$ is the methane fraction in landfill gas (%), ρ is the density of methane (0.66 kg/m^3), ω is the calorific value of methane (55.53 MJ/kg), and $\eta_{el.}$ is the electrical efficiency of the gas engine (%).

3.4. Determination of k Value for Waste Degradation in Karachi

The k value is the biodecomposition half-life value in a year (year^{-1}) for landfilled waste, and is influenced by waste depth, density, pH, and other environmental conditions [77,84]. Several authors have [84–86] reported precipitation as the most significant parameter in the estimation of k value, because a higher moisture content results in faster biodegradation of waste. Thus, for the estimation of k value considering the local precipitation regime, the following Equation (5) provided by [87] and reported by [84] can be used

$$k = (3.2 \times 10^{-5} \times annual\ precipitation\ in\ mm) - 0.01 \quad (5)$$

3.5. Estimation of Liquid Required for Bioreactor Landfill

The degradation rate constant (k) value (0.3/year) considered for the proposed bioreactor landfill was taken from the literature [43,77] and is shown in Table 5. In the case of a bioreactor landfill where additional liquids are introduced into the landfill, the amount of additional liquid should be determined and added to the amount of precipitation, as suggested by Alberta Environment [87]. In this case, the equation for k value would be

$$k = 3.2 \times 10^{-5} \times (AP + AL) + 0.01 \quad (6)$$

where AP is the annual precipitation rate in mm and AL is the amount of additional liquid required.

Table 5. Data used for estimation of designing a bioreactor landfill for Karachi.

Parameter	Value	Unit	Reference
Waste tipping	3700	[tonnes/day]	
Waste compaction	0.8	[tonnes/m^3]	
Landfill height	30	[meters]	
Total *DOC* in the waste	155	kg/tonne FM	
DOC loss in pre-treatment	10	[%]	
DOC in the waste disposed in bioreactor landfill	139.9	kg/tonne FM	
Landfill gas collection efficiency	50	[%]	[88]
k value for bioreactor landfill	0.3		[43,77]
Density of methane	0.66	[kg/m^3]	
Methane fraction in landfill gas	64	[%]	Average CH_4 concentration in LFG in simulating bioreactor landfill conditions in Karachi [19]
1 kWh	3.6	[MJ]	
Electric efficiency (η)	30	[%]	[78]
Calorific value of methane	55.53	[MJ/tonne]	
LHV of methane	36.48	[kJ/m^3]	

4. Proposal for Development of Bioreactor Landfills in Karachi

The conventional sanitary waste landfill method (dry tomb) is not a long-term sustainable solution and has negative impacts on the environment and urban sustainability [52,89]. According to one study [11], two major obstacles are associated with conventional (dry tomb) sanitary landfills; the first is slow gas production, and the second is that the use of low-permeability daily/intermediate cover layers hinders the free flow of gas during extraction. Hence, conventional sanitary landfills are not compatible for the landfill gas recovery and utilization approach and only serve as places for perpetual storage of waste, occupying valuable land resources [46].

In the development of new landfill sites in Karachi, a hybrid form of the bioreactor landfill approach can be applied for rapid gas production and waste stabilization. This approach can be more environmentally sustainable when bioreactor landfill facilities are planned with aftercare measures (in situ aeration) taken into account and followed by a decrease in landfill gas production rate. Furthermore, in [56] the hybrid bioreactor concept is demonstrated to be an efficient technique for enhancing methane production and achieving landfill completion in a 25–35% shorter time compared to traditional (dry tomb) anaerobic landfill systems. Furthermore, the results of a study [19] conducted by simulating bioreactor landfill conditions in the situation of Karachi (MSW composition and climatic conditions) reported that a bioreactor landfill with post-aeration (a hybrid bioreactor) showed accelerated methane production higher than that of a conventional sanitary landfill.

In this context, the present paper proposes a more advanced and environmentally sustainable solid waste landfill approach, a hybrid bioreactor landfill for future landfills development in the city. Under this approach, the waste placed in the landfill would be subjected to aerobic oxidization by means of in situ aeration after completion of an anaerobic phase when the landfill gas production will be significantly reduced. The post-aeration phase is intended to accelerate the degradation of the remaining hardly-degradable organic material and shorten the aftercare period of the landfill, as reported by various authors [35,36].

The bioreactor landfill approach without the aftercare option is only better than existing waste disposal sites in Karachi regarding its environmental performance and landfill gas generation. Under this approach there will be significant risks to the environment, such as long-term residual gas emissions even after power generation from landfill gas can no longer be economically feasible. As the MSW generated in Karachi contains a high organic fraction and as most of the recyclable material in the solid waste is collected by waste pickers (or can be systematically collected through the establishment of material recovery facilities), leftover organic material can be valorised by methane production through application of the bioreactor landfill approach. Later, when the gas production reaches minimal levels, the landfill could be aerated.

Karachi has two major official solid waste landfill sites, known as Jam Chakro (N = 25°01.675′, E = 67°01.61′) and Gond Pass (N = 25°00.634′, E = 66°55.263′), located north-west and west of the city, respectively [19,90]. Presently, there is no landfill for the disposal of solid waste generated in the eastern parts of the city. The absence of an official designated waste disposal site on the eastern side of the city leads to mismanagement of waste, and provides reasons to the public for open disposal of the waste on street sides, vacant plots, drainage channels, and in the Malir river.

Furthermore, transportation of solid waste from the eastern side to officially designated landfill sites located on the northern side of the city has high costs in terms of both fuel consumption and time. The Sindh Solid Waste Management Board (SSWMB) is planning to establish a new sanitary landfill to serve the waste disposal needs of the eastern parts of the city, for which 3000 acres (1214 hectares) of land have been allocated at the Dhabeji site (N = 24°48.804′, E = 67°30.567′) [91]. The locations of Jam Chakro and Gond Pass landfill sites and the future landfill site at Dhabeji are shown in Figure 1.

Figure 1. Locations of Jam Chakro, Gond Pass, and Dhabeji landfill sites in Karachi (Google maps).

According to SSWMB, both the Jam Chakro and Gond Pass waste disposal sites are operated as controlled landfills, and sanitary landfills are under planning for future waste disposal [92].The proposed concept for transformation of waste disposal strategies from open dumps to sustainable waste disposal in Karachi is illustrated in Figure 2.

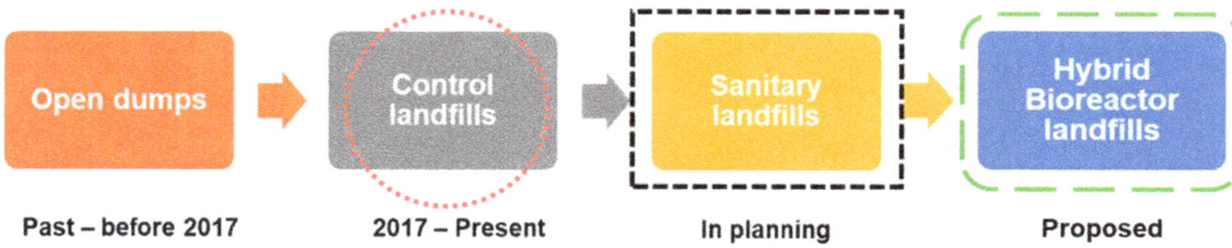

Figure 2. Evolution of solid waste landfilling approach in Karachi.

The following sections provide details regarding the proposed approach.

4.1. Estimation of Methane Emissions from Waste Disposal Sites in Karachi

The methane emission potential is estimated using Equation (1), with three different values of the degradation rate constant (k) considered: 0.095/year, 0.05/year, and 0.04/year. The results with all three values of the degradation rate constant (k) were similar, ranging from 438.6 to 446.9 m^3/tonne fresh waste (FW) for the estimation of landfill gas over 100 years. For the modelling of methane emissions from waste disposal sites in Karachi, the middle value of the degradation constant (k) 0.05 suggested by the IPCC (2000) [76] was used. The comparison of landfill gas emissions with three different values of the degradation rate constant (k) for LFG emissions over 100 years is shown in Figure 3.

A significant amount of solid waste generated in Karachi is openly burned, either at community bins in the city or at the landfill sites [93,94]. Therefore, as a result of burning, the biodegradable fraction (*DOC*) in the waste can be significantly reduced.

Therefore, the estimations of methane production for waste disposed at landfill sites in Karachi are made at four different *DOC* ranges here: first, considering theoretical (100%) *DOC* in solid waste with no loss of *DOC*; second, at 75% *DOC* (25% *DOC* loss); third at 50% *DOC*; and fourth at 25% *DOC* (75% *DOC* loss). The estimated cumulative methane production from waste disposal sites over 100 years at different *DOC* fractions in the solid waste disposed of is shown in Figure 4.

Considering the latest waste disposal quantity and data in Table 4, the theoretical global warming potential (GWP) over time of 100 years for the solid waste annually disposed (about 4 million tonnes/year) at dumpsites in Karachi is estimated as 7.3 $MtCO_2$-eq., with a specific GWP of 1.83 tCO_2-eq/t fresh mass (FM). Furthermore, at the *DOC* levels of 75%, 50% and 25%, the GWP of the waste quantity disposed annually at dumpsites is estimated as 5.5 $MtCO_2$-eq (1.4 tCO_2-eq/t FM), 3.7 $MtCO_2$-eq (0.9 tCO_2-eq/t FM), and 1.8 $MtCO_2$-eq (0.5 tCO_2-eq/t FM), respectively, over a time of 100 years.

According to the results obtained from this study and modelling of the methane emission potential of landfilled solid waste, it is evident that the existing dumpsites in Karachi are causing significant GHG emissions. These waste disposal sites can be transformed into sanitary landfill facilities and sources of renewable energy generation by extracting methane-rich landfill gas. Later, captured methane can be utilized for power generation, transportation, and industrial purposes. After reaching a point where power generation from produced landfill gas is no longer economically feasible, the waste placed in landfills can be rapidly stabilized by employing in situ aeration as a landfill aftercare approach. Given this idea, a sustainable approach is proposed for the development of new landfills in Karachi.

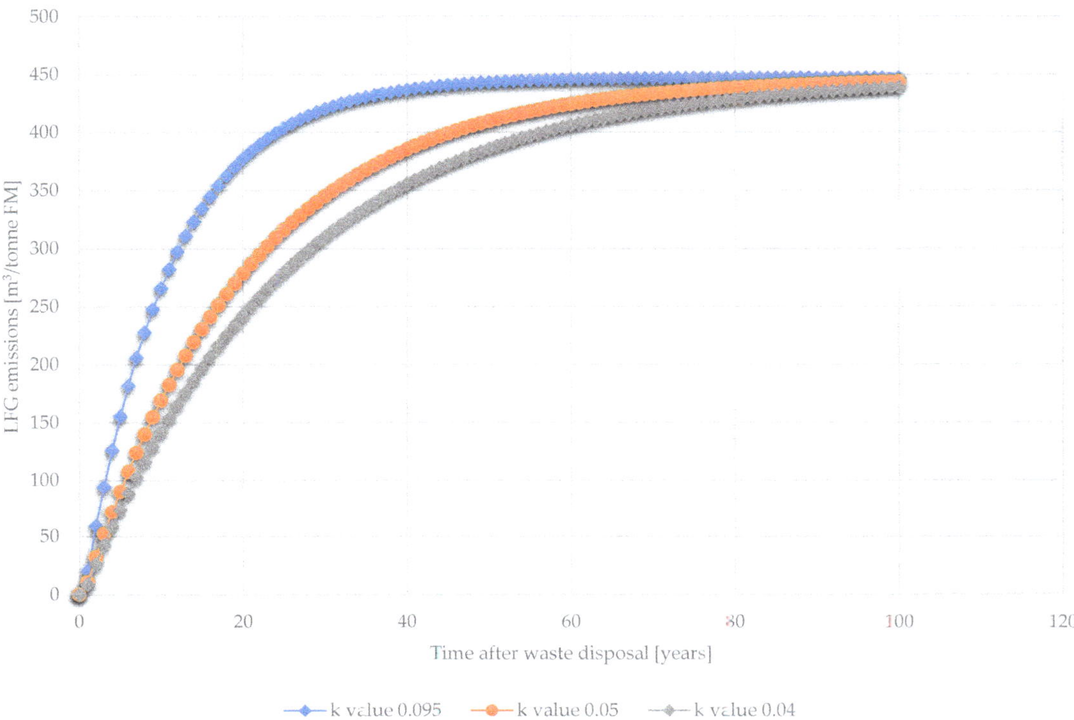

Figure 3. Comparison of different degradation constant (k) values for LFG emissions.

4.2. Bioreactor Landfill Operations

The operation of solid waste tipping in a new bioreactor landfill is proposed in ten phases, and the duration of each phase is assumed to be one year. After ten years of waste deposition, the whole landfill could be closed and capped by providing a final cover. Furthermore, in order to minimize leachate generation and initial operational costs the waste footprint is divided into cells, as recommended by [95].

Hence, the placement of the solid waste at the landfill site is planned by tipping in small daily cells of one larger cell of landfill and providing a daily cover, as recommended by the authors of [96]. The proposed operational layout of the new bioreactor landfill for Karachi is presented in Figure 5.

Solid waste arriving at the landfill would be weighed first at the entrance of the site, and material would be collected at the material recovery and treatment facility where the recyclable waste fraction would be separated from the organic and non-recyclable fractions of the waste. Furthermore, it is proposed that before tipping into the daily cell, waste material should be pre-treated by means of shredding and in situ aeration to reduce readily degradable organics and enhance landfill gas production, as recommended by the authors of [36,97]. A study by Ali et al. [98] recommended at least 27% reduction of volatile solids (VS) during aerobic pre-treatment in order to realize an early start to methanogenesis and increase LFG generation in the anaerobic phase.

Later, the anaerobic phase of landfill operation would be initiated to establish favourable conditions for methane production, with a landfill gas (LFG) capture and utilisation approach for power generation. At the point in time when the rate of LFG production would significantly decrease and the methane recovered would not be economically/technically feasible to utilise for power generation, the aftercare phase (in situ post-aeration) would be started for accelerated biodegradation of the remaining (mostly hardly-degradable) organics in the waste, as proposed by various studies [39,64,99]. The concept of transforming waste disposal strategies from open dumps to sustainable waste disposal is illustrated in Figure 6.

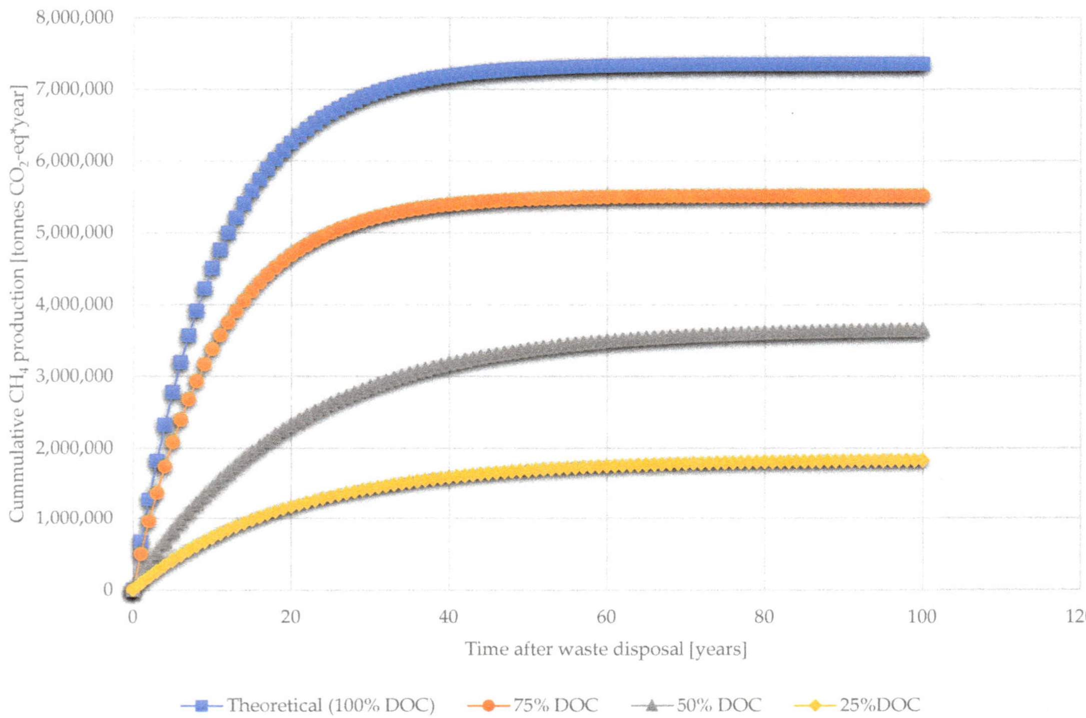

Figure 4. Estimated cumulative methane emissions from waste disposal sites in Karachi.

4.3. Estimation of Land Requirement for Bioreactor Landfill

It is assumed that the solid waste will be collected in closed community bins without initial segregation of recyclables by waste pickers and directly transported to the landfill site. Taking the total quantity of solid waste coming to the landfill site as 5000 tonnes/day, 25% (about 1300 tonnes/day) of recyclable material will be collected by establishing a material recovery facility at the landfill site, and only the organic fraction of MSW will be deposited in the daily cell of the landfill. Furthermore, it is assumed that each daily cell will receive about 3700 tonnes of solid waste on daily basis. Hence, each phase will be completed and covered after one year of waste tipping with a total capacity of 1.3 million tonnes. Overall, it is assumed that 13.5 million tonnes of waste would be accepted at the landfill facility.

The waste height for a sanitary landfill ranges between 15 and 30 meters (m) [82]. A similar study assumed a waste height of 22 m for the determination of the required landfill area [82]. However, if the waste height decreases, the area required for waste disposal increases [82]. In this proposal, the mean height of waste in the bioreactor landfill is assumed as 30 m, excluding intermediate and final covers. According to one study [52], the compaction density of the waste achieved by moderate compaction may range from 0.5–0.85 tonnes/m^3. This study assumed the specific density of waste to be placed in the landfill as 0.8 tonnes/m^3, the commonly-used value for compacted waste in sanitary landfills [32,82]. The land required for landfill construction was determined using Equation (3).

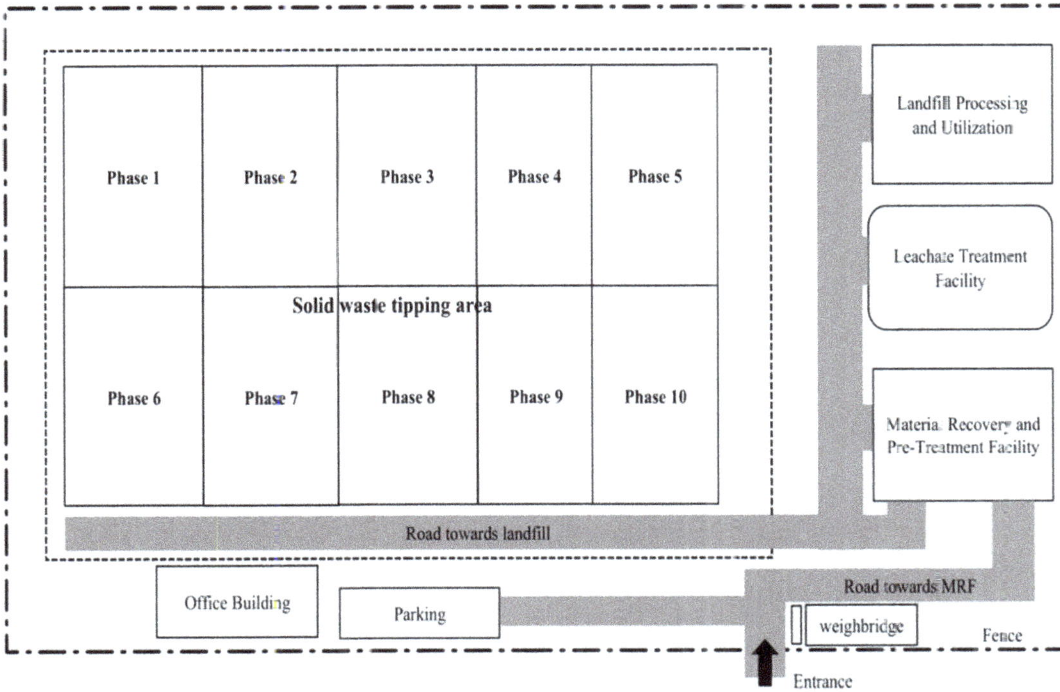

Figure 5. Proposed operational layout of new bioreactor landfill for Karachi.

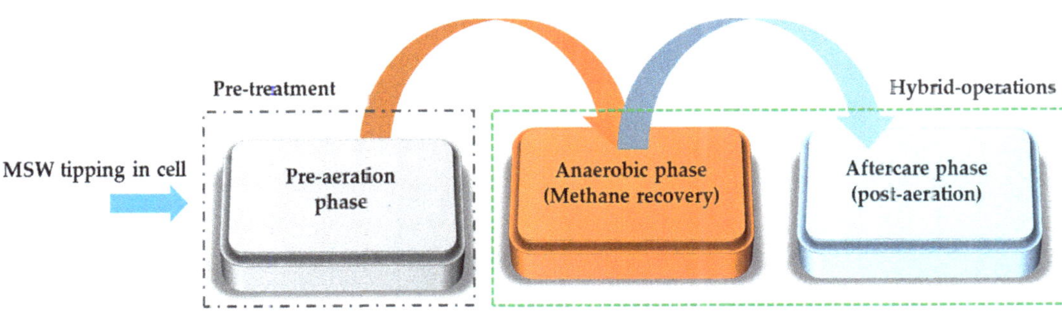

Figure 6. Proposed sequence of bioreactor landfill operations.

By assuming the total waste deposition capacity of a phase as 1.35 million tonnes of waste, it can be determined that an area of 5.6 ha (56,271 m^2) will be required for construction of each landfill phase, excluding the area required by daily and final covers and other landfill facilities. This area should be increased by 10–15% for the placement of daily and final covers [82]. An additional 40–50% area will be required for other facilities such as a receiving area, treatment facilities, and administration buildings [32,82]. The land area required by cover material (daily and final covers) was determined to be 0.84 ha (8440 m^2) by assuming 15% of the waste landfilling area. Hence, the total area required for waste placement is 6.5 ha for each phase of the landfill. Overall, 65 ha of land area will be required for tipping the waste during the landfill's operational life.

Similarly, the area required by the establishment of other facilities (leachate treatment, LFG process and utilization, road construction, office buildings, etc.) at the landfill site was determined to be 19.4 ha by assuming 30% of the total area required for waste placement and cover material. The total land area required to establish a new bioreactor landfill in Karachi, including waste tipping, cover material, and development of other facilities, is estimated to be 84 ha.

4.4. Estimation of Methane Production and Power Generation from Bioreactor Landfill

As discussed above in Section 3.2, landfill operation is divided into ten phases; therefore, methane production and power generation are estimated from the total amount of waste assumed to be disposed during the landfill during its life (ten years). The estimation of methane production from the biodegradation of the organic fraction of the waste in the landfill was calculated based on the LFG production model using Equation (1). The data considered for modelling methane production in the bioreactor, such as the k value, LFG collection rate, methane fraction in LFG, etc., are provided in Table 5. Landfill gas (methane) production from landfilled waste is modelled by considering a 10% loss in initial *DOC* content of 155 kg/tonne FM solid waste material during the pre-treatment phase.

The efficiency of the landfill gas collection system ranges from 13%–80%, with an average of 50% [88]. In this study, landfill gas collection efficiency is considered to be 50%, as shown in Table 5. Based on the modelled methane production and collection rate, the anaerobic phase is supposed to be prolonged until 23 years pass due to a significant reduction in LFG (methane) recovery rate, reaching about 52 m^3/h, as shown in Figure 6. The estimated methane recovery from bioreactor landfill starts from 2572 m^3/h in the first year of anaerobic operation and reaches a maximum rate of 9429 m^3/h in ten years of waste disposal. After closure of landfill, the methane recovery rate gradually decreases to 52 m^3/h thirteen years after closure (23 years of landfill anaerobic operation). The prognosis of the methane recovery rate during the anaerobic phase is illustrated in Figure 7.

The electric power generation potential of a bioreactor landfill could range from 7.8 MW to a maximum of 28.7 MW during the disposal period, and would be reduced to 0.16 MW until the 23rd year of landfill anaerobic operation. The estimated power generation from a bioreactor during the anaerobic operation period is provided in Figure 8.

Moreover, through estimating the specific global warming potential of fresh MSW disposed at landfill sites in Karachi, the reduction in global warming potential by waste deposition at each phase of bioreactor landfill operation is estimated as being in the range of 2.5 $MtCO_2$-eq to 0.6 $MtCO_2$-eq (with different *DOC* levels, 100% to 25%, in solid waste) through methane collection and sustainable utilisation via power generation or flaring. Overall, approximately 25 $MtCO_2$-eq to 6 $MtCO_2$-eq of methane emissions can be controlled by total waste deposition during the ten year period of bioreactor landfill operation.

4.5. Determination of k Value for Waste Degradation in Karachi

Various researchers have found that the k value increases with higher moisture content and higher temperature [84,100,101]. The degradation rate constant (k) value for the biodegradation of the organic fraction of MSW under the climatic conditions (annual rainfall) of Karachi is determined by understanding the waste decomposition dynamics using Equation (5) and considering the total annual rainfall, 176 mm, as reported by [19,102].

It can be determined that with the moisture received through rainfall, the k value for waste degradation at landfill sites in Karachi is 0.016/year. This lower k value is due to low annual rainfall rates in Karachi. However, a study by authors Amini et al. [77] reported a k value of 0.1/year for wet cells and 0.08/year for a traditional landfill due to fact that the study was carried out on a landfill located in Florida, which has relatively high annual rainfall rates, therefore resulting in a higher k value.

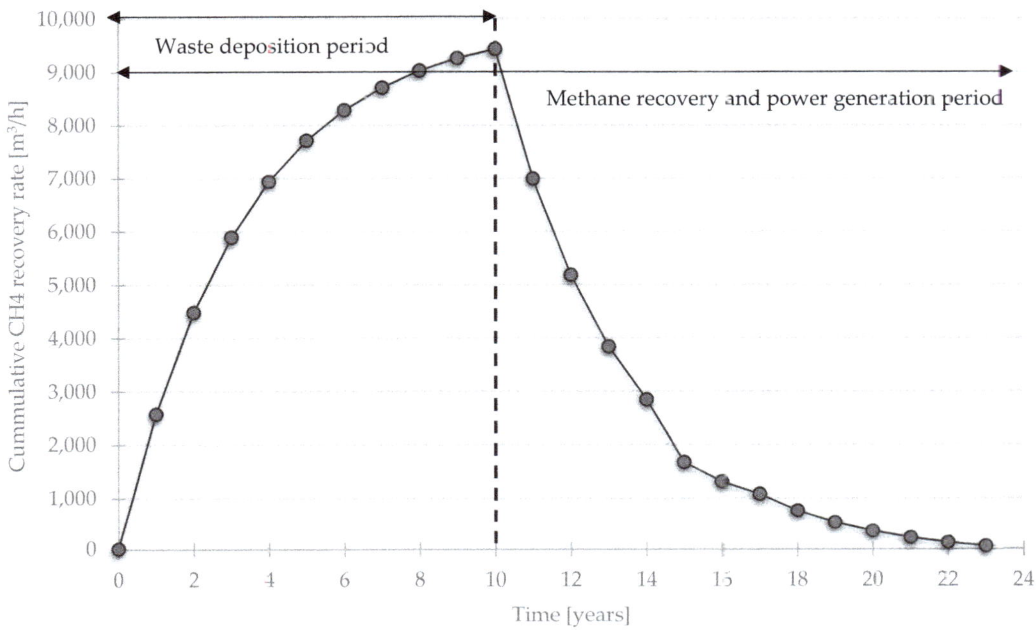

Figure 7. Prognosis of centralised methane recovery rate from bioreactor landfill.

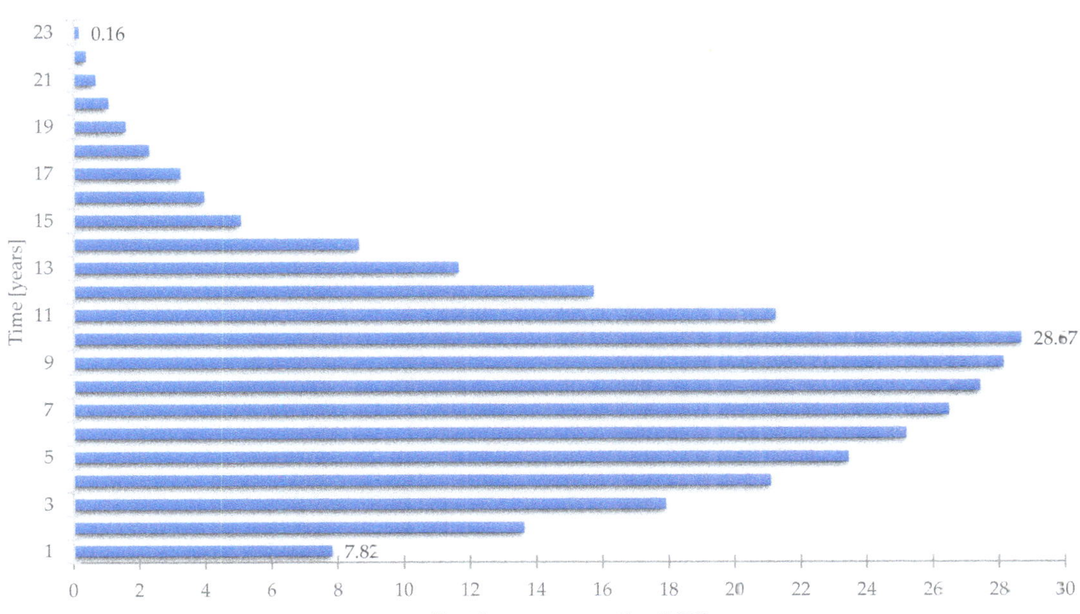

Figure 8. Estimated electric power generation during anaerobic operation of bioreactor landfill.

4.6. Estimation of Liquid Requirement for Bioreactor Landfill

The sources of liquid addition in a landfill may include storm water, groundwater, infiltrating rainfall, or leachate [47]. In the case of a lower annual precipitation rate, as in Karachi, additional liquid would be required to introduce moisture and leachate generation in the bioreactor landfill. Later, leachate can be collected and reintroduced into the landfill after pre-treatment (nitrification).

As the k value for the bioreactor landfill (0.3/year) and precipitation rate in Karachi (176 mm) are known, the additional liquid required to maintain the considered k value in the landfill can be estimated using Equation (6). Similarly, the additional liquid required for a bioreactor landfill has previously been estimated in [103]. By taking their recommendations and integrating Equation (6), it is estimated that 9063 mm of liquid will be required annually in order to maintain a k value of 0.3/year in the bioreactor landfill considering the local precipitation regime in Karachi.

Hence, in order to maintain the considered degradation rate, it is determined that 434.2 L/tonne waste liquid will be required for the daily amount of waste disposed in a cell of the bioreactor landfill at a rate of 67 m^3/h. Furthermore, it is determined that given the annual rainfall rate of the city, only 31 m^3/day water can be expected to be available as run-off from the area allocated to every phase of landfill. However, in the monsoon season (June-September), the run-off collection rate could increase to 77 m^3/day due to the higher precipitation rate in that period. Overall, due to the low annual rainfall rate in the city, almost all (98%) of the required liquid will have to be supplied.

4.7. Design Componetnts of Bioreactor Landfill

To operate a bioreactor landfill effectively, careful construction and operation of infrastructure is required beyond what is necessary in a conventional landfill [55]. The major infrastructure for an engineered landfill facility, includes bottom liner, daily covers, top cover, landfill leachate and gas collection system, embankments, berms, and monitoring systems, and the service life of this infrastructure is assumed to be up to 100 years [12]. Moreover, the design components of a bioreactor landfill include a leachate recirculation system, air injection system, intermediate covers, and final cap [47,104,105]. The leachate recirculation includes the collection of leachate from the bottom of the landfill cell for pumping back into the landfill waste mass [52]. The leachate recirculation system may consist of horizontal distribution pipes/trenches at different depths inside the landfill cell [52].

The landfill gas collection infrastructure consists of gas extraction wells, including a transmission pipe network and condensate knockout system [104]. The key component of the landfill gas collection system is a horizontal pipe network installed during the placement of the waste [104]. However, according to [106] the horizontal gas extraction pipe network is vulnerable to damage by overburdened pressure from the waste, and is easily clogged by leachate components. The vertical extraction wells are most commonly used; while these are easy to install and operate, they are mostly installed after the closure of a landfill cell [52].

Alternatively, during the aerobic operation phase an existing LFG extraction system can be utilized for air injection landfill waste mass [89]. The most commonly used bottom liner system in bioreactor landfills is the composite liner system, which includes a compacted clay liner (CCL) and a flexible membrane liner (FML) [52]. The preliminary design concept of the bioreactor landfill development in Karachi is presented in Figure 9.

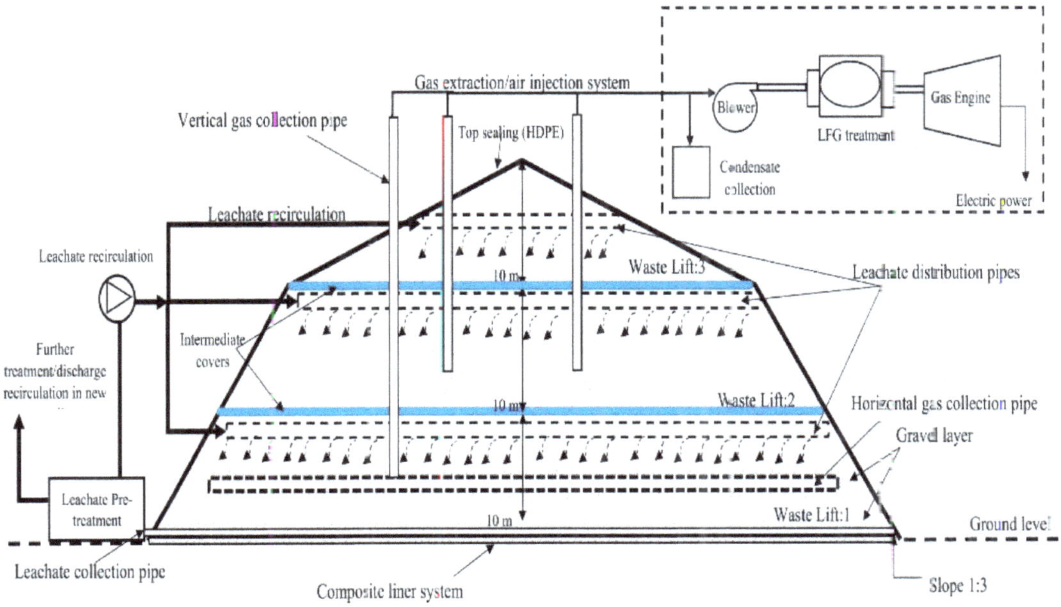

Figure 9. The preliminary design concept of the bioreactor landfill proposed for Karachi (adopted from [107]).

5. Conclusions and Recommendations

Population growth and increasing commercial activities are cumulatively increasing the amount of waste generated in Karachi. Additionally, the waste disposal sites in the city are reaching their saturation point, causing a continual degradation of environment and public health. Therefore, there is an immediate necessity for development of new sanitary landfills for sustainable disposal of a huge amount of waste while minimizing the negative implications associated with its uncontrolled disposal.

Both the former (City District Government of Karachi, CDGK) and present (SSWMB) authorities responsible for solid waste management have been planning the development of a new sanitary landfill to serve the solid waste disposal needs of the eastern side of the city beginning in 2007; however, the planning is at the very initial stage. This delay in the execution of the planned landfill project can be associated with various political, administrative, technical, and financial reasons.

Considering the recent progress in sanitary landfill development for Karachi from SSWMB, this study proposes the approach of a hybrid bioreactor with post-aeration for aftercare for the development of new sustainable landfills in the city based on the concept of energy recovery from municipal solid waste. All estimations made here (such as the quantity and organic fraction of MSW arriving at the landfill, land requirements, timing of each operation phase, methane production and power generation, etc.) as well as the proposed design for the development of the bioreactor landfill in the study area are based on carefully considered assumptions and an extensive literature survey related to the concept. In conclusion, a comprehensive feasibility study shall be conducted by developing a pilot-scale bioreactor on the proposed landfill sites in Karachi to confirm the assumptions taken in this study.

Furthermore, in order to improve the solid waste management situation and optimize the GHG mitigation potential of landfills, this study recommends the adoption of an integrated solid waste management approach in Karachi, with full financial, legal administrative and institutional support. The valorisation of the organic fraction of MSW generated in Karachi should be enhanced through separate collection and utilization for energy generation.

Author Contributions: Conceptualization I.S. and M.R.; methodology, I.S., M.R., M.S. and M.F.; formal analysis, I.S., M.S. and M.F.; investigation, I.S., M.S. and M.F.; data curation, I.S., M.S. and M.F.; writing—original draft, I.S., M.S. and M.F.; writing—review and editing, I.S., M.R., M.S. and M.F.; validation, I.S., M.R., M.S. and M.F.; supervision, M.R. and K.K.; visualization, I.S., M.R., M.S., M.F. and K.K.; project administration, I.S., M.R. and K.K.; resources, K.K.; supervision, M.R. and K.K.; funding acquisition K.K. All authors have read and agreed to the published version of the manuscript.

Funding: The PhD study of Ihsanullah Sohoo (I.S.) was funded by a scholarship from the Higher Education Commission–HEC, Pakistan incorporation with German Academic Exchange Service–DAAD, Germany. Publishing fees funded by the Deutsche Forschungsgemeinschaft (DFG, German Research Foundation)—project number 491268466 and the Hamburg University of Technology (TUHH) in the funding program *Open Access Publishing*.

Institutional Review Board Statement: Not applicable.

Informed Consent Statement: Not applicable.

Data Availability Statement: Not applicable.

Acknowledgments: Publishing fees funded by the Deutsche Forschungsgemeinschaft (DFG, German Research Foundation)—project number 491268466 and the Hamburg University of Technology (TUHH) in the funding program *Open Access Publishing*.

Conflicts of Interest: The authors declare no conflict of interest.

References

1. Liu, A.; Ren, F.; Lin, W.Y.; Wang, J.-Y. A review of municipal solid waste environmental standards with a focus on incinerator residues. *Int. J. Sustain. Built Environ.* **2015**, *4*, 165–188. [CrossRef]
2. Zia, U.U.R.; Rashid, T.U.; Ali, M.; Awan, W.N. Techno-economic assessment of energy generation through municipal solid waste: A case study for small/medium size districts in Pakistan. *Waste Dispos. Sustain. Energy* **2020**, *2*, 337–350. [CrossRef]
3. Wilson, D.C.; Velis, C.A. Waste management—Still a global challenge in the 21st century: An evidence-based call for action. *Waste Manag. Res.* **2015**, *33*, 1049–1051. [CrossRef] [PubMed]
4. Gautam, M.; Agrawal, M. Greenhouse Gas Emissions from Municipal Solid Waste Management: A Review of Global Scenario. Available online: https://link.springer.com/chapter/10.1007/978-981-15-9577-6_5 (accessed on 6 February 2022).
5. Kaza, S.; Yao, L.; Bhada-Tata, P.; Van Woerden, F. *What a Waste 2.0: A Global Snapshot of Solid Waste Management to 2050*; World Bank: Washington, DC, USA, 2018.
6. Reinhart, D.; Bolyard, S.C.; Berge, N. Grand Challenges—Management of municipal solid waste. *Waste Manag.* **2016**, *49*, 1–2. [CrossRef] [PubMed]
7. Das, S.; Lee, S.H.; Kumar, P.; Kim, K.H.; Lee, S.S.; Bhattacharya, S.S. Solid waste management: Scope and the challenge of sustainability. *J. Clean. Prod.* **2019**, *228*, 658–678. [CrossRef]
8. Marshall, R.E.; Farahbakhsh, K. Systems approaches to integrated solid waste management in developing countries. *Waste Manag.* **2013**, *33*, 988–1003. [CrossRef]
9. Ritzkowski, M.; Stegmann, R. Landfill aeration worldwide: Concepts, indications and findings. *Waste Manag.* **2012**, *32*, 1411–1419. [CrossRef]
10. Jouhara, H.; Czajczyńska, D.; Ghazal, H.; Krzyżyńska, R.; Anguilano, L.; Reynolds, A.J.; Spencer, N. Municipal waste management systems for domestic use. *Energy* **2017**, *139*, 485–506. [CrossRef]
11. Hettiaratchi, J.P.A.; Jayasinghe, P.A.; Yarandy, T.A.; Attalage, D.; Jalilzadeh, H.; Pokhrel, D.; Bartholameuz, E.; Hunte, C. Innovative Practices to Maximize Resource Recovery and Minimize Greenhouse Gas Emissions from Landfill Waste Cells: Historical and Recent Developments. *J. Indian Inst. Sci.* **2021**, *101*, 1–20. [CrossRef]
12. Fei, X.; Fang, M.; Wang, Y. Climate change affects land-disposed waste. *Nat. Clim. Chang.* **2021**, *32*, 1411–1419. [CrossRef]
13. Cossu, R. Technical evolution of landfilling. *Waste Manag.* **2010**, *30*, 947–948. [CrossRef] [PubMed]
14. Mir, I.S.; Cheema, P.P.S.; Singh, S.P. Implementation analysis of solid waste management in Ludhiana city of Punjab. *Environ. Challenges* **2021**, *2*, 100023. [CrossRef]
15. Cossu, R. *Multibarrier Principles in Landfilling*; Elsevier Inc.: Amsterdam, The Netherlands, 2018.
16. Raheem, A.; Hassan, M.Y.; Shakoor, R. Bioenergy from anaerobic digestion in Pakistan: Potential, development and prospects. *Renew. Sustain. Energy Rev.* **2016**, *59*, 264–275. [CrossRef]
17. Aslam, S.; Ali, F.; Naseer, A.; Sheikh, Z. Application of material flow analysis for the assessment of current municipal solid waste management in Karachi, Pakistan. *Waste Manag. Res.* **2021**, *40*, 0734242X211000427. [CrossRef] [PubMed]

18. Ricci-Jürgensen, M.; Gilbert, J.; Ramola, A. *Global Assessment of Municipal Organic Waste Production and Recycling*; ISWA: Rotterdam, The Netherlands, 2020.
19. Sohoo, I.; Ritzkowski, M.; Guo, J.; Sohoo, K.; Kuchta, K. Municipal Solid Waste Management through Sustainable Landfilling: In View of the Situation in Karachi, Pakistan. *Int. J. Environ. Res. Public Health* **2022**, *19*, 773. [CrossRef] [PubMed]
20. Lavagnolo, M.C. Landfilling in developing countries. In *Solid Waste Landfilling Concepts, Processes, Technology*; Cossu, R., Stegmann, R., Eds.; Elsevier Inc.: Amsterdam, The Netherlands, 2019; pp. 773–796.
21. Dos Muchangos, L.S.; Tokai, A. Greenhouse gas emission analysis of upgrading from an open dump to a semi-aerobic landfill in Mozambique—the case of Hulene dumpsite. *Sci. Afr.* **2020**, *10*, e00638. [CrossRef]
22. Sharma, A.; Gupta, A.K.; Ganguly, R. Impact of open dumping of municipal solid waste on soil properties in mountainous region. *J. Rock Mech. Geotech. Eng.* **2018**, *10*, 725–739. [CrossRef]
23. Amadi, A.N.; Olasehinde, P.I.; Okosun, E.A.; Okoye, N.O.; Okunlola, I.A.; Alkali, Y.B.; Dan-Hassan, M.A. A Comparative Study on the Impact of Avu and Ihie Dumpsites on Soil Quality in Southeastern Nigeria. *Am. J. Chem.* **2012**, *2*, 17–23. [CrossRef]
24. Babu, R.; Prieto Veramendi, P.M.; Rene, E.R. Strategies for resource recovery from the organic fraction of municipal solid waste. *Case Stud. Chem. Environ. Eng.* **2021**, *3*, 100098. [CrossRef]
25. Krecl, P.; de Lima, C.H.; Dal Bosco, T.C.; Targino, A.C.; Hashimoto, E.M.; Oukawa, G.Y. Open waste burning causes fast and sharp changes in particulate concentrations in peripheral neighborhoods. *Sci. Total Environ.* **2021**, *765*, 142736. [CrossRef]
26. Warith, M.A. Solid waste management: New trends in landfill design. *Emirates J. Eng. Res.* **2003**, *8*, 61–70.
27. Tonini, D.; Manfredi, S.; Bakas, I.; Kai-Sørensen Brogaard, L.; Damgaard, A. Life Cycle Assessment of Landfilling. In *Solid Waste Landfilling*; Elsevier: Amsterdam, The Netherlands, 2018; pp. 955–972. [CrossRef]
28. Yang, R.; Xu, Z.; Chai, J. A review of characteristics of landfilled municipal solid waste in several countries: Physical composition, unit weight, and permeability coefficient. *Polish J. Environ. Stud.* **2018**, *27*, 2425–2435. [CrossRef]
29. Slezak, R.; Krzystek, L.; Ledakowicz, S. Simulation of aerobic landfill in laboratory scale lysimeters—Effect of aeration rate. *Chem. Pap.* **2010**, *64*, 223–229. [CrossRef]
30. Heyer, K.-U.; Hupe, K.; Stegmann, R. Landfill Aftercare—Scope for Actions, Duration, Costs and Quantitative Criteria for the Completion. In Proceedings of the Sardinia Tenth International Waste Management and Landfill Symposium S. Margherita di Pula, Cagliari, Italy, 3–7 October 2005.
31. Matsufuji, Y.; Tanaka, A.; Cossu, R. Semiaerobic Landfilling. In *Solid Waste Landfilling*; Elsevier: Amsterdam, The Netherlands, 2018; pp. 807–830. [CrossRef]
32. Theng, L.C.; Matsufuji, Y.; Mohd, N.H. Implementation of the semi-aerobic landfill system (Fukuoka method) in developing countries: A Malaysia cost analysis. *Waste Manag.* **2005**, *25*, 702–711. [CrossRef]
33. Matsuto, T.; Zhang, X.; Matsuo, T.; Yamada, S. Onsite survey on the mechanism of passive aeration and air flow path in a semi-aerobic landfill. *Waste Manag.* **2015**, *36*, 204–212. [CrossRef] [PubMed]
34. Yanase, R.; Matsufuji, Y.; Tashiro, T.; Nakatomi, S. Study on the Gas Flow of Semi-Aerobic Landfill. In Proceedings of the 21th Annual Conference of Japan Society of Material Cycles and Waste Management, Takaokamachi, Japan, 7 November 2010; pp. 39–540.
35. Ritzkowski, M.; Stegmann, R. Landfill In Situ Aeration. In *Solid Waste Landfilling*; Elsevier: Amsterdam, The Netherlands, 2018; pp. 899–914. [CrossRef]
36. Cossu, R.; Grossule, V. Landfill Bioreactors. In *Solid Waste Landfilling*; Elsevier: Amsterdam, The Netherlands, 2018; pp. 831–841. [CrossRef]
37. Ritzkowski, M.; Heyer, K.U.; Stegmann, R. Fundamental processes and implications during in situ aeration of old landfills. *Waste Manag.* **2006**, *26*, 356–372. [CrossRef]
38. Ritzkowski, M.; Heerenklage, J.; Stegmann, R. An overview on techniques and regulations of mechanical-biological pre-treatment of municipal solid waste *. *Clean Air* **2006**, *2*, 57–68.
39. Ritzkowski, M.; Stegmann, R. Landfill aeration within the scope of post-closure care and its completion. *Waste Manag.* **2013**, *33*, 2074–2082. [CrossRef]
40. Berge, N.D.; Reinhart, D.R.; Dietz, J.; Townsend, T. In situ ammonia removal in bioreactor landfill leachate. *Waste Manag.* **2006**, *26*, 334–343. [CrossRef]
41. Shao, L.M.; He, P.J.; Li, G.J. In situ nitrogen removal from leachate by bioreactor landfill with limited aeration. *Waste Manag.* **2008**, *28*, 1000–1007. [CrossRef]
42. Raga, R.; Cossu, R.; Heerenklage, J.; Pivato, A.; Ritzkowski, M. Landfill aeration for emission control before and during landfill mining. *Waste Manag.* **2015**, *46*, 420–429. [CrossRef]
43. Nwaokorie, K.J.; Bareither, C.A.; Mantell, S.C.; Leclaire, D.J. The influence of moisture enhancement on landfill gas generation in a full-scale landfill. *Waste Manag.* **2018**, *79*, 647–657. [CrossRef]
44. Erses, A.S.; Onay, T.T.; Yenigun, O. Comparison of aerobic and anaerobic degradation of municipal solid waste in bioreactor landfills. *Bioresour. Technol.* **2008**, *99*, 5418–5426. [CrossRef] [PubMed]
45. Nanda, S.; Berruti, F. Municipal solid waste management and landfilling technologies: A review. *Environ. Chem. Lett.* **2020**, *19*, 1433–1456. [CrossRef]
46. Meegoda, J.N.; Bhuvaneshwari, S.; Hettiaratchi, P.A.; Hettiarachchi, H. *A Comprehensive Model for Anaerobic Degradation in Bio-Reactor Landfills*; Scholars' Mine: Chicago, IL, USA, 2013; pp. 1–7.

47. Kumar, S.; Chiemchaisri, C.; Mudhoo, A. Bioreactor landfill technology in municipal solid waste treatment: An overview. *Crit. Rev. Biotechnol.* **2011**, *31*, 77–97. [CrossRef] [PubMed]
48. Morello, L.; Cossu, R.; Raga, R.; Pivato, A.; Lavagnolo, M.C. Recirculation of reverse osmosis concentrate in lab-scale anaerobic and aerobic landfill simulation reactors. *Waste Manag.* **2016**, *56*, 262–270. [CrossRef] [PubMed]
49. Frank, R.R.; Davies, S.; Wagland, S.T.; Villa, R.; Trois, C.; Coulon, F. Evaluating leachate recirculation with cellulase addition to enhance waste biostabilisation and landfill gas production. *Waste Manag.* **2016**, *55*, 61–70. [CrossRef]
50. Liu, L.; Xiong, H.; Ma, J.; Ge, S.; Yu, X.; Zeng, G. Leachate Recirculation for Enhancing Methane Generation within Field Site in China. *J. Chem.* **2018**, *2018*, 9056561. [CrossRef]
51. Sang, N.N.; Soda, S.; Ishigaki, T.; Ike, M. Microorganisms in landfill bioreactors for accelerated stabilization of solid wastes. *J. Biosci. Bioeng.* **2012**, *114*, 243–250. [CrossRef]
52. Hettiaratchi, P.; Jayasinghe, P.; Tay, J.H.; Yadav, S. Recent advances of biomass waste to gas using landfill bioreactor technology-a review. *Curr. Org. Chem.* **2015**, *19*, 413–422. [CrossRef]
53. Grossule, V.; Morello, L.; Cossu, R.; Lavagnolo, M.C. Bioreactor landfills: Comparison and kinetics of the different systems. *Detritus* **2018**, *3*, 100–113. [CrossRef]
54. Hater, G.R.; Young, K.E.; Simpson, C.; Harris, J.M. Economics of Eight Scenarios for Landfill Bioreactors as Compared to a Base Case Subtitle D Landfill. In Proceedings of the Waste Tech 2001, San Diego, CA, USA, 13 February 2001.
55. Berge, N.D.; Reinhart, D.R.; Batarseh, E.S. An assessment of bioreactor landfill costs and benefits. *Waste Manag.* **2009**, *29*, 1558–1567. [CrossRef] [PubMed]
56. Morello, L.; Raga, R.; Lavagnolo, M.C.; Pivato, A.; Ali, M.; Yue, D.; Cossu, R. The S.An.A.®concept: Semi-aerobic, Anaerobic, Aerated bioreactor landfill. *Waste Manag.* **2017**, *67*, 193–202. [CrossRef]
57. Valencia, R.; Van der Zon, W.; Woelders, H.; Lubberding, H.J.; Gijzen, H.J. Achieving "Final Storage Quality" of municipal solid waste in pilot scale bioreactor landfills. *Waste Manag.* **2009**, *29*, 78–85. [CrossRef] [PubMed]
58. Price, G.A.; Barlaz, M.A.; Hater, G.R. Nitrogen management in bioreactor landfills. *Waste Manag.* **2003**, *23*, 675–688. [CrossRef]
59. Omar, H.; Rohani, S. Treatment of landfill waste, leachate and landfill gas: A review. *Front. Chem. Sci. Eng.* **2015**, *9*, 15–32. [CrossRef]
60. Read, A.D.; Hudgins, M.; Harper, S.; Phillips, P.; Morris, J. The successful demonstration of aerobic landfilling. The potential for a more sustainable solid waste management approach? *Resour. Conserv. Recycl.* **2001**, *32*, 115–146. [CrossRef]
61. Jain, P.; Powell, J.; Townsend, T.G.; Reinhart, D.R. Air Permeability of Waste in a Municipal Solid Waste Landfill. *J. Environ. Eng.* **2005**, *131*, 1565–1573. [CrossRef]
62. Yazdani, R.; Mostafid, M.E.; Han, B.; Imhoff, P.T.; Chiu, P.; Augenstein, D.; Kayhanian, M.; Tchobanoglous, G. Quantifying factors limiting aerobic degradation during aerobic bioreactor landfilling. *Environ. Sci. Technol.* **2010**, *44*, 6215–6220. [CrossRef]
63. Wu, C.; Shimaoka, T.; Nakayama, H.; Komiya, T.; Chai, X. Stimulation of waste decomposition in an old landfill by air injection. *Bioresour. Technol.* **2016**, *222*, 66–74. [CrossRef]
64. Sohoo, I.; Ritzkowski, M.; Kuchta, K. Influence of moisture content and leachate recirculation on oxygen consumption and waste stabilization in post aeration phase of landfill operation. *Sci. Total Environ.* **2021**, *773*, 145584. [CrossRef]
65. Yazdani, R. *Quantifying Factors Limiting Aerobic Degradation During Aerobic Bioreactor Landfilling and Performance Evaluation of a Landfill-Based Anaerobic Composting Digester for Energy Recovery and Compost Production*; University of California: Davis, CA, USA, 2010; ISBN 1124354654.
66. Žgajnar Gotvajn, A.; Pavko, A. Perspectives on Biological Treatment of Sanitary Landfill Leachate. In *Wastewater Treatment Engineering*; InTech: Rijeka, Croatia, 2015; Volume 13, pp. 31–39.
67. Ko, J.H.; Xu, Q.; Jang, Y.-C. Emissions and control of hydrogen sulfide at landfills: A review. *Crit. Rev. Environ. Sci. Technol.* **2015**, *45*, 2043–2083. [CrossRef]
68. Grossule, V.; Stegmann, R. Problems in traditional landfilling and proposals for solutions based on sustainability. *Detritus* **2020**, *12*, 78–91. [CrossRef]
69. Tabasaran, O.; Rettenberger, G. Grundlagen zur Planung von Entgasungsanlagen. *Hösel Schenkel Schurer. Müll-Handbuch. E. Schmidt Berl. Bd* **1987**, *3*, 5000–8061.
70. Ritzkowski, M.; Stegmann, R. Controlling greenhouse gas emissions through landfill in situ aeration. *Int. J. Greenh. Gas Control* **2007**, *1*, 281–288. [CrossRef]
71. Towprayoon, S. CHAPTER 3: Solid Waste Disposal. In *2019 Refinement to the 2006 IPCC Guidelines for National Greenhouse Gas Inventory*; Intergovernmental Panel on Climate Change—IPCC: Geneva, Switzerland, 2019; pp. 1–25.
72. Sohoo, I.; Ritzkowski, M.; Kuchta, K.; Cinar, S.Ö. Environmental Sustainability Enhancement of Waste Disposal Sites in Developing Countries through Controlling Greenhouse Gas Emissions. *Sustainability* **2021**, *13*, 151. [CrossRef]
73. Shahid, M.; Nergis, Y.; Siddiqui, S.A.; Farooq Choudhry, A. Environmental impact of municipal solid waste in Karachi city. *World Appl. Sci. J.* **2014**, *29*, 1516–1526. [CrossRef]
74. Sohoo, I.; Ritzkowski, M.; Kuchta, K. Evaluation of behavior of waste disposal sites in Karachi, Pakistan and effects of enhanced leaching on their emission potential. *Detritus* **2019**, *7*, 96–103. [CrossRef]
75. IPCC. Good Practice Guidance and Uncertainty Management in National Greenhouse Gas Inventories. 2001. Available online: https://www.ipcc-nggip.iges.or.jp/public/gp/english/gpgaum_en.html (accessed on 12 December 2021).

76. IPCC. Conceptual Basis for Uncertainty Analysis. In *Good Practice Guidance and Uncertainty Management in National Greenhouse Gas Inventories*; 2000. Available online: https://www.ipcc-nggip.iges.or.jp/public/gp/english/A1_Conceptual.pdf. (accessed on 12 December 2021).
77. Amini, H.R.; Reinhart, D.R.; Mackie, K.R. Determination of first-order landfill gas modeling parameters and uncertainties. *Waste Manag.* **2012**, *32*, 305–316. [CrossRef]
78. Zuberi, M.J.S.; Ali, S.F. Greenhouse effect reduction by recovering energy from waste landfills in Pakistan. *Renew. Sustain. Energy Rev.* **2015**, *44*, 117–131. [CrossRef]
79. Yu, X.; Ye, X.; Lin, H.; Feng, N.; Gao, S.; Zhang, X.; Wang, Y.; Yu, H.; Deng, X.; Qian, B. Knockdown of long non-coding RNA LCPAT1 inhibits autophagy in lung cancer. *Cancer Biol. Med.* **2018**, *15*, 228–237. [CrossRef] [PubMed]
80. IPCC. *Climate Change 2013: The Physical Science Basis: Working Group I Contribution to the Fifth Assessment Report of the Intergovernmental Panel on Climate Change*; Cambridge University Press: Cambridge, UK, 2013; ISBN 110705799X.
81. Salukele, F.M. Innovative Landfill Bioreactor Systems for Municipal Solid Waste Treatment in East Africa Aimed at Optimal Energy Recovery and Minimal Greenhouse Gas Emissions. Ph.D. Thesis, Wageningen University, Wageningen, The Netherlands, 2013. Available online: https://edepot.wur.nl/264686 (accessed on 8 November 2021).
82. Younes, M.K.; Nopiah, Z.M.; Basri, N.E.A.; Basri, H.; Abushammala, M.F.M.; Younes, M.Y. Landfill area estimation based on integrated waste disposal options and solid waste forecasting using modified ANFIS model. *Waste Manag.* **2016**. [CrossRef] [PubMed]
83. Yedla, S. Modified landfill design for sustainable waste management. *Int. J. Glob. Energy Issues* **2005**, *23*, 93–105. [CrossRef]
84. Park, J.K.; Chong, Y.G.; Tameda, K.; Lee, N.H. Methods for determining the methane generation potential and methane generation rate constant for the FOD model: A review. *Waste Manag. Res.* **2018**, *36*, 200–220. [CrossRef]
85. Garg, A.; Achari, G.; Joshi, R.C. A model to estimate the methane generation rate constant in sanitary landfills using fuzzy synthetic evaluation. *Waste Manag. Res.* **2006**, *24*, 363–375. [CrossRef]
86. Machado, S.L.; Carvalho, M.F.; Gourc, J.P.; Vilar, O.M.; do Nascimento, J.C.F. Methane generation in tropical landfills: Simplified methods and field results. *Waste Manag.* **2009**, *29*, 153–161. [CrossRef]
87. Alberta Environment Technical Guidance for the Quantification of Specified Gas Emissions from Landfills (Version 1.2). Available online: https://open.alberta.ca/publications/9780778576792 (accessed on 4 July 2021).
88. Duan, Z.; Kjeldsen, P.; Scheutz, C. Efficiency of gas collection systems at Danish landfills and implications for regulations. *Waste Manag.* **2022**, *139*, 269–278. [CrossRef]
89. Rahman, N. *Sustainable Waste Management through Operating Landfill as Biocell*; The University of Texas at Arlington: Arlington, TX, USA, 2018.
90. SSWMB. *Solid Waste Emergency and Efficiency Project (SWEEP) Sindh Solid Waste Management Board Government of Sindh*; Sindh Solid Waste Management Board (SSWMB): Karachi, Pakistan, 2020.
91. SSWMB Tenders. Available online: http://sswmb.gos.pk/cms/?page_id=1617 (accessed on 28 April 2021).
92. SSWMB. *Government of Sindh Bidding Document Operation/Maintenance of Sswmb Landfill Site at DEH Procuring Agency*; Solid Waste Management Board (SSWMB): Karachi, Pakistan, 2020.
93. Sohoo, I.; Ritzkowski, M.; Heerenklage, J.; Kuchta, K. Biochemical methane potential assessment of municipal solid waste generated in Asian cities: A case study of Karachi, Pakistan. *Renew. Sustain. Energy Rev.* **2021**, *135*, 110175. [CrossRef]
94. Korai, M.S.; Mahar, R.B.; Uqaili, M.A. The feasibility of municipal solid waste for energy generation and its existing management practices in Pakistan. *Renew. Sustain. Energy Rev.* **2017**, *72*, 338–353. [CrossRef]
95. Meegoda, J.N.; Hettiarachchi, H.; Hettiaratchi, P. Landfill design and operation. *Sustain. Solid Waste Manag.* **2016**, *18*, 577–604.
96. Pazoki, M.; Ghasemzadeh, R. Landfilling. In *Municipal Landfill Leachate Management*; Springer: Berlin/Heidelberg, Germany, 2020; pp. 21–100.
97. Ni, Z.; Liu, J.; Girotto, F.; Cossu, R.; Qi, G. Targeted modification of organic components of municipal solid waste by short-term pre-aeration and its enhancement on anaerobic degradation in simulated landfill bioreactors. *Bioresour. Technol.* **2016**, *216*, 250–259. [CrossRef]
98. Ali, M.; Zhang, J.; Raga, R.; Lavagnolo, M.C.; Pivato, A.; Wang, X.; Zhang, Y.; Cossu, R.; Yue, D. Effectiveness of aerobic pretreatment of municipal solid waste for accelerating biogas generation during simulated landfilling. *Front. Environ. Sci. Eng.* **2018**, *12*, 1–9. [CrossRef]
99. Ritzkowski, M.; Walker, B.; Kuchta, K.; Raga, R.; Stegmann, R. Aeration of the teuftal landfill: Field scale concept and lab scale simulation. *Waste Manag.* **2016**, *55*, 99–107. [CrossRef] [PubMed]
100. Ishii, K.; Furuichi, T. Estimation of methane emission rate changes using age-defined waste in a landfill site. *Waste Manag.* **2013**, *33*, 1861–1869. [CrossRef]
101. Baldwin, T.D.; Stinson, J.; Ham, R.K. Decomposition of Specific Materials Buried within Sanitary Landfills. *J. Environ. Eng.* **1998**, *124*, 1193–1202. [CrossRef]
102. World Meteorological Organization. World Weather Information Service. Available online: http://worldweather.wmo.int/en/city.html?cityId=892 (accessed on 23 May 2017).
103. Hunte, C.A. *Performance of a Full-Scale Bioreactor Landfill*; University of Calgary: Calgary, AB, Canada, 2010; Volume 72.
104. Meegoda, J.N.; Soliman, A.; Hettiaratchi, P.A.; Agbakpe, M. Resource mining for a bioreactor landfill. *Curr. Environ. Eng.* **2019**, *6*, 17–34. [CrossRef]

105. Pacey, J.; Augenstein, D.; Morck, R.; Reinhart, D.; Yazdani, R. The bioreactor landfill-an innovation in solid waste management. *MSW Manag.* **1999**, *1*, 53–60.
106. Hettiarachchi, H.; Hettiaratchi, J.P.A.; Hunte, C.A.; Meegoda, J.N. Operation of a landfill bioreactor in a cold climate: Early results and lessons learned. *J. Hazard. Toxic Radioact. Waste* **2013**, *17*, 307–316. [CrossRef]
107. Sohoo, I.; Ritzkowski, M.; Sohu, Z.A.; Cinar, S.Ö.; Chong, Z.K.; Kuchta, K. Estimation of methane production and electrical energy generation from municipal solid waste disposal sites in Pakistan. *Energies* **2021**, *14*, 2444. [CrossRef]

sustainability

Article

Quality Evaluation of Lemon Cordial Stored at Different Times with Microwave Heating (Pasteurization)

Faiqa Malik [1,*], Muhammad Nadeem [1,*], Ammara Ainee [1], Rabia Kanwal [2], Muhammad Sultan [3,*], Aqsa Iqbal [1], Samy F. Mahmoud [4], Garsa Ali Alshehry [5], Huda Abdalrahman Al-Jumayi [5] and Eman Hassan Ahmed Algarni [5]

1 Institute of Food Science and Nutrition, University of Sargodha, Sargodha 40100, Pakistan; ammara.ainee@uos.edu.pk (A.A.); aqsa0057@gmail.com (A.I.)
2 Post-Harvest Research Centre, Ayub Agriculture Research Institute, Jhang Road, Faisalabad 38350, Pakistan; rabiak_018@yahoo.com
3 Department of Agricultural Engineering, Bahauddin Zakariya University, Bosan Road, Multan 60800, Pakistan
4 Department of Biotechnology, College of Science, Taif University, P.O. Box 11099, Taif 21944, Saudi Arabia; s.farouk@tu.edu.sa
5 Department of Food Science and Nutrition, College of Science, Taif University, Taif 21944, Saudi Arabia; garsa.a@tu.edu.sa (G.A.A.); huda.a@tu.edu.sa (H.A.A.-J.); eman1400@tu.edu.sa (E.H.A.A.)
* Correspondence: faiqamalikft@gmail.com (F.M.); nadeem.abdul@uos.edu.pk (M.N.); muhammadsultan@bzu.edu.pk (M.S.); Tel.: +92-300-604-9890 (M.N.)

Abstract: Consumer interest in food quality and safety has shifted over time, as consumers increasingly prefer minimally processed items. As a result, numerous non-thermal approaches have been implemented due to their potential to preserve the nutritional profile of products along with lengthening their storability. Microwaving, a green processing technique, volumetrically heats the product because of the interactions developed between charged ions, polar water molecules of foodstuff and the incoming electromagnetic waves. The study was mapped out to investigate the effect of microwave exposure time (60, 90 and 120 s) at fixed power (1000 W) and frequency (2450 MHz) on physicochemical properties, phytochemical constituents, antioxidant potential and microbial counts of lemon cordial stored at refrigerated temperature (4 ± 2 °C). The mentioned parameters were analyzed after an interval of 30–90 days. Statistical findings illustrated a highly significant ($p \leq 0.01$) impact of microwave treatment and storage on titratable acidity, pH, total soluble solids, total phenolic contents, total flavonoids contents, antioxidant potential and total plate count. Sample microwaved for 120 s showed the highest pH values (2.45 ± 0.050), total soluble solids (56.68 ± 2.512 °B) and antioxidant activity (1212.03 ± 716.5 µg—equivalent of ascorbic acid per 100 mL of cordial); meanwhile, it exhibited the lowest total plate counts (1.75 ± 0.144 Log 10 CFU/mL). Therefore, microwaving can be suggested as a suitable alternate to traditional pasteurization techniques as well as to chemical preservatives.

Keywords: lemon cordial; microwave; preservation; green processing; antioxidant potential

Citation: Malik, F.; Nadeem, M.; Ainee, A.; Kanwal, R.; Sultan, M.; Iqbal, A.; Mahmoud, S.F.; Alshehry, G.A.; Al-Jumayi, H.A.; Algarni, E.H.A. Quality Evaluation of Lemon Cordial Stored at Different Times with Microwave Heating (Pasteurization). *Sustainability* **2022**, *14*, 1953. https://doi.org/10.3390/su14041953

Academic Editor: Rajeev Bhat

Received: 10 January 2022
Accepted: 3 February 2022
Published: 9 February 2022

1. Introduction

Prevention of various diseases is possible by including fruits in our diet as they are an excellent source of minerals, vitamins, antioxidant components and other phytoconstituents [1]. Generation of free radicals, which are triggering factors for several acute and chronic diseases, can be prohibited by the antioxidant potential of the fruits, thus promoting a healthier life [2]. Elevated levels of plasma carotenoids and vitamin C are associated with the increased intake of fruits, which ultimately reduces the probability of diabetes, cardiovascular diseases, neurological disorders and cancer [3,4].

Citrus limon (L.) Burm. f., most commonly known as lemon, a yellow-colored edible fruit, is the third most widely produced representative of the Rutaceae family and hybrid of genus *Citrus*—just after orange and mandarin—worldwide [5]. *Citrus limon* can either be consumed as a fresh fruit, as beverage, as cooking material or for preservation purposes. Because of its tart flavor, it is most often used in manufacturing of beverages, desserts, ice creams, salad dressings, jams, jellies, pickles and in several kinds of vegetable and meat dishes [6,7].

Citrus limon is also a good supplier of a variety of phytoconstituents, including phenolics. Among these phenolics, eriocitrin, coumarins, flavonoids and limonoid glycosides are present in adequate concentrations [8]. Consumption of *Citrus limon* has shown reduction in the risk of several types of cancers, along with cardiovascular disorders [7]. Organic compounds present in lemons are effective against asthma and can serve as antidepressants and stress relievers. Moreover, these also stimulate digestion and are effective in case of flu, fever, boils and in several kinds of ulcers, particularly mouth ulcers [9]. In the past, when vitamin C was not discovered, the juice extracted from *Citrus limon* fruit was used for treatment of scurvy [10]. Additionally, the juice has also served as traditional medicine for the cure of hypertension, common cold, sore throat, chest pain and rheumatism [11].

Cordial can be defined as sparkling, clear, or syrup concentrated fruit juice formed by the complete removal of pulp and other suspended particles, and needs to be diluted upon consumption [12] (Yusof and Chiong). *Citrus limon* is also cultivated in Pakistan and it occupies 6th position in terms of area and production [13]. Due to more production and less utilization, the fruit obtain wasted. In order to overcome the losses, there is a need to convert them to some value-added products, such as squashes and cordials. The study focused on making cordial from lemons in order to meet the rising consumer demands regarding new value-added products.

During the last couple of years, consumers' interest throughout the entire world has been changed regarding quality and safety of the food product [14]. They demand minimally processed products that are not only healthy, but are also processed by means of safe preservation techniques, so that their quality and nutritional profile are not affected [15].

In general, fruit juices are preserved by heating them near the boiling point of water, or slightly below it, for a set amount of time, to kill or inactivate deterioration-causing microbes and enzymes. [16]. Although traditional heating methods ensure the safety and stability of the juices, they greatly affect the phytochemical profiles of the fruit juices, along with causing a decline in the physical and chemical properties, the nutritional profiling and the volatile compounds [14,17]. Similarly, chemical preservatives are other means to extend the shelf life of products, but these chemicals also produce several health complications in humans, particularly cancer, neurological dysfunction, asthma, hyperactivity and hypersensitivity, dermatitis, allergies and gastrointestinal and respiratory disorders [18–20].

Due to disadvantages of thermal techniques and chemical preservatives, several novel non-thermal techniques have been in practice to enhance the shelf stability of processed products [21,22]. Microwaving, a novel technique, utilizes electromagnetic waves that heat the product by means of molecular interaction, as generated by the electromagnetic field. In this technique, there is a direct interaction of food particles with that of the incoming waves. As a result of this direct contact, penetration of heat is easier within the food, and thus volumetric heating takes place [23,24]. Microwaving also preserves the nutritional contents of products to a greater extent, such as the retention of vitamins, thus enhancing the quality and safety of products through the inactivation of microbes and enzymes within a short duration [25,26].

Keeping the above benefits in mind, the present research aimed to develop lemon cordial and to investigate the effect of the microwave technique for preserving lemon cordial's physicochemical properties and microbial counts, and to determine the effect of microwave treatment on the shelf stability of lemon cordial at refrigerated temperature for a period of 90 days.

2. Materials and Methods

2.1. Collection of Fruit

Lemons were purchased from a local farm and sorted to separate damaged and diseased fruits from healthy fruit. Subsequently, the fruits were washed using tap water to remove dirt and dust.

2.2. Chemicals and Reagents

All the chemicals used in analysis were of analytical grade and purchased from Sigma-Aldrich (Gillingham, UK), available in the local market.

2.3. Preparation of Lemon Cordial

Extraction of the juice from the fresh fruit was carried out using a manual juice extractor. After extraction, the juice was filtered through 4 folds of muslin cloth to remove seeds and juice vesicles to obtain clear filtrate. The remaining ingredients, such as sugar, water, citric acid and lemon-yellow color, were added to the cleared juice (Table 1) with constant stirring to obtain lemon cordial (45 °brix), following the procedure of [27], with some modifications.

Table 1. Formulation of lemon cordial.

Ingredients	Quantity
Clarified lemon juice	1 L
Sugar	1.5 Kg
Water	500 ml
Citric acid	1 g/1 L
Carboxymethyl cellulose	1 g/1 L
Lemon-yellow coloring	0.1 g/1 L

2.4. Microwave Processing of Lemon Cordial

The cordial was subjected to microwave treatment with a domestic microwave processor (Model No: DW- 131A operating at 1000 W power and frequency of 2450 MHz) for 60, 90 and 120 s. Microwaving of cordial samples (200 mL) was carried out in sterilized beakers (500 mL). Immediately after pasteurization, the product was transferred and packed in presterilized 250 mL plastic bottles that were immersed in ice cold water to prevent shrinkage of the bottles and to cool the product instantly, as the temperature after pasteurization for 120 s reaches 90 ± 2 °C. The treated lemon cordial was later stored at refrigerated temperature (4 ± 2 °C) for further study (Figure 1).

2.5. Chemical Preservative

In treatment, T_{0+} potassium metabisulphite (KMS) was used as preservative to compare the cordial with other treatments, as well as with the control (T_{0-}), without adding chemical preservatives and microwave application. The treatment plan is depicted in (Table 2).

Table 2. Treatment plan of lemon cordial.

Treatments	Microwave Time (Seconds)/Preservatives	Storage Conditions
T_{0+}	0.1% (KMS)	Temperature (4 ± 2 °C)
T_{0-}	-	Temperature (4 ± 2 °C)
T_1	60 s	Temperature (4 ± 2 °C)
T_2	90 s	Temperature (4 ± 2 °C)
T_3	120 s	Temperature (4 ± 2 °C)

Figure 1. Processing of lemon cordial (1st row); preservation of lemon cordial at refrigerated temperature (4 ± 2 °C) (2nd row); different analysis of lemon cordial (3rd row).

2.6. Physicochemical Analysis

Acidity was accessed through procedure no. 942.15, as mentioned in [28]. pH was evaluated by using a digital pH meter (AD 1040 Benchtop meter, Adwa, Hungary), as per method no. 981.12 of [28]. Degree brix were measured using a hand refractometer (Atago, Japan), according to method no 932.12, as explained by [28].

2.7. Determination of Total Phenolic Contents

Total phenolics of lemon cordial were accessed through a modified Folin–Ciocalteu reagent method, as explained by [29]. A diluted lemon cordial sample (0.5 mL, or 500 μL) was used for analysis and absorbance was measured by spectrophotometer at 760 nm. Gallic acid (in ethanol) was taken as standard, and the findings of total phenolic contents were indicated as mg of gallic acid equivalents (GAE) per 100 mL of lemon cordial.

2.8. Determination of Total Flavonoids Contents

The total flavonoids of the lemon cordial were accessed by using aluminum chloride reagent through procedure described by [30]. Diluted lemon cordial samples (0.25 mL or 250 μL) were used for analysis and absorbance was measured by spectrophotometer at 510 nm wavelength. Catechin (in ethanol) was taken as standard, and the findings of total flavonoids were demonstrated as mg of (+)—catechin equivalent (CE) per 100 mL of lemon cordial.

2.9. Determination of Total Antioxidant Activity

Total antioxidant activity of all the diluted lemon cordial samples was accessed through the procedure described by [31]. Diluted lemon cordial samples (0.4 mL or 400 μL) were used for analysis and absorbance was probed through a spectrophotometer at 695 nm wavelength. Standard calibration curves were made by using ascorbic acid. Findings of total antioxidant activity were demonstrated as μg ascorbic acid equivalent (AAE) per mL sample.

2.10. Microbiological Analysis

Total plate count of all lemon cordial samples was accessed through the [32] standard method of the *Bacteriological Analytical Manual* [33]. A measure of 1 mL of lemon cordial sample was poured in presterilized test tubes with the aid of sterilized pipette. To these test tubes, 9 mL of normal saline solution was added to make dilutions. Serial dilutions were made by transferring 1 mL of previously generated dilution to a test tube, containing 9 mL of normal saline as diluent. The prepared dilutions were shaken well to prevent the settling of suspended materials. Subsequently, the prepared dilutions were sprinkled upon the control plates for each series of samples. After the media and poured dilutions had completely solidified, the Petri dishes were inverted and placed in an incubator at 35 °C for 48 h. Plates with colonies between 30 and 300 were compounded with dilution factor based on the number of colonies that appeared on the plates. The arithmetic average was calculated as the total plate count per mL.

2.11. Statistical Analysis

Findings of each parameter were statistically analyzed through statistics software. ANOVA technique and Tukey's HSD test were employed to determine difference among means by having the level of significance at 5% confidence interval, according to the method illustrated by [34].

3. Results and Discussion

The impact of microwave treatments and storage on titratable acidity of lemon cordial retained at refrigerated temperature is illustrated in (Table 3). Microwaving produced a highly significant impact ($p < 0.01$) on titratable acidity, leading to a gradual reduction in the acidity of lemon cordial. The highest value for acidity was exhibited by T_{0-}, proclaiming a mean value of 0.38 ± 0.038%, while the lowest was observed in T_3, exhibiting a mean value of 0.32 ± 0.045%. The observations are in conformance with those obtained by Pandiselvam et al. [35] and Bozkir et al. [36], who reported the same decreasing tendency in the acidity after the application of microwave treatment on kalparasa and apple juice. They articulated that increased time and temperature during microwave treatment resulted in destruction of fermenting microbes, due to which the production of organic acids declines. Considering the impact of storage on the titratable acidity, it was noticed that titratable acidity increased gradually throughout the storage duration. The highest mean value of 0.41 ± 0.014% was observed at 90 days, while the lowest mean value of 0.29 ± 0.024% was examined at 0 days. These results are supported by the findings of Pandiselvam et al. [35], who reported an increase in titratable acidity of microwave-treated kalparasa (coconut inflorescence sap) upon refrigeration storage and claimed that an increased production of organic acids during anaerobic fermentation is responsible for an increase in acidity. Similarly, Palanisamy et al. [37] reported that increase in the acidity of noni fruit juice blended squash during storage was caused by the soluble proteins hydrolyzing to free amino acids, which ultimately resulted in an interlinkage of citric acid in squash. An increase in acidity during storage in mixed fruit squash and microwaved apple puree was also communicated by Jothi et al. [38] and Picouet et al. [39], respectively.

Microwave treatment and storage duration both produced a highly significant impact ($p \leq 0.01$) upon the pH of the lemon cordial (Table 4). Gradual increase in the pH of the lemon cordial upon microwaving was observed. The highest pH value was expressed by T_3, with a mean value of 2.45 ± 0.050, while the lowest was observed in T_{0-}, exhibiting a mean value of 2.38 ± 0.046. A similar trend for pH was reported by Pandiselvam et al. [35] in microwave-treated kalparasa, who documented that reduction in the contents of organic acids of kalparasa, as result of heating, increased the pH. Picouet et al. [39] articulated that the pH of apple puree, when subjected to microwaving for 35 s at 652 W power, slightly enhanced from 3.2 to 3.3. Considering the impact of storage on the pH, it was noticed that it decreased throughout the storage duration. The highest mean value of 2.46 ± 0.027 was noticed at start of study while the lowest mean value of 2.35 ± 0.024 was observed at 90 days. The observation of the conducted study correlates with those proposed by Palanisamy et al. [37], Yadav et al. [40] and Jothi et al. [38]. They claimed that the pH of noni fruit juice blended squash, guava mango RTS and squash and mixed fruit squash made from banana, papaya and carrot juice decreased during storage, which was attributed to the accumulation of acidic compounds—particularly lactic acid and acetic acid—because of the activity of microbes during natural fermentation process.

Table 3. Titratable acidity (%) of lemon cordial.

Treatments	Storage (Days)				Means
	0	30	60	90	
T_{0-}	0.33 $^{e–g}$ ± 0.007	0.35 ^{ef} ± 0.001	0.39 ^{bc} ± 0.004	0.43 ^{a} ± 0.003	0.38 ^{A} ± 0.038
T_{0+}	0.30 ^{i} ± 0.013	0.34 ^{ef} ± 0.004	0.37 ^{d} ± 0.010	0.41 ^{a} ± 0.004	0.36 ^{B} ± 0.044
T_1	0.29 ^{ij} ± 0.010	0.34 $^{e–g}$ ± 0.004	0.35 ^{de} ± 0.006	0.41 ^{ab} ± 0.006	0.35 ^{C} ± 0.045
T_2	0.28 ^{j} ± 0.004	0.32 ^{gh} ± 0.006	0.34 ^{ef} ± 0.004	0.39 ^{bc} ± 0.004	0.33 ^{D} ± 0.044
T_3	0.27 ^{j} ± 0.004	0.30 ^{hi} ± 0.007	0.33 ^{fg} ± 0.004	0.39 ^{c} ± 0.004	0.32 ^{E} ± 0.045
Means	0.29 ^{D} ± 0.024	0.33 ^{C} ± 0.017	0.36 ^{B} ± 0.024	0.41 ^{A} ± 0.014	

Values exhibiting similar alphabets are statistically non-significant. T_{0-}—control (lemon cordial with no preservative or treatment); T_{0+}—control (lemon cordial preserved with KMS); T_1—microwave treatment (60 s); T_2—microwave treatment (90 s); T_3—microwave treatment (120 s).

Table 4. pH of lemon cordial during storage.

Treatments	Storage (Days)				Means
	0	30	60	90	
T_{0-}	2.43 $^{c–e}$ ± 0.015	2.40 $^{e–g}$ ± 0.015	2.36 $^{g–i}$ ± 0.023	2.32 ^{j} ± 0.010	2.38 ^{D} ± 0.046
T_{0+}	2.45 $^{b-d}$ ± 0.012	2.42 $^{d–f}$ ± 0.015	2.38 $^{g–i}$ ± 0.021	2.34 ^{ij} ± 0.010	2.40 ^{C} ± 0.044
T_1	2.46 ^{bc} ± 0.012	2.43 $^{c–e}$ ± 0.010	2.38 $^{f–h}$ ± 0.015	2.35 $^{h–j}$ ± 0.010	2.41 ^{BC} ± 0.044
T_2	2.46 ^{bc} ± 0.015	2.44 $^{b–d}$ ± 0.006	2.40 $^{e–g}$ ± 0.006	2.37 $^{g–i}$ ± 0.010	2.42 ^{B} ± 0.040
T_3	2.50 ^{a} ± 0.006	2.48 ^{ab} ± 0.006	2.42 $^{d–f}$ ± 0.006	2.38 $^{f–h}$ ± 0.006	2.45 ^{A} ± 0.050
Means	2.46 ^{A} ± 0.027	2.43 ^{B} ± 0.029	2.39 ^{C} ± 0.023	2.35 ^{D} ± 0.024	

Values exhibiting similar small alphabets are statistically non-significant ($p > 0.05$) whereas similar capital alphabets represent statistically non-significant ($p > 0.05$) difference among overall means values. T_{0-}—control (lemon cordial with no preservative and treatment); T_{0+}—control (lemon cordial preserved with KMS); T_1—microwave treatment (60 s); T_2—microwave treatment (90 s); T_3—microwave treatment (120 s).

The influences of microwave treatment and storage on the total soluble solids (°brix) of lemon cordial at refrigerated temperature are illustrated in (Table 5). The presented data clearly depicts that microwave as well as storage duration produced a highly significant impact ($p \leq 0.01$) upon the °brix of lemon cordial. A gradual increase in the °brix of lemon cordial was observed upon microwaving. The highest °brix were exhibited by T_3 expressing a mean value of 56.68 ± 2.612 °B while the lowest were observed in T_{0-}, exhibiting a mean value of 50.83 ± 0.755 °B. The observations of the study are supported by the findings of Sattar et al. [41], Song et al. [42] and Fazaeli et al. [43], who stated an increment in total soluble solids of functional peach beverage, apple and black mulberry juice upon microwaving. They figured out that during microwaving, evaporation of water from the product takes place, and this phenomenon is directly linked to the increase in the internal temperature of juice. It leads to juice concentration and increase in total soluble solid contents. Considering the impact of storage, it was noticed that °brix of lemon cordial decreased throughout the storage duration. The highest mean value of 55.28 ± 3.499 °B was noticed at 0 days while the lowest mean value of 51.05 ± 1.746 °B was observed at 90 days. These findings are similar to those reported by Pandiselvam et al. [35] and Yusof and Chiong [12]. During storage sugars are utilized by the fermenting microbes, particularly yeast and acetic acid bacteria, as a source of nutrient, thus converting them into their respective acids and alcohol, along with the liberation of CO_2.

Table 5. °Brix of lemon cordial during storage.

Treatments	Storage (Days)				Means
	0	30	60	90	
T_{0-}	51.83 ^{g} ± 0.208	51.13 ^{h} ± 0.058	50.33 ^{i} ± 0.153	50.00 ^{i} ± 0.100	50.83 ^{E} ± 0.755
T_{0+}	51.73 ^{g} ± 0.252	51.13 ^{h} ± 0.058	50.97 ^{h} ± 0.058	50.30 ^{i} ± 0.200	51.03 ^{D} ± 0.552
T_1	55.17 ^{d} ± 0.289	51.97 ^{g} ± 0.058	50.07 ^{i} ± 0.058	49.13 ^{j} ± 0.058	51.58 ^{C} ± 2.413
T_2	56.93 ^{b} ± 0.115	53.90 ^{e} ± 0.100	53.17 ^{f} ± 0.058	51.90 ^{g} ± 0.100	53.98 ^{B} ± 1.936
T_3	60.73 ^{a} ± 0.252	56.10 ^{c} ± 0.100	56.00 ^{c} ± 0.100	53.90 ^{e} ± 0.100	56.68 ^{A} ± 2.612
Means	55.28 ^{A} ± 3.499	52.85 ^{B} ± 1.983	52.11 ^{C} ± 2.311	51.05 ^{D} ± 1.746	

Values exhibiting similar small alphabets are statistically non-significant ($p > 0.05$) whereas similar capital alphabets represent statistically non-significant ($p > 0.05$) difference among overall means values. T_{0-}—control (lemon cordial with no preservative and treatment); T_{0+}—control (lemon cordial preserved with KMS); T_1—microwave treatment (60 s); T_2—microwave treatment (90 s); T_3—microwave treatment (120 s).

Data demonstrating the influence of microwave treatment and storage on total phenolic contents of lemon cordial during storage at refrigeration temperature is depicted in (Table 6). Microwave treatment as well as storage duration produced a highly significant impact ($p \leq 0.01$) upon total phenolic contents of lemon cordial. Reduction in total phenolic contents of lemon cordial was observed upon the application of microwave. The highest values of total phenolics were exhibited by T_{0+}, expressing a mean value of 399.08 ± 67.745 mg GAE per 100 mL of cordial, while the lowest were observed in T_3, exhibiting a mean value of 346.42 ± 64.387 mg GAE per 100 mL of cordial. Findings of the conducted study were in line with those reported by Cheng et al. [44] and Igual et al. [45]. They observed reduction in total phenolics contents of *Citrus* unshiu juice and grapefruit juice, respectively. According to de Souza et al. [46] and Rasoulian et al. [47], electromagnetic radiation generated by microwaves first liberates bound phenolics from their binding sites, then causes their cleavage due to increasing temperature, thus changing their composition and stability. In contrast to the microwave results, the storage data indicate that the total phenolic content increased considerably during storage. At 90 days, the highest mean value of 451.20 ± 26.940 mg GAE per 100 mL of cordial was recorded, whereas the lowest value of 281.00 ± 24.474 mg GAE per 100 mL of cordial was observed at 0 days. Increase in total phenolic contents of black mulberry juice was observed during storage at 3 different temperatures—5, 15 and 25 °C—by Jiang et al. [48]. Pandiselvam et al. [35] and Piljac-Zegarac et al. [49] explained that there is a possibility some of the peroxidase may survive the pasteurization treatment, which might promote oxidation, thus elevating the total phenolic contents. Barba et al. [50] figure out that during storage some products are formed because of Maillard reactions. These chemicals have antioxidant properties and, when combined with the Folin reagent, increase the concentration of total phenols.

Table 6. Total phenolic contents (mg GAE per 100 mL) of lemon cordial.

Treatments	Storage (Days)				Means
	0	30	60	90	
T_{0-}	293.00 ^{l} ± 5.196	368.33 ^{gh} ± 2.887	410.33 ^{de} ± 4.619	460.00 ^{b} ± 5.000	382.92 ^{B} ± 64.057
T_{0+}	318.33 ^{k} ± 2.887	369.67 ^{gh} ± 4.041	413.33 ^{de} ± 2.887	495.00 ^{a} ± 5.000	399.08 ^{A} ± 67.745
T_1	280.00 ^{m} ± 3.464	358.33 ^{hi} ± 2.887	405.00 ^{ef} ± 5.196	448.33 ^{b} ± 2.887	372.92 ^{C} ± 65.228
T_2	261.00 ^{n} ± 3.464	352.33 ^{i} ± 2.887	393.67 ^{f} ± 5.774	433.33 ^{c} ± 2.887	360.08 ^{D} ± 66.904
T_3	252.67 ^{n} ± 4.619	336.33 ^{j} ± 2.887	377.33 ^{g} ± 4.619	419.33 ^{d} ± 2.309	346.42 ^{E} ± 64.387
Means	281.00 ^{D} ± 24.474	357.00 ^{C} ± 12.867	399.93 ^{B} ± 14.175	451.20 ^{A} ± 26.940	

Values exhibiting similar small alphabets are statistically non-significant ($p > 0.05$) whereas similar capital alphabets represent statistically non-significant ($p > 0.05$) difference among overall means values. T_{0-}—control (lemon cordial with no preservative and treatment); T_{0+}—control (lemon cordial preserved with KMS); T_1—microwave treatment (60 s); T_2—microwave treatment (90 s); T_3—microwave treatment (120 s).

Findings regarding the influence of microwave treatment and storage on the total flavonoids content of lemon cordial at refrigerated temperature are displayed in (Table 7), which clearly depicts that the treatments and storage periods had a highly significant impact ($p \leq 0.01$) upon total flavonoid contents. Reduction in total flavonoid contents of lemon cordial were observed upon the application of microwave treatment. The highest values of total flavonoids were exhibited by T_{0+}, 178.70 ± 24.472 mg CE per 100 mL of cordial, while the lowest values were observed in T_3, exhibiting a mean value of 118.21 ± 24.078 mg CE per 100 mL of cordial. Decrease in total flavonoids after microwaving was also reported by Papoutsis et al. [51], and Igual et al. [45] in *Citrus limon* L. pomace and grapefruit juice, respectively. They stated that higher power and long exposure time led to degradation of heat sensitive polyphenols. Cheng et al. [44] communicated that water serves as a very important medium for absorbing microwave energy. At higher temperatures and long exposure time, evaporation of water becomes rapid, thus its ability to preserve the bioactive compounds is reduced, which, ultimately, drops their concentration. Therefore, microwave power, time, frequency and temperature play very critical roles in determining the concentration of these polyphenolic constituents. Storage also caused reduction in the total flavonoid contents of lemon cordial. The highest total flavonoid contents with mean value of 174.81 ± 23.349 mg CE per 100 mL of cordial were noticed at 0 days, while the lowest contents with mean value of 117.04 ± 21.678 mg CE per 100 mL of cordial was observed at 90 days. When total flavonoid contents were compared to total phenolic contents during storage, a discrepancy in the data was noticed, indicating that total flavonoid contents began to decline during storage. A 57% reduction in total flavonoid contents was observed when comparison was carried out at beginning and at the end of storage duration. Total flavonoid contents of microwave-treated functional peach beverage and sugarcane juice also declined during storage. Sattar et al. [41] and Zia et al. [24] stated that, during storage, generation of free radicals takes place to a greater extent, which ultimately reduces the total flavonoids contents.

Table 7. Total flavonoid contents (mg CE per 100 mL) of lemon cordial.

Treatments	Storage (Days)				Means
	0	30	60	90	
T_{0-}	187.65 ^{b} ± 5.658	164.20 $^{c–e}$ ± 2.14	151.85 $^{e–g}$ ± 3.70	128.40 ^{jk} ± 2.138	158.02 ^{B} ± 22.575
T_{0+}	209.88 ^{a} ± 5.658	188.89 ^{b} ± 6.415	167.90 ^{cd} ± 4.277	148.15 $^{f–h}$ ± 3.70	178.70 ^{A} ± 24.472
T_1	171.60 ^{c} ± 4.277	155.56 $^{d–g}$ ± 3.7	137.04 $^{h–j}$ ± 3.70	116.05 ^{kl} ± 4.277	145.06 ^{C} ± 21.930
T_2	158.02 $^{d–f}$ ± 5.66	143.21 $^{g–i}$ ± 2.14	124.69 ^{jk} ± 2.138	106.17 ^{l} ± 2.138	133.02 ^{D} ± 20.558
T_3	146.91 $^{f–h}$ ± 2.14	130.86 ^{ij} ± 4.277	108.64 ^{l} ± 4.277	86.42 ^{m} ± 4.277	118.21 ^{E} ± 24.078
Means	174.81 ^{A} ± 23.349	156.54 ^{B} ± 20.691	138.02 ^{C} ± 21.572	117.04 ^{D} ± 21.678	

Values exhibiting similar small alphabets are statistically non-significant ($p > 0.05$) whereas similar capital alphabets represent statistically non-significant ($p > 0.05$) difference among overall means values. T_{0-}—control (lemon cordial with no preservative and treatment); T_{0+}—control (lemon cordial preserved with KMS); T_1—microwave treatment (60 s); T_2—microwave treatment (90 s); T_3—microwave treatment (120 s).

Statistical results indicated highly significant impact ($p \leq 0.01$) of microwave treatment, as well as storage, on the antioxidant activity of lemon cordial stored at refrigerated temperature (Table 8). Increment in total antioxidant activity of lemon cordial was observed upon the application of microwave. The highest value for total antioxidants was expressed by T_3, depicting a mean value of 1212.03 ± 716.5 µg equivalent of ascorbic acid per 100 mL of cordial, while the lowest was observed in T_{0-}, exhibiting a mean value of 837.73 ± 427.1 µg equivalent of ascorbic acid per 100 mL of cordial. Results of the conducted study coincide with the findings of Martins et al. [52] and Zia et al. [24], who proposed that total antioxidant potential of orange juicemilk beverage and sugarcane juice increased with application of microwaves, as the enzymes responsible for promoting oxidation are inhibited. Besides, microwaving also leads to greater extraction of antioxidant compounds from the product to which they are applied. Storage results revealed reduction in the total antioxidant activity of lemon cordial. The highest mean value of 1685.22 ± 264.73 µg equivalent of ascorbic acid per 100 mL of cordial for total antioxidants was noticed at 0 days, while the lowest mean value of 337.76 ± 27.03 µg equivalent of ascorbic acid per 100 mL of cordial was observed at 90 days. During storage, a huge reduction in the total antioxidant activity was observed. At the end of storage, a fourfold reduction in the antioxidant activity was recorded. Results are in conformance with the findings of Jiang et al. [48], who stated that, during storage, loss in vitamin C and flavonoid contents triggers the reduction in the antioxidant activity of black mulberry juice. Additionally, onset of complex chemical reactions, particularly polymerization reactions, limits the availability of free hydroxyl groups, leading to diminished antioxidant potential.

Table 8. Total antioxidant activity (µg equivalent of ascorbic acid per 100 mL) of lemon cordial.

Treatments	Storage (Days)				Means
	0	30	60	90	
T_{0-}	1378.02 ^{e} ± 1.56	1046.49 ^{h} ± 0.10	625.76 ^{m} ± 1.57	300.55 ^{r} ± 0.59	837.73 ^{E} ± 427.1
T_{0+}	1441.08 ^{d} ± 2.70	1120.36 ^{g} ± 1.56	659.10 ^{l} ± 1.56	317.15 ^{q} ± 3.25	884.42 ^{D} ± 448.7
T_1	1683.42 ^{c} ± 1.56	1113.15 ^{g} ± 1.56	685.52 ^{k} ± 4.99	341.11 ^{p} ± 4.63	955.80 ^{C} ± 523.6
T_2	1861.80 ^{b} ± 1.56	1316.76 ^{f} ± 0.10	703.24 ^{j} ± 0.10	359.10 ^{o} ± 6.24	1060.22 ^{B} ± 601.7
T_3	2061.80 ^{a} ± 9.49	1683.42 ^{c} ± 1.56	732.07 ^{i} ± 1.56	370.81 ^{n} ± 0.10	1212.03 ^{A} ± 716.5
Means	1685.22 ^{A} ± 264.73	1256.04 ^{B} ± 240.13	681.14 ^{C} ± 37.82	337.76 ^{D} ± 27.03	

Values exhibiting similar small alphabets are statistically non-significant ($p > 0.05$) whereas similar capital alphabets represent statistically non-significant ($p > 0.05$) difference among overall means values. T_{0-}—control (lemon cordial with no preservative and treatment); T_{0+}—control (lemon cordial preserved with KMS); T_1—microwave treatment (60 s); T_2—microwave treatment (90 s); T_3—microwave treatment (120 s).

At refrigerated temperature, the influence of microwave and storage on the total plate count of lemon cordial is presented in (Table 9), which clearly depicts a highly significant impact ($p \leq 0.01$). Reduction in total plate count of lemon cordial was observed, upon the application of microwave treatment. Lowest values for microbial populations were expressed by T_{0+} and T_3, exhibiting mean values of 1.72 ± 0.131 Log $_{10}$ CFU/mL and 1.75 ± 0.144 Log $_{10}$ CFU/mL of cordial, respectively, while highest were observed in T_{0-}, exhibiting a mean value of 1.83 ± 0.125 Log $_{10}$ CFU/mL of cordial. Observations of the conducted study correlate with investigations of Adulvitayakorn et al. [53] and Li et al. [54], who emphasized that microwaving causes cell lysis of bacteria due to coupling of electromagnetic energy. Storage results revealed an increment in the total plate count of lemon cordial. Lowest mean value of 1.58 ± 0.052 Log $_{10}$ CFU/mL of cordial for total plate count was noticed at 0 days, while highest mean value of 1.93 ± 0.039 Log $_{10}$ CFU/mL of cordial was observed at 90 days. Storage results contradict in a way that total plate count increased throughout storage. Pradhan et al. [55] claimed that, during storage of microwave-treated sugarcane juice, enhancement in total plate count was noted.

Table 9. Total plate count Log_{10} CFU/mL of lemon cordial.

Treatments	Storage (Days)				Means
	0	30	60	90	
T_{0-}	1.65 ^{ij} ± 0.023	1.80 ^{f} ± 0.017	1.89 $^{b-d}$ ± 0.012	1.97 ^{a} ± 0.012	1.83 ^{A} ± 0.125
T_{0+}	1.53 ^{m} ± 0.029	1.71 ^{hi} ± 0.023	1.78 ^{fg} ± 0.006	1.86 ^{de} ± 0.006	1.72 ^{D} ± 0.131
T_1	1.61 ^{jk} ± 0.023	1.79 ^{f} ± 0.017	1.89 ^{cd} ± 0.012	1.95 ^{ab} ± 0.012	1.81 ^{B} ± 0.132
T_2	1.59 ^{kl} ± 0.023	1.77 ^{fg} ± 0.017	1.87 ^{de} ± 0.012	1.93 $^{a-c}$ ± 0.012	1.79 ^{B} ± 0.137
T_3	1.54 ^{lm} ± 0.029	1.73 ^{gh} ± 0.017	1.82 ^{ef} ± 0.012	1.91 $^{b-d}$ ± 0.012	1.75 ^{C} ± 0.144
Means	1.58 ^{D} ± 0.052	1.76 ^{C} ± 0.040	1.85 ^{B} ± 0.044	1.93 ^{A} ± 0.039	

Values exhibiting similar small alphabets are statistically non-significant ($p > 0.05$) whereas similar capital alphabets represent statistically non-significant ($p > 0.05$) difference among overall means values. T_{0-}—control (lemon cordial with no preservative and treatment); T_{0+}—control (lemon cordial preserved with KMS); T_1—microwave treatment (60 s); T_2—microwave treatment (90 s); T_3—microwave treatment (120 s).

4. Conclusions

It was observed that microwaving resulted in a nutritious and shelf-stable product, as it exhibited great antioxidant potential and low microbial counts in formulated lemon cordial. Furthermore, microwave treatment also enhanced the physicochemical profile of the constituted lemon cordial; however, at greater exposure time, there was a reduction in phytochemical constituents. Treatment T_3, microwaved for 120 s, showed better results in terms of physicochemical attributes, antioxidant capacity and microbial counts. It is thus suggested that microwaving should be used at both laboratory and industrial scales as an alternative to chemical preservatives and thermal methods, owing to its cost effectiveness and ease in processing. Additionally, it produces minimally processed products with great shelf stability and high nutritional value.

Author Contributions: Conceptualization, F.M.; data curation, F.M. and A.I.; formal analysis, F.M. and A.I.; funding acquisition, S.F.M., G.A.A., H.A.A.-J. and E.H.A.A.; investigation, F.M. and A.I.; methodology, F.M. and A.A.; resources, M.N. and A.A.; software, M.N. and A.A.; validation, M.N., A.A. and R.K.; visualization, M.N., A.A., R.K., M.S., S.F.M., G.A.A., H.A.A.-J. and E.H.A.A.; writing—original draft preparation, F.M. and A.I.; writing—review and editing, M.N., A.A., R.K., M.S., S.F.M., G.A.A., H.A.A.-J. and E.H.A.A. All authors have read and agreed to the published version of the manuscript.

Funding: The authors would like to acknowledge the financial support of Taif University Researchers Supporting Project number (TURSP-2020/138), Taif University, Taif, Saudi Arabia.

Institutional Review Board Statement: Not applicable.

Informed Consent Statement: Not applicable.

Data Availability Statement: Data are contained within the article.

Acknowledgments: The authors would like to acknowledge the financial support of Taif University Researchers Supporting Project number (TURSP-2020/138), Taif University, Taif, Saudi Arabia.

Conflicts of Interest: The authors declare no conflict of interest.

References

1. Jabbar, S.; Abid, M.; Hu, B.; Wu, T.; Hashim, M.M.; Lei, S.; Zhu, X.; Zeng, X. Quality of Carrot Juice as Influenced by Blanching and Sonication Treatments. *LWT-Food Sci. Technol.* **2014**, *55*, 16–21. [CrossRef]
2. Miller, H.E.; Rigelhof, F.; Marquart, L.; Prakash, A.; Kanter, M. Antioxidant Content of Whole Grain Breakfast Cereals, Fruits and Vegetables. *J. Am. Coll. Nutr.* **2000**, *19*, 312S–319S. [CrossRef]
3. Pomerleau, J.; Lock, K.; McKee, M. The Burden of Cardiovascular Disease and Cancer Attributable to Low Fruit and Vegetable Intake in the European Union: Differences between Old and New Member States. *Public Health Nutr.* **2006**, *9*, 575–583. [CrossRef] [PubMed]
4. Broekmans, W.M.R.; Klö Pping-Ketelaars, I.A.A.; Schuurman, C.R.W.C.; Verhagen, H.; van den Berg, H.; Kok, F.J.; van Poppel, G. Fruits and Vegetables Increase Plasma Carotenoids and Vitamins and Decrease Homocysteine in Humans. *J. Nutr.* **2000**, *1*, 1578–1583. [CrossRef] [PubMed]
5. Lücker, J.; el Tamer, M.K.; Schwab, W.; Verstappen, F.W.A.; van der Plas, L.H.W.; Bouwmeester, H.J.; Verhoeven, H.A. Monoterpene Biosynthesis in Lemon (*Citrus Limon*) CDNA Isolation and Functional Analysis of Four Monoterpene Synthases. *Eur. J. Biochem.* **2002**, *269*, 3160–3171. [CrossRef]
6. Sarker, N.I.; Barman, S.C.; Islam, M.; Chakma, A.S. Role of Lemon (*Citrus Limon*) Production on Livelihoods of Rural People in Bangladesh. *J. Agric. Econ. Rural. Dev.* **2017**, *3*, 167–175.
7. Xi, W.; Lu, J.; Qun, J.; Jiao, B. Characterization of Phenolic Profile and Antioxidant Capacity of Different Fruit Part from Lemon (*Citrus Limon* Burm.) Cultivars. *J. Food Sci. Technol.* **2017**, *54*, 1108–1118. [CrossRef]
8. Del Río, J.A.; Fuster, M.D.; Gómez, P.; Porras, I.; García-Lidón, A.; Ortuño, A. *Citrus Limon*: A Source of Flavonoids of Pharmaceutical Interest. *Food Chem.* **2004**, *84*, 457–461. [CrossRef]
9. Saleem, F.; Sarkar, D.; Ankolekar, C.; Shetty, K. Phenolic Bioactives and Associated Antioxidant and Anti-Hyperglycemic Functions of Select Species of Apiaceae Family Targeting for Type 2 Diabetes Relevant Nutraceuticals. *Ind. Crops Prod.* **2017**, *107*, 518–525. [CrossRef]
10. Mabberley, D.J. *Citrus* (*Rutaceae*): A Review of Recent Advances in Etymology, Systematics and Medical Applications. *Blumea J. Plant Taxon. Plant Geogr.* **2004**, *49*, 481–498. [CrossRef]

11. Cañumir, J.A.; Celis, J.E.; de Bruijn, J.; Vidal, L.V. Pasteurisation of Apple Juice by Using Microwaves. *LWT-Food Sci. Technol.* **2002**, *35*, 389–392. [CrossRef]
12. Yusof, S.; Kok Chiong, L. Effects of Brix, Processing Techniques and Storage Temperature on the Quality of Carambola Fruit Cordial. *Food Chem.* **1997**, *59*, 27–32. [CrossRef]
13. Siddique, M.I.; Gamevska, E. Citrus value chain (s): A survey of Pakistan citrus industry. *Agric. Value Chain* **2018**, 37–53. [CrossRef]
14. Fazaeli, M.; Yousefi, S.; Emam-Djomeh, Z. Investigation on the Effects of Microwave and Conventional Heating Methods on the Phytochemicals of Pomegranate (*Punica Granatum* L.) and Black Mulberry Juices. *Food Res. Int.* **2013**, *50*, 568–573. [CrossRef]
15. Shams Najafabadi, N.; Sahari, M.A.; Barzegar, M.; Hamidi Esfahani, Z. Effects of Concentration Method and Storage Time on Some Bioactive Compounds and Color of Jujube (*Ziziphus Jujuba* Var *Vulgaris*) Concentrate. *J. Food Sci. Technol.* **2017**, *54*, 2947–2955. [CrossRef]
16. Renge, M.M.; Suryawanshi, H.M. Five-Level Diode Clamped Inverter to Eliminate Common Mode Voltage and Reduce Dv/Dt in Medium Voltage Rating Induction Motor Drives. *IEEE Trans. Power Electron.* **2008**, *23*, 1598–1607. [CrossRef]
17. Maskan, M. Production of Pomegranate (*Punica Granatum* L.) Juice Concentrate by Various Heating Methods: Colour Degradation and Kinetics. *J. Food Eng.* **2006**, *72*, 218–224. [CrossRef]
18. Anand, S.; Sati, N. Artificial Preservatives and Their Harmful Effects: Looking Toward Nature for Safer Alternatives. *Int. J. Pharm. Sci. Res.* **2013**, *4*, 2496. [CrossRef]
19. Carocho, M.; Barreiro, M.F.; Morales, P.; Ferreira, I.C.F.R. Adding Molecules to Food, Pros and Cons-A Review on Synthetic and Natural Food Additives. *Compr. Rev. Food Sci. Food Saf.* **2014**, *13*, 377–399. [CrossRef] [PubMed]
20. Anyasi, T.A.; Jideani, A.I.O.; Mchau, G.R.A. Effects of Organic Acid Pretreatment on Microstructure, Functional and Thermal Properties of Unripe Banana Flour. *J. Food Meas. Charact.* **2017**, *11*, 99–110. [CrossRef]
21. Fundo, J.F.; Miller, F.A.; Tremarin, A.; Garcia, E.; Brandão, T.R.S.; Silva, C.L.M. Quality Assessment of Cantaloupe Melon Juice under Ozone Processing. *Innov. Food Sci. Emerg. Technol.* **2018**, *47*, 461–466. [CrossRef]
22. Shah Sulaiman, N.N.A.K.; Sidek, A.; Supian, N.S.M. Quality Assessment of Ozone-Treated Citrus Fruit Juices. *Int. Food Res. J.* **2019**, *26*, 1405–1415.
23. Benlloch-Tinoco, M.; Igual, M.; Rodrigo, D.; Martínez-Navarrete, N. Superiority of Microwaves over Conventional Heating to Preserve Shelf-Life and Quality of Kiwifruit Puree. *Food Control.* **2015**, *50*, 620–629. [CrossRef]
24. Zia, S.; Khan, M.R.; Zeng, X.A.; Sehrish Shabbir, M.A.; Aadil, R.M. Combined Effect of Microwave and Ultrasonication Treatments on the Quality and Stability of Sugarcane Juice during Cold Storage. *Int. J. Food Sci. Technol.* **2019**, *54*, 2563–2569. [CrossRef]
25. Math, R.G.; Nayani, S. Continuous Microwave Processing and Preservation of Acidic and Non Acidic Juice Blends. *IJAFST Int. J. Agric. Food Sci. Technol.* **2014**, *5*, 81–90.
26. Stratakos, A.C.; Delgado-Pando, G.; Linton, M.; Patterson, M.F.; Koidis, A. Industrial Scale Microwave Processing of Tomato Juice Using a Novel Continuous Microwave System. *Food Chem.* **2016**, *190*, 622–628. [CrossRef] [PubMed]
27. Helali, M.O.H.; Ibrahim, M.; Shafique, M.Z.; Rahman, M.M.; Biswas, S.K.; Islam, M.S. Formulation, preparation and preservation of lemon (*Citrus Limon* l.) cordial. *J. Bio-Sci.* **2008**, *16*, 125–127. [CrossRef]
28. AOAC. *Official Methods of Analysis*, 18th ed.; The Association of Official Analytical Chemists: Gaithersburg, MD, USA, 2011.
29. Saeeduddin, M.; Abid, M.; Jabbar, S.; Hu, B.; Hashim, M.M.; Khan, M.A.; Xie, M.; Wu, T.; Zeng, X. Physicochemical Parameters, Bioactive Compounds and Microbial Quality of Sonicated Pear Juice. *Int. J. Food Sci. Technol.* **2016**, *51*, 1552–1559. [CrossRef]
30. Abid, M.; Jabbar, S.; Hu, B.; Hashim, M.M.; Wu, T.; Wu, Z.; Khan, M.A.; Zeng, X. Synergistic Impact of Sonication and High Hydrostatic Pressure on Microbial and Enzymatic Inactivation of Apple Juice. *LWT-Food Sci. Technol.* **2014**, *59*, 70–76. [CrossRef]
31. Prieto, P.; Pineda, M.; Aguilar, M. Spectrophotometric Quantitation of Antioxidant Capacity through the Formation of a Phosphomolybdenum Complex: Specific Application to the Determination of Vitamin E. *Anal. Biochem.* **1999**, *269*, 337–341. [CrossRef]
32. FDA. *U.S. Food and Drug Administration Bacteriological Analytical Manual*, 8th ed.; FDA Center for Food Safety and Applied Nutrition: Silver Spring, MD, USA, 2001
33. Maturin, L.; Peeler, J.T. *BAM: Aerobic Plate Count*; US Food and Drug Administration: Silver Spring, MD, USA, 2001.
34. Steel, R.G.D.; Torrie, J.H.; Dicky, D.A. *Principles and Procedures of Statistics: A Biometrical Approaches*, 3rd ed.; McGraw Hill Book Co., Inc.: Singapore, 1997; pp. 204–227.
35. Pandiselvam, R.; Hebbar, K.B.; Manikantan, M.R.; Prashanth, B.K.; Beegum, S.; Ramesh, S.V. Microwave Treatment of Coconut Inflorescence Sap (Kalparasa®): A Panacea to Preserve Quality Attributes. *Sugar Tech* **2020**, *22*, 718–726. [CrossRef]
36. Bozkir, H.; Baysal, T. Concentration of Apple Juice Using a Vacuum Microwave Evaporator as a Novel Technique: Determination of Quality Characteristics. *J. Food Process Eng.* **2017**, *40*, e12535. [CrossRef]
37. Thirukkumar, S.; Vennila, P.; Kanchana, S. Physico-Chemical Characteristics of Noni Fruit Juice Blended Squashes during Storage. *Int. J. Chem. Stud.* **2018**, *6*, 449–455.
38. Jothi, J.S.; Karmoker, P.; Sarower, K. Quality Assessment of Mixed Fruit Squash: Physico-Chemical Analysis, Sensory Evaluation and Storage Studies. *J. Bangladesh Agric. Univ.* **2014**, *12*, 195–201. [CrossRef]
39. Picouet, P.A.; Landl, A.; Abadias, M.; Castellari, M.; Viñas, I. Minimal Processing of a Granny Smith Apple Purée by Microwave Heating. *Innov. Food Sci. Emerg. Technol.* **2009**, *10*, 545–550. [CrossRef]

40. Yadav, S.R.; Gehlot, S.; Siddiqui, G.R.B. Changes in chemical constituents and overall acceptability of guavamango ready-to-serve (RTS) drink and squash. *Beverage Food World* **2014**, *41*, 30–33.
41. Sattar, S.; Ahmad, T.; Mahr-un, N.; Imran, M.; Holmes, M.; Maycock, J.; Nadeem, M.; Khan, M.K. Microwave Processing Impact on Physicochemical and Bioactive Attributes of Optimized Peach Functional Beverage. *J. Food Process. Preserv.* **2019**, *43*, e13952. [CrossRef]
42. Song, W.J.; Sung, H.J.; Kang, D.H. Inactivation of *Escherichia Coli* O157: H7 and *Salmonella Typhimurium* in Apple Juices with Different Soluble Solids Content by Combining Ozone Treatment with Mild Heat. *J. Appl. Microbiol.* **2015**, *118*, 112–122. [CrossRef]
43. Fazaeli, M.; Hojjatpanah, G.; Emam-Djomeh, Z. Effects of Heating Method and Conditions on the Evaporation Rate and Quality Attributes of Black Mulberry (*Morus Nigra*) Juice Concentrate. *J. Food Sci. Technol.* **2013**, *50*, 35–43. [CrossRef]
44. Cheng, C.-X.; Jia, M.; Gui, Y.; Ma, Y. Comparison of the Effects of Novel Processing Technologies and Conventional Thermal Pasteurisation on the Nutritional Quality and Aroma of Mandarin (*Citrus Unshiu*) Juice. *Innov. Food Sci. Emerg. Technol.* **2020**, *64*, 102425. [CrossRef]
45. Igual, M.; García-Martínez, E.; Camacho, M.M.; Martínez-Navarrete, N. Effect of Thermal Treatment and Storage on the Stability of Organic Acids and the Functional Value of Grapefruit Juice. *Food Chem.* **2010**, *118*, 291–299. [CrossRef]
46. De Souza Comapa, S.; Carvalho, L.M.S.; Lamarão, C.V.; das Chagas do Amaral Souza, F.; Aguiar, J.P.L.; da Silva, L.S.; Mar, J.M.; Sanches, E.A.; dos Santos, F.F.; de Araújo Bezerra, J.; et al. Microwave Processing of Camu-Camu Juices: Physicochemical and Microbiological Parameters. *J. Food Process. Preserv.* **2019**, *43*, e13989. [CrossRef]
47. Rasoulian, S.S.; Aminifar, M.; Rashidi, L. Investigating the effect of two different thermal methods of rotary evaporation and microwave on the phenolic content, antioxidant activity and color of grapefruit juice concentrate. *Iran. J. Nutr. Sci. Food Technol.* **2017**, *12*, 47–54.
48. Jiang, B.; Mantri, N.; Hu, Y.; Lu, J.; Jiang, W.; Lu, H. Evaluation of Bioactive Compounds of Black Mulberry Juice after Thermal, Microwave, Ultrasonic Processing, and Storage at Different Temperatures. *Food Sci. Technol. Int.* **2015**, *21*, 392–399. [CrossRef]
49. Piljac-Žegarac, J.; Valek, L.; Martinez, S.; Belščak, A. Fluctuations in the Phenolic Content and Antioxidant Capacity of Dark Fruit Juices in Refrigerated Storage. *Food Chem.* **2009**, *113*, 394–400. [CrossRef]
50. Barba, F.J.; Esteve, M.J.; Frígola, A. High Pressure Treatment Effect on Physicochemical and Nutritional Properties of Fluid Foods During Storage: A Review. *Compr. Rev. Food Sci. Food Saf.* **2012**, *11*, 307–322. [CrossRef]
51. Papoutsis, K.; Pristijono, P.; Golding, J.B.; Stathopoulos, C.E.; Bowyer, M.C.; Scarlett, C.J.; Vuong, Q.V. Enhancement of the total phenolic compounds andantioxidant activity of aqueous *Citrus Limon* L. pomace extract using microwavepretreatment on the dry powder. *J. Food Process. Preserv.* **2017**, *41*, e13152. [CrossRef]
52. Martins, C.C.P.; Cavalcanti, R.N.; Cardozo, F.T.S.; Couto, S.M.; Esmerino, A.; Guimarães, J.T.; Balthazar, C.F.; Rocha, R.S.; Pimentel, T.C.; Freitas, M.Q.; et al. Effects of Microwave Heating on the Chemical Composition and Bioactivity of Orange Juice-Milk Beverages. *Food Chem.* **2021**, *345*, 128746. [CrossRef] [PubMed]
53. Adulvitayakorn, S.; Azhari, S.H.; Hasan, H. The Effects of Conventional Thermal, Microwave Heating, and Thermosonication Treatments on the Quality of Sugarcane Juice. *J. Food Process. Preserv.* **2020**, *44*, e14322. [CrossRef]
54. Li, T.; Li, J.; Hu, W.; Zhang, X.; Li, X.; Zhao, J. Shelf-Life Extension of Crucian Carp (*Carassius Auratus*) Using Natural Preservatives during Chilled Storage. *Food Chem.* **2012**, *135*, 140–145. [CrossRef]
55. Pradhan, N.; Kumar, D.; Singh, P.; Pisalkar, P.S. Sensory and Microbial Evaluation of Microwave Treated Sugarcane Juice. *Int. J. Curr. Microbiol. Appl. Sci.* **2020**, *9*, 1313–1320. [CrossRef]

sustainability

MDPI

Article

Determining the Factors Affecting the Boiling Heat Transfer Coefficient of Sintered Coated Porous Surfaces

Uzair Sajjad [1,*], Imtiyaz Hussain [2], Muhammad Sultan [3,*], Sadaf Mehdi [4], Chi-Chuan Wang [1], Kashif Rasool [5], Sayed M. Saleh [6], Ashraf Y. Elnaggar [7] and Enas E. Hussein [8]

1 Department of Mechanical Engineering, National Yang Ming Chiao Tung University, 1001 University Road, Hsinchu 300, Taiwan; ccwang@nctu.edu.tw
2 Department of Power Mechanical Engineering, National Tsing Hua University, No. 101, Section 2, Guangfu Road, East District, Hsinchu 300, Taiwan; imtiyazkou@yahoo.com
3 Department of Agricultural Engineering, Bahauddin Zakariya University, Bosan Road, Multan 60800, Pakistan
4 Department of Mechanical Engineering, Wichita State University, Wichita, KS 67260, USA; mehdi.sadaf96@gmail.com
5 Qatar Environment and Energy Research Institute, Hamad Bin Khalifa University (HBKU), Qatar Foundation, Doha P.O. Box 5825, Qatar; krasool@hbku.edu.qa
6 Department of Chemistry, College of Science, Qassim University, Buraidah 51452, Saudi Arabia; e.saleh@qu.edu.sa
7 Department of Food Nutrition Science, College of Science, Taif University, P.O. Box 11099, Taif 21944, Saudi Arabia; aynaggar@tu.edu.sa
8 National Water Research Center, P.O. Box 74, Shubra El-Kheima 13411, Egypt; enas_el-sayed@nwrc.gov.eg
* Correspondence: energyengineer01@gmail.com (U.S.); muhammadsultan@bzu.edu.pk (M.S.)

Citation: Sajjad, U.; Hussain, I.; Sultan, M.; Mehdi, S.; Wang, C.-C.; Rasool, K.; Saleh, S.M.; Elnaggar, A.Y.; Hussein, E.E. Determining the Factors Affecting the Boiling Heat Transfer Coefficient of Sintered Coated Porous Surfaces. *Sustainability* **2021**, *13*, 12631. https://doi.org/10.3390/su132212631

Academic Editor: Cinzia Buratti

Received: 29 September 2021
Accepted: 4 November 2021
Published: 16 November 2021

Publisher's Note: MDPI stays neutral with regard to jurisdictional claims in published maps and institutional affiliations.

Abstract: The boiling heat transfer performance of porous surfaces greatly depends on the morphological parameters, liquid thermophysical properties, and pool boiling conditions. Hence, to develop a predictive model valid for diverse working fluids, it is necessary to incorporate the effects of the most influential parameters into the architecture of the model. In this regard, two Bayesian optimization algorithms including Gaussian process regression (GPR) and gradient boosting regression trees (GBRT) are used for tuning the hyper-parameters (number of input and dense nodes, number of dense layers, activation function, batch size, Adam decay, and learning rate) of the deep neural network. The optimized model is then employed to perform sensitivity analysis for finding the most influential parameters in the boiling heat transfer assessment of sintered coated porous surfaces on copper substrate subjected to a variety of high- and low-wetting working fluids, including water, dielectric fluids, and refrigerants, under saturated pool boiling conditions and different surface inclination angles of the heater surface. The model with all the surface morphological features, liquid thermophysical properties, and pool boiling testing parameters demonstrates the highest correlation coefficient, $R^2 = 0.985$, for HTC prediction. The superheated wall is noted to have the maximum effect on the predictive accuracy of the boiling heat transfer coefficient. For example, if the wall superheat is dropped from the modeling parameters, the lowest prediction of R^2 (0.893) is achieved. The surface morphological features show relatively less influence compared to the liquid thermophysical properties. The proposed methodology is effective in determining the highly influencing surface and liquid parameters for the boiling heat transfer assessment of porous surfaces.

Keywords: pool boiling heat transfer coefficient; sintered coated porous surfaces; deep neural network; Bayesian optimization; gaussian process; gradient boosting regression trees

1. Introduction

Due to rapid advancements in the machining industry, microelectronics devices have gained popularity. These devices produce large amounts of heat. For the system safety and health of these devices, the removal of high heat fluxes in minimal space has become a challenge. These high heat fluxes cannot be handled by single-phase heat transfer. With the exploitation of latent heat, two-phase techniques are providing improved results, and several phase change phenomena are under investigation, for instance pool boiling, gas-assisted evaporative cooling, and spray cooling. Boiling heat transfer is a ubiquitous phenomenon because of its large heat-removing ability. Its applications are diverse, from renewable energy systems to refrigeration industries, desalination, nuclear reactors, and waste heat recovery plants. Similarly, if there is an uplift in the performance of the evaporator, it will be reflected in the efficiency of the heat pump. In the past decade, much research has been carried out to understand the boiling phenomenon.

Boiling heat transfer mainly depends on the surface and liquid thermophysical properties, and usually there are some constraints on the working fluid [1]. So, the best way to enhance the heat transfer is to modify the surface geometry [2]. Enhanced surfaces' morphologies can be changed by using various methods, broadly classified as active and passive techniques [2]. Surface engineering is performed to encourage rewetting, increase the nucleation sites, and improve the effective boiling surface area [3]. Advanced additive manufacturing technique are providing a great deal of flexibility in getting an optimized porous geometry, and many studies have been conducted on porous surfaces. As reported by researchers, the boiling performance of enhanced surfaces is improved due to the presence of a large number of bubble seeding cavities (nucleation sites), facilitating earlier and more numerous bubble detachment [4,5]. At the same time, porous substrates have also been able to extend the CHF limit by interrupting the bubbles' coalescence because of their having separate liquid–vapor paths, facilitating sustainable liquid replenishment [4]. Leonardo et al. [6] reported on two porous metal foams' (Ni and Cu) performances with respect to the heat flux. They found that copper foam showed significant improvement throughout the boiling curve as compared to the simple surface. They attributed this enhancement to the improved thermal conductivity of copper. Through visualization, they found that Ni foam successfully generated many small bubbles at low heat flux and reduced the ONB; however, at high heat flux, it resisted the early removal of large bubbles, which deteriorated the heat transfer.

Among various coated surfaces, sintered surfaces have shown promising results with higher stability [7]. Pastsuzko et al. [8] prepared sintered microporous surfaces and conducted visualization experimentation during the pool boiling of water and FC-72. They observed 130% and 75% enhancements in heat flux for water and Fc-72, respectively, with respect to the plain sample. Furthermore, they also developed a simple semi-analytical model to predict the boiling heat transfer. Halon et al. [9] tested the samples manufactured by Pastsuzko et al. [8] and analyzed the boiling behavior under sub-atmospheric conditions using water as the working fluid. They also observed enhancement in heat transfer. However, the best-performing sample under atmospheric conditions was the worst-performing under sub-atmospheric conditions. Arvind and Satish [10] created three boiling surfaces to analyze the microchannel performance with varied microporous coatings. During their tests, the maximum degree of enhancement was achieved by the sample with fully sintered microchannels as compared to the other two surfaces (only fin tops and channel walls). Xu et al. [11] investigated the role of the shape of the pore opening, coating thickness and thermal conductivity of the material during the pool boiling of DI water on open-celled metallic foam-sintered surfaces. They found that the sintered foam sample performed better than the grooved-shape surface because of the high pore density of the sintered foam structure. Moreover, the larger grooved samples' performance was poorer than the narrow-grooved one's because of the low capillary force of wide grooves. Jun et al. [12] conducted a parametric study of the pool boiling of water on sintered copper surfaces. They studied the particle sizes of 10 μm, 25 μm and 67 μm with different coating thicknesses. They observed a plausible increase in heat transfer coefficient (HTC) and critical heat flux (CHF). As compared to the plain sample, the maximum enhancement in HTC was 8× at a particle size of 67 μm with a coating thickness of 296 μm, and the highest reported improvement in CHF was 2× at the particle size of 67 μm and coating thickness of 428 μm. Within their tested range, both HTC and CHF increased as the particle size or coating thickness increased. Another research group [13] prepared honeycomb porous structures with and without radial gradient by electrodeposition and sintering in a reducing environment. They tested the samples in the pool boiling of DI water and found that the samples with radial gradient, and smaller pores at the center than near the edges, showed better heat transfer than the sample with uniform structure. They attributed this enhancement to the higher K/Reff, and as per their investigation, a radial gradient from edges to center assisted in quicker rewetting. Furthermore, through visualization experimentation, they also found that the radially distributed porous sample generated larger bubbles at a faster rate, which helped in removing more heat than in other samples. Recently, Pastsuzko et al. [14] investigated the boiling heat transfer of water, FC-72 and Novec-649 on microporous sintered surfaces with and without mesh. They postulated that mesh coverings on the sintered micro-finned surfaces are only effective at low heat flux for liquids with low surface tension, because at the high heat flux (50 kW/m^2 for FC-72 and 100 kW/m^2 for ethanol), maximum performance was recorded in the sample with no mesh or covering.

Aim and Motivation of the Study

As reported by researchers, porous surfaces' performance greatly depends on the morphological parameters, and these parameters are affected by changes in the working fluid. So, to develop a predictive model valid for diverse working fluids, it is necessary to incorporate the effect of these parameters and the thermophysical properties of the working fluid. The predictability of the empirical correlations is usually hampered by these parameters, because these correlations were developed on a limited database. To cover a wide range of data, an artificial intelligence (AI) model, based on advance algorithms and new libraries, can be developed with high accuracy [1,15]. Keeping this in view, the objective of this study is to develop Bayesian optimized deep neural network models to perform a sensitivity analysis for finding the most influential parameters in the boiling heat transfer of sintered coated porous surfaces fabricated on copper substrate, subjected to a variety of high- and low-wetting working fluids, including water, dielectric fluids, and refrigerants, under saturated pool boiling conditions and different surface inclination angles of the heater surface. The detailed impacts of combined and individual surfaces, liquids, and boiling condition parameters have been assessed for a range of tested data. In line with this, the most impactful parameters are highlighted by representing their predictions. The proposed method can help to assess the strong morphological parameters of sintered coated porous surfaces subjected to a range of working fluids and pool boiling conditions. The proposed methodology is effective in determining the highly influential surface and liquid parameters for boiling heat transfer assessment.

2. Materials and Methods

2.1. Experimental Data Collection

In the present investigation, 380 data points have been collected from the saturated pool boiling experiments of microporous coated surfaces for different high- and low-surface tension working fluids including water, dielectric liquids, and refrigerants [11,12,14,16–34]. Most of the tested porous surfaces are manufactured by sintering techniques. The considered porous surfaces have a range of morphological parameters in terms of particle diameter, surface roughness, coating thickness, and porosity. The considered data include the saturated pool boiling results for a variety of surface inclination angles (0–180°). The investigated wall superheat is 0–40 K and the heat transfer coefficient ranges between 0.5 and 476 kW/m^2 K. The experimental data range of the studied parameters can be seen in Table 1.

Table 1. Investigated experimental data range.

Parameter	Value	Unit
Wall Superheat	0.5–38	K
HTC	0.45–476	$kW\ m^{-2}\ K^{-1}$
Heat Flux	0.3–18088	$kW\ m^{-2}$
Surface Inclination	0–180	°
Saturation Temperature	56–100	°C
Liquid Density	997–1680	$Kg\ m^{-3}$
Heat of Vaporization	88–2257	$kJ\ kg^{-1}$
Specific Heat	1100–4180	$J\ Kg^{-1}\ K^{-1}$
Surface Tension	10–72	$m\ Nm^{-1}$
Thermal Conductivity of the Working Fluid	0.057–0.608	$W\ m^{-1}\ K^{-1}$
Porosity	39–65	%
Particle Diameter	11.2–1000	μm
Coating Thickness	250–590	μm

2.2. Methodology

Training and testing data are divided 80–20% as is commonly practiced in the literature [35,36]. Firstly, two models were developed by considering the surface parameters, liquid thermophysical features, and surface inclination angle. In one of the models, other than the aforementioned parameters, heat flux was taken as the added input parameter, while in the second model, wall superheat was taken as the added input parameter. It was noticed that the wall superheat model showed much better prediction ability compared to the heat flux model. Based on these findings, we decided to consider the wall superheat model for the further assessment of boiling heat transfer coefficient.

2.2.1. Bayesian Optimization (BO)

Bayesian optimization is used to select the hyper-parameters for evaluation in the true objective function by building a probability model of the objective function. This employs the Bayes theorem for determining the maximum and minimum of an objective function. There are different techniques available based on the BO. Here, the two most common BO methods, GPR and GBRT, are considered for hyper-parameter optimization.

2.2.2. Gaussian Process Regression (GPR)

The GPR model is an ML framework employed for classification and regression problems. This is a nonparametric Bayesian method for regression. GPR can work well for small databanks. The GPR model predicts by including prior knowledge, and provides predictions uncertainty measures.

2.2.3. Gradient Boosting Regression Trees (GBRT)

Decision trees are machine learning algorithms that are most famous for their feature selection. These can be used for classification and regression problems. Decision trees for regression problems are known as regression trees. The model starts learning with an increase in the iterations. The training process stops when the hyper-parameters are triggered. GBRTs are iterative algorithms in which each tree takes account of the error in the previous one, and their final outcome is the mean of all trees' predictions.

These algorithms require comparatively less effort in preprocessing data, and can process incomplete data. However, they are expensive in terms of the computational training time.

In order to achieve the optimized deep learning model, the hyper-parameters were tuned by using the Bayesian optimization method. Here, two different Bayesian optimization methods, namely, Gaussian process regression (GPR) and gradient boosting regression trees (GBRT), were used for hyper-parameter tuning. For the models' performance evaluation, different error metrics, such as the correlation coefficient (R^2), mean absolute error (MAE), and absolute average relative deviation (AARD), were used.

Figure 1a,b represents the procedure and flow charts for fine-tuning the hyper-parameters by using two different Bayesian optimization methods, such as GPR and GBRT.

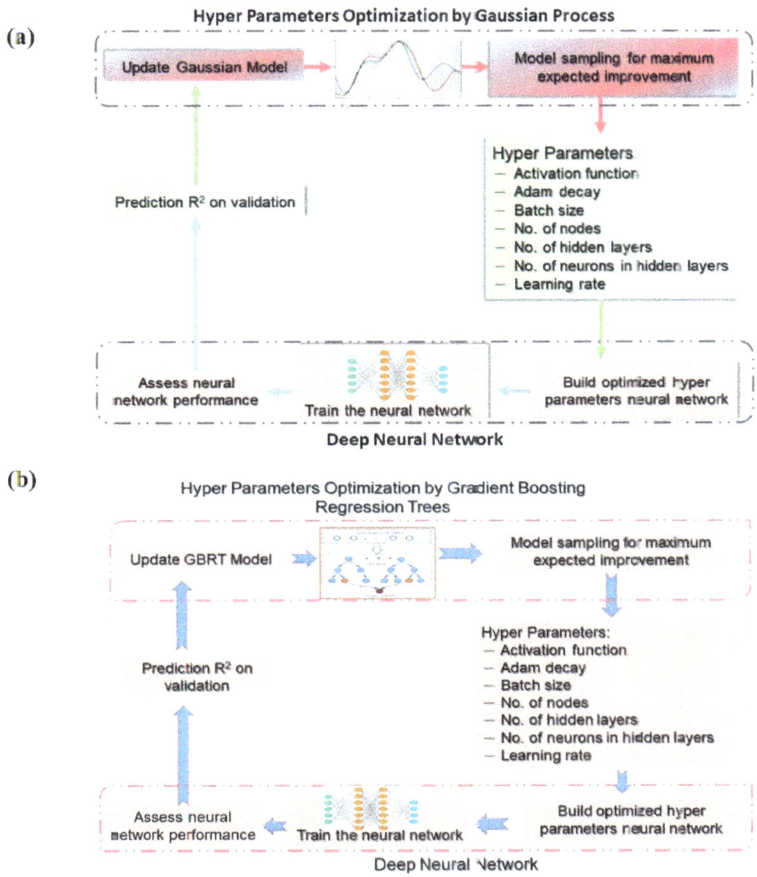

Figure 1. Procedure of hyper-parameter tuning by (a) Gaussian process and (b) gradient boosting regression trees.

Once the hyper-parameters are optimized, then a deep neural network was developed based on the acquired information of the optimized hyper-parameters.

2.2.4. Hyper-Parameters

These are the parameters whose values are used in controlling the learning process. In order to achieve an optimal model with a short computational time, the hyper-parameters' tuning should be optimized, which is a major goal of the BO. In the developed model, the hyper-parameters are activation function, learning rate, Adam decay, number of nodes, no. of hidden layers, no. of neurons in each hidden layer, and batch size, as shown in Figure 1. A pair plot showing the fine-tuned hyper-parameter can be seen in Figure 2. The red circles represent the optimal value for each hyper-parameter. In Figure 2, the yellow region shows a strong relationship (those values can be selected for the hyper-parameter) while the green region represents a weak relationship (those values are not recommended for selection of that hyper-parameter). The optimal hyper-parameters in terms of the partial dependence are highlighted in Figure 2. It can be seen that the no. of dense layers should be 6, the activation function should be ReLU, and so on. The range of hyper-parameters considered in this study can be witnessed in Table 2.

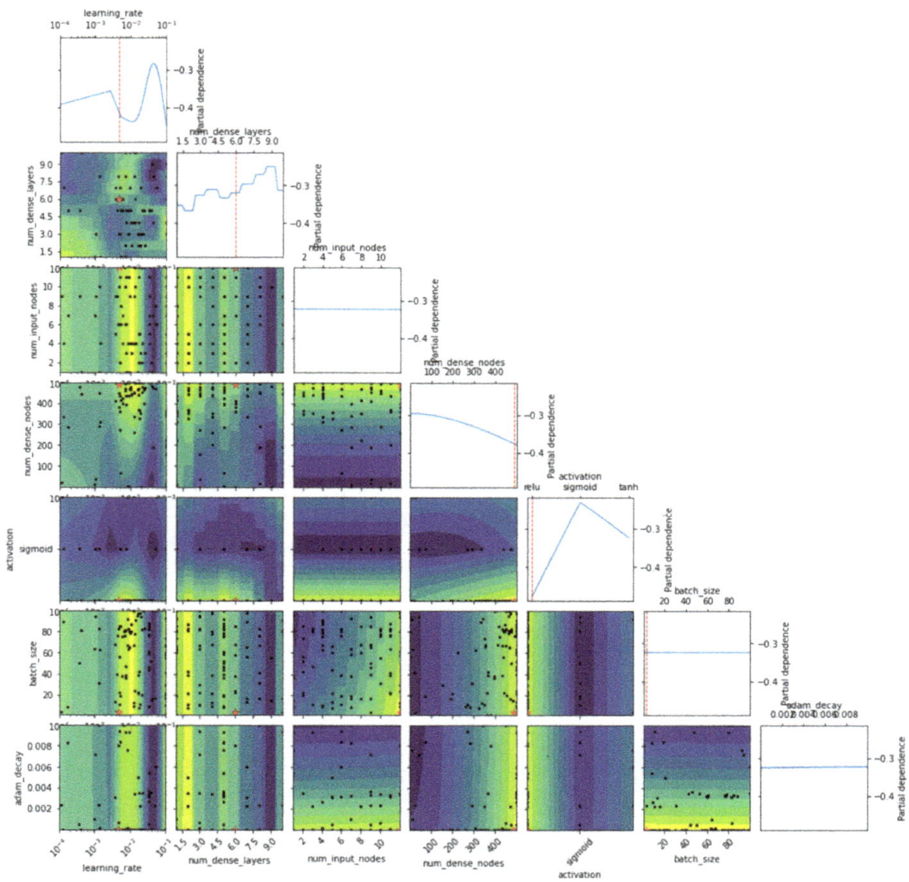

Figure 2. Pair plot showing the fine-tuned hyper-parameters.

Table 2. Range of hyper-parameters considered in this study.

Hyper-Parameter	Range
Learning rate	0.0001 to 0.1
Adam decay	0.000001 to 0.01
Input nodes	1 to 12
Dense layers	1 to 10
Dense nodes	1 to 500
Batch size	1 to 100
Activation function	ReLU, Sigmoid, tanh

The method for achieving the best solution and minimizing the error with respect to the number of calls can be seen in Figure 3. It is obvious that the error reaches a minimum level after 40 calls, although the full convergence occurs at the 18th iteration. This represents the potential of the Bayesian optimization technique when considering a large number of input variables for a wide range of data. For instance, timely convergence was achieved for an optimal solution. The architecture of the optimized models is provided in Table 3, which clearly shows that both of the Bayesian optimization techniques, GPR and GBRT, yield the same values of correlation coefficient (R^2). Other details regarding learning rate, no. of input and dense nodes, activation function, batch size, Adam decay, and no. of dense layers for both of the algorithms, can be found in Table 3.

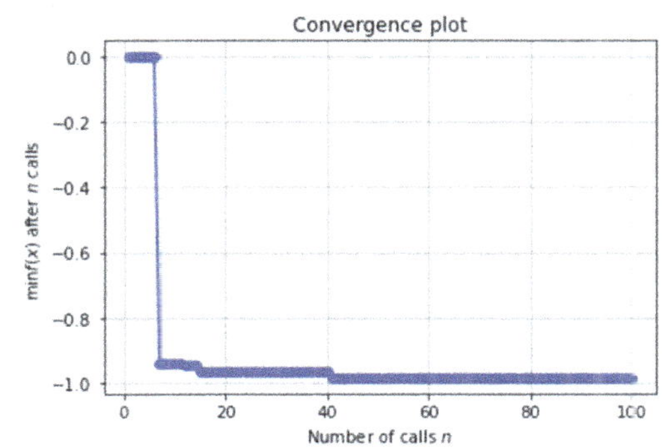

Figure 3. Error convergence plot.

Table 3. Architecture of the optimized models.

Bayesian Optimization Method	Gradient Boosting Regression Trees	Gaussian Process
Learning rate	0.00787965	0.004673739
No. of hidden layers	6	6
No. of neurons in input layer	12	12
No. of neurons in each hidden layer	467	489
Activation function	ReLU	ReLU
Batch size	19	3
Adam decay	0.0048345	0.000001
No. of neurons in output layer	1	1
Correlation coefficient (R^2)	98.48	98.48

3. Results and Discussions

Factors Affecting the Boiling Heat Transfer Coefficient

Pool boiling heat transfer mainly relies on the surface morphological features, liquid thermophysical properties, and pool boiling testing conditions (see Figure 4).

Generally, the porosity of a material is defined by the ratio of void volume to total volume, and various hypotheses about the influence of porosity on the boiling surface have been reported, for instance it can influence the heat transfer through the thickness of the porous coating, the diameter of the coating particles, the number of pores, and pore connectivity. By increasing the porosity, the wetted area is increased, and more nucleation sites become active as it becomes easier to supply the liquid to the nucleation sites, and this increases the number of bubbles and the bubble generation frequency, which result in heat transfer enhancement. At low heat fluxes, boiling performance is improved with the addition of the porous coated layer because of the formation of a large number of bubbles. However, at high heat flux, increases in coating thickness may result in the formation of a vapor blanket and heat transfer deterioration, and this is why the ratio of the coating thickness to particle diameter is calculated and an optimal value is reported [37]. Furthermore, the heat transfer coefficient can also be augmented by increasing the roughness of the boiling surface. This effect was reported for the first time by Jackob in 1931 [38]. A surface is roughened to create bubbles' seeding cavities in order to produce a large number of bubbles, and due to this, the heat transfer coefficient is improved [39,40].

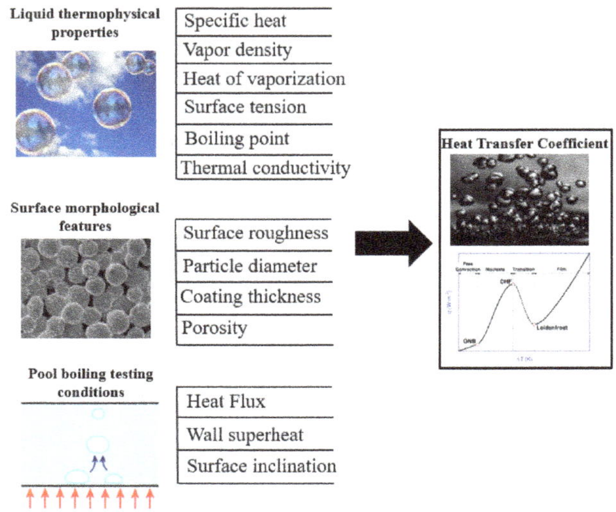

Figure 4. Parameters affecting the boiling heat transfer coefficient.

Some of the most influential liquid thermophysical properties in boiling heat transfer are liquid density, specific heat, vapor density, the latent heat of vaporization, surface tension, boiling point, and liquid thermal conductivity. Similarly, for microporous coated surfaces, the morphological parameters of porosity, coating thickness, particle diameter, and surface roughness have a strong influence on the boiling heat transfer coefficient.

The surface morphological parameters affecting the boiling heat transfer coefficient or porous surfaces are porosity, coating thickness, particle diameter, and surface roughness. The impact of individual morphological parameter on the boiling heat transfer assessment of various working fluids is provided in Figure 5, followed by the error density analysis in Figure 6. The impact of liquid thermophysical properties on the BHTC prediction along with error density analysis is given in Figures 7 and 8, while Figure 9 shows the sensitivity analysis of surface inclination and wall superheat.

Table 4 presents the individual and combined impacts of surface morphological features, liquid thermophysical properties, and pool boiling conditions on the assessment of heat transfer coefficient. Apparently, the model with all the surface morphological features, liquid thermophysical properties, and pool boiling testing parameters demonstrated the highest accuracy for HTC prediction. The wall superheat is noted to have the maximum impact on the predictive accuracy of the boiling heat transfer coefficient. More specifically, if the wall superheat is dropped from the modeling parameters, the lowest predictive R^2 (0.893%) is achieved. Among the surface morphological parameters, the particle diameter exhibits the strongest influence on the heat transfer coefficient. On the contrary, the surface roughness and coating thickness do not seem to have a strong impact on the pool boiling data of sintered coated porous surfaces for water, refrigerants, and dielectric liquids. In general, the surface features show relatively less influence compared to the liquid thermophysical properties. This is because liquids with totally different thermophysical properties result in very different boiling phenomena. However, liquid thermal conductivity and specific heat cause a noticeable impact on the boiling heat transfer coefficient of porous surfaces. From the above results, it can be stated that liquid thermophysical properties have much more of an effect on the pool boiling phenomenon of sintered coated porous surfaces compared to the morphology of the heater surfaces. In addition, the BHTC is strongly influenced by the surface inclination angle of the heater surface.

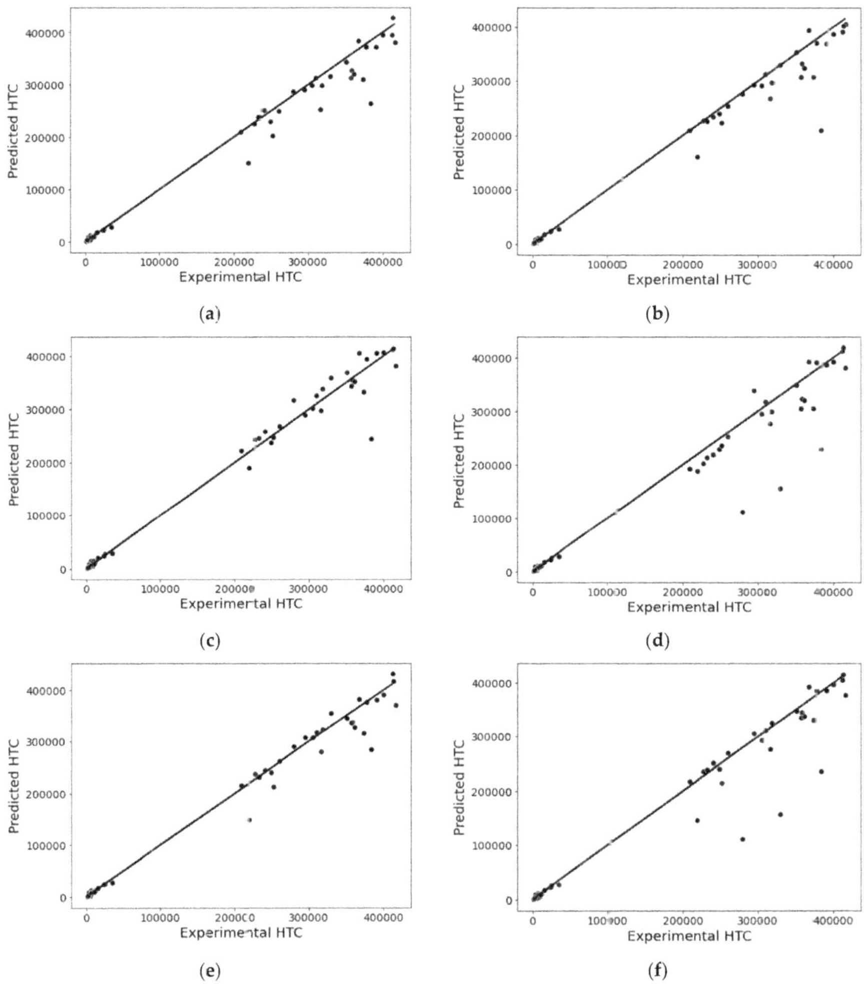

Figure 5. Impact of surface morphological features on the BHTC prediction. (**a**) Roughness dropped. (**b**) Porosity dropped. (**c**) Coating thickness dropped. (**d**) Particle diameter dropped. (**e**) Original with all features. (**f**) No surface feature.

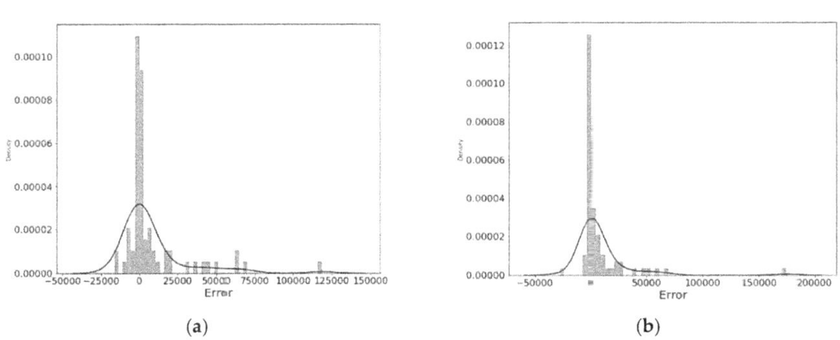

Figure 6. *Cont.*

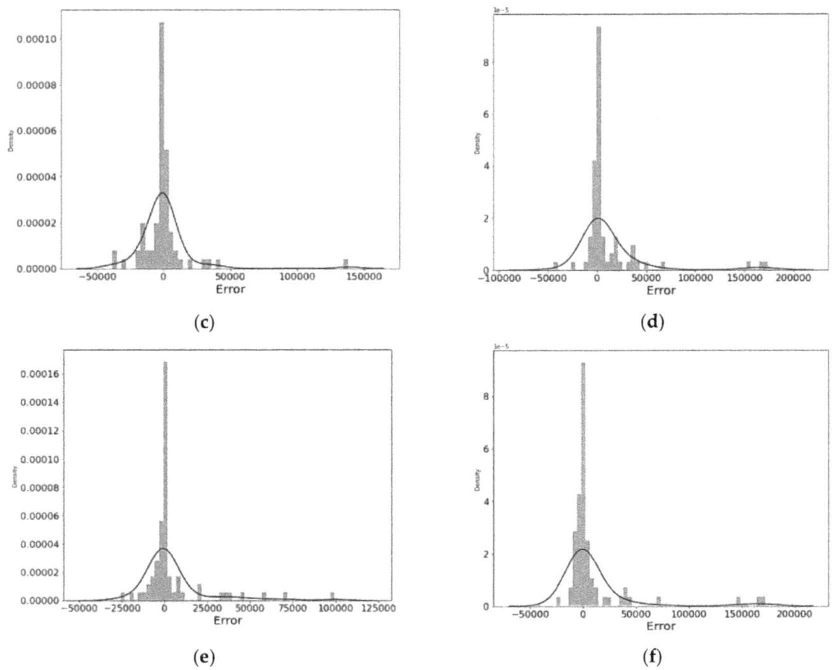

Figure 6. Error density analysis for surface morphological features in the BHTC prediction. (**a**) Roughness dropped. (**b**) Porosity dropped. (**c**) Coating thickness dropped. (**d**) Particle diameter dropped. (**e**) Original with all features. (**f**) All surface features dropped.

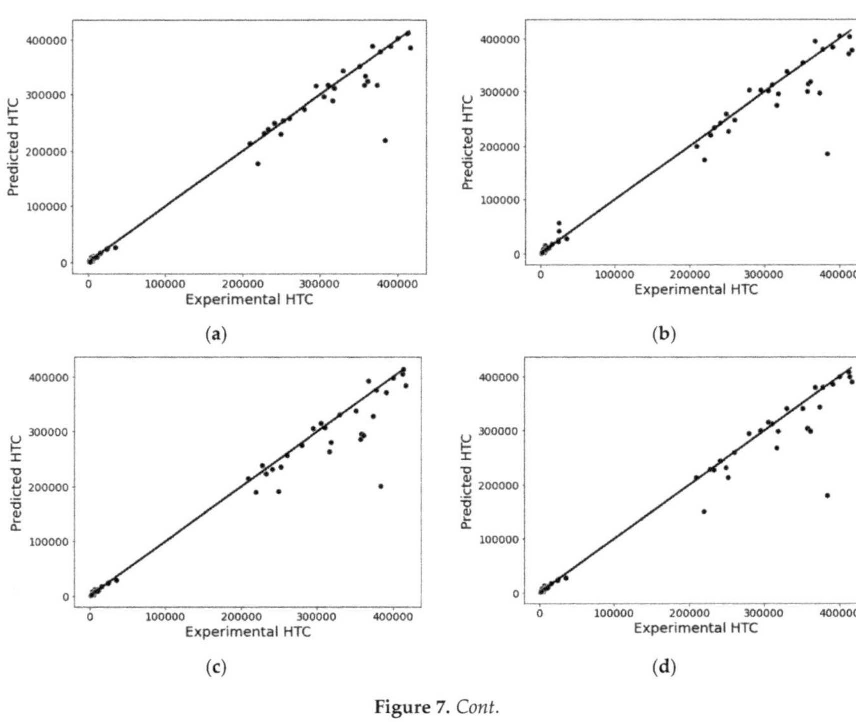

Figure 7. *Cont.*

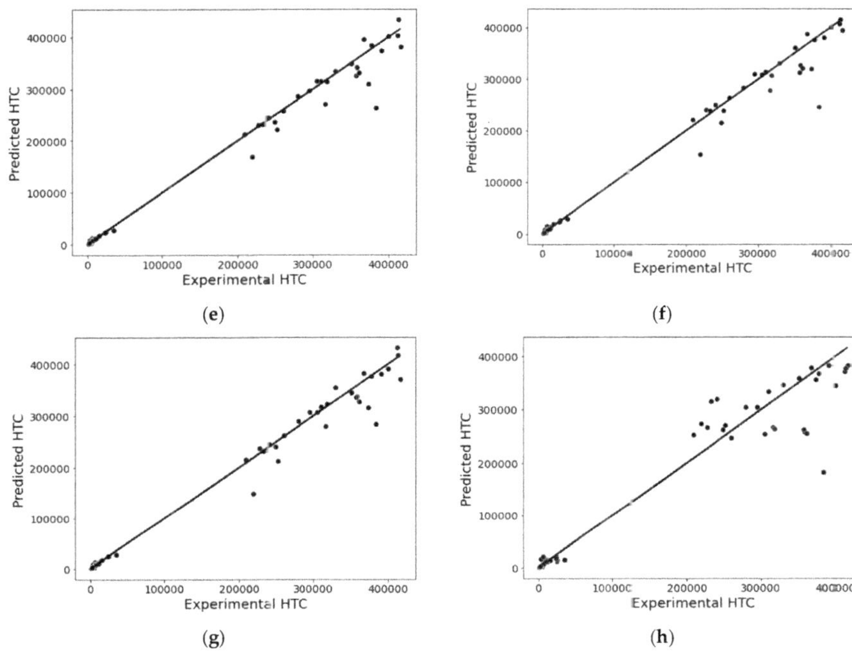

Figure 7. Impact of liquid thermophysical properties on the BHTC prediction. (**a**) Surface tension dropped. (**b**) Boiling point dropped. (**c**) Liquid density dropped. (**d**) Heat of vaporization dropped. (**e**) Specific heat dropped. (**f**) Liquid thermal conductivity dropped. (**g**) Original with all features. (**h**) No liquid properties.

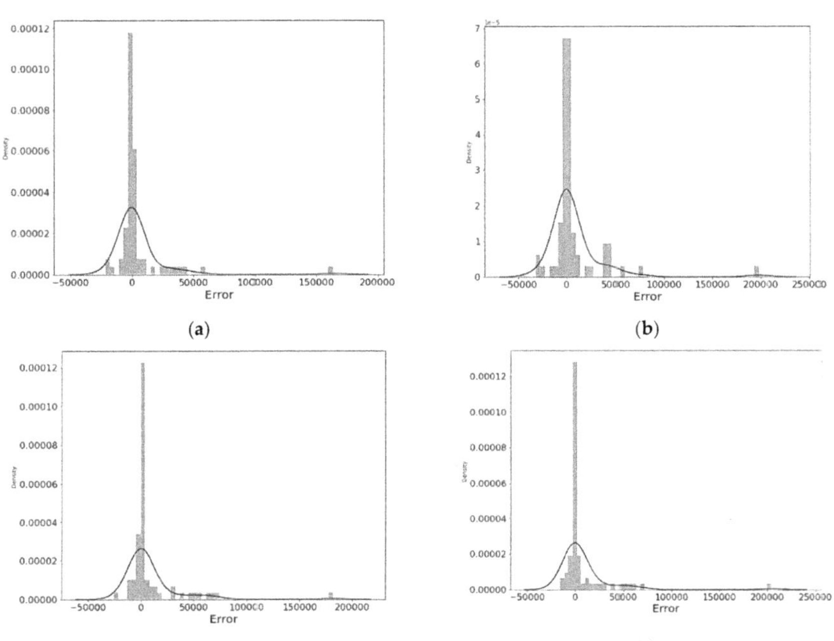

Figure 8. *Cont.*

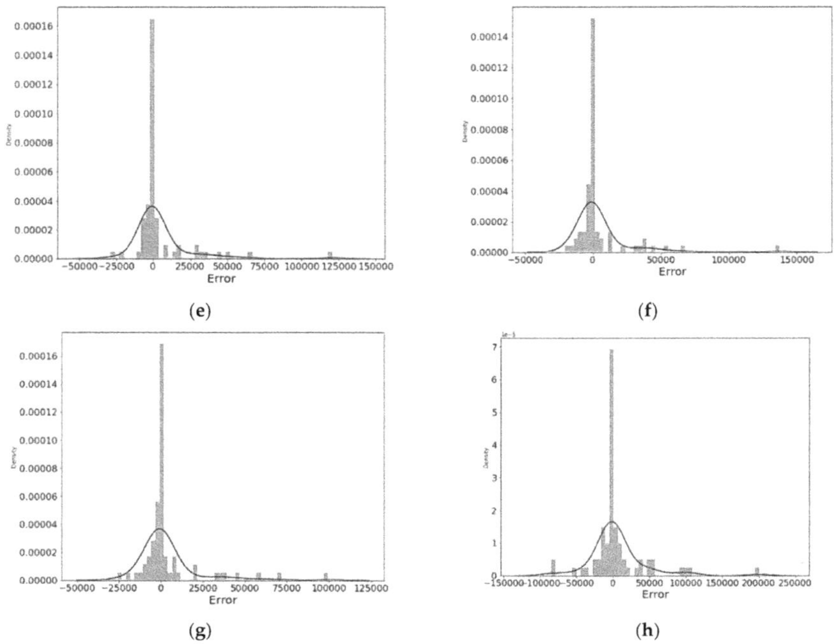

Figure 8. Error density analysis of liquid thermophysical properties in the BHTC prediction. (**a**) Surface tension dropped. (**b**) Boiling point dropped. (**c**) Liquid density dropped. (**d**) Heat of vaporization dropped. (**e**) Specific heat dropped. (**f**) Liquid thermal conductivity dropped. (**g**) Original with all features. (**h**) All liquid properties dropped.

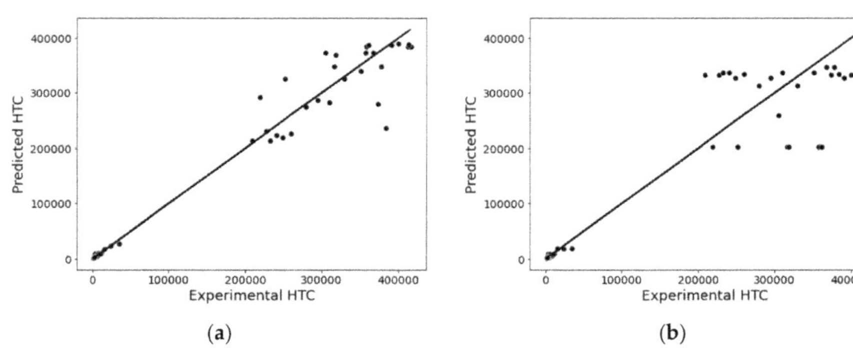

Figure 9. *Cont.*

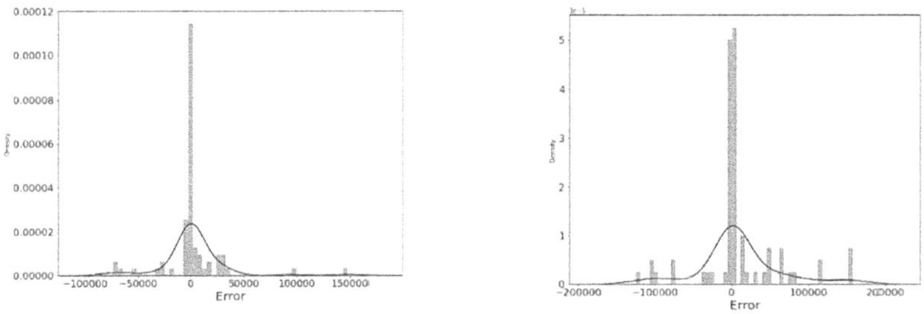

(c) (d)

Figure 9. Impact of (**a**) surface inclination, (**b**) wall superheat and error density analysis of (**c**) surface inclination and (**d**) wall superheat in the BHTC prediction.

Table 4. Input parameters of the developed models along with their performances.

Model	Input Parameters	R^2	AARD
Original model with all features	porosity coating thickness particle diameter surface roughness liquid density specific heat latent heat of vaporization surface tension boiling point liquid thermal conductivity surface inclination wall superheat	0.9855	17.127
Porosity dropped	coating thickness particle diameter surface roughness liquid density specific heat latent heat of vaporization surface tension boiling point liquid thermal conductivity surface inclination wall superheat	0.975	16.301
Coating thickness dropped	porosity particle diameter surface roughness liquid density specific heat latent heat of vaporization surface tension boiling point liquid thermal conductivity surface inclination wall superheat	0.983	19.25

Table 4. *Cont.*

Model	Input Parameters	R^2	AARD
Particle diameter dropped	porosity coating thickness surface roughness liquid density specific heat latent heat of vaporization surface tension boiling point liquid thermal conductivity surface inclination wall superheat	0.948	19.41
Surface roughness dropped	porosity coating thickness particle diameter liquid density specific heat latent heat of vaporization surface tension boiling point liquid thermal conductivity surface inclination wall superheat	0.981	22.088
All surface features dropped	liquid density specific heat latent heat of vaporization surface tension boiling point liquid thermal conductivity surface inclination wall superheat	0.951	25.541
Liquid density dropped	porosity coating thickness particle diameter surface roughness specific heat latent heat of vaporization surface tension boiling point liquid thermal conductivity surface inclination wall superheat	0.969	15.678
Specific heat dropped	porosity coating thickness particle diameter surface roughness liquid density latent heat of vaporization surface tension boiling point liquid thermal conductivity surface inclination wall superheat	0.984	15.981

Table 4. *Cont.*

Model	Input Parameters	R^2	AARD
Heat of vaporization dropped	porosity coating thickness particle diameter surface roughness liquid density specific heat surface tension boiling point liquid thermal conductivity surface inclination wall superheat	0.967	18.039
Surface tension dropped	porosity coating thickness particle diameter surface roughness liquid density specific heat latent heat of vaporization boiling point liquid thermal conductivity surface inclination wall superheat	0.979	16.932
Boiling point dropped	porosity coating thickness particle diameter surface roughness liquid density specific heat latent heat of vaporization surface tension liquid thermal conductivity surface inclination wall superheat	0.966	18.643
liquid thermal conductivity dropped	porosity coating thickness particle diameter surface roughness liquid density specific heat latent heat of vaporization surface tension boiling point surface inclination wall superheat	0.981	17.019
All liquid features dropped	porosity coating thickness particle diameter surface roughness surface inclination wall superheat	0.94	35.6

Table 4. *Cont.*

Model	Input Parameters	R^2	AARD
Surface inclination dropped	porosity coating thickness particle diameter surface roughness liquid density specific heat latent heat of vaporization surface tension boiling point liquid thermal conductivity wall superheat	0.967	20.408
Wall superheat dropped	porosity coating thickness particle diameter surface roughness liquid density specific heat latent heat of vaporization surface tension boiling point liquid thermal conductivity surface inclination	0.893	30.079

The choice of the pool boiling parameters is a very important aspect of this study. For instance, to predict the boiling heat transfer coefficient of engineered surfaces subjected to different working fluids of completely different thermophysical properties for a wide range of pool boiling conditions, it is extremely important to include the most influential parameters in the architecture of the model. This is the major limitation of the existing correlations in the literature—that these correlations cannot account for all of the important surface, liquid, and testing parameters, resulting in a very poor predictability for different surface liquid combinations and testing ranges. Other than considering the important surface and liquid features, the proposed model accounts for important pool boiling testing parameters. For example, in the proposed model, wall superheat and surface inclination angle were considered as the inputs. However, the predictions can be done by replacing the wall superheat with the applied heat flux. However, for the investigated data, the predictability of the model with the heat flux was relatively lower than the model with the wall superheat. Hence, wall superheat was chosen as the input parameter.

With the help of the proposed methodology, a highly accurate BHTC can be estimated by incorporating the strongly influencing liquid, surface, and pool boiling testing parameters into the Bayesian optimization-based neural network model. Owing to improved heat transfer performance, microporous surfaces are the prime candidate to remove the high heat fluxes generated through various applications. For instance, 3M copper powder is widely used in industry to form a boiling coated surface. These boiling surfaces can be used to cool microprocessors, LEDs, power electronics, MOSFETs and IGBTs. Other than that, sintered coated surfaces can be used to manufacture high-flux surfaces, employed to reduce plant size by decreasing the number of reboilers required in petrochemical plants.

4. Conclusions

The objective of this study is to develop Bayesian optimized deep neural network models to perform a sensitivity analysis for finding the most influential parameters in the boiling heat transfer assessment of sintered coated porous surfaces subjected to a variety of high- and low-wetting working fluids, including water, dielectric fluids, and refrigerants, under saturated pool boiling conditions and different surface inclination angles of the heater surface. Some of very specific conclusions are provided as follows.

- The model with all the surface morphological features, liquid thermophysical properties, and pool boiling testing parameters demonstrates the highest $R^2 = 0.985$ for HTC prediction.
- The wall superheat is noted to have the maximum impact on the predictive accuracy of the boiling heat transfer coefficient. For example, if the wall superheat is dropped from the modeling parameters, the lowest prediction of R^2 (0.893) is achieved.
- The surface morphological features show relatively less influence compared to the liquid thermophysical properties, e.g., liquid thermophysical properties are much more sensitive to the pool boiling phenomenon of sintered coated porous surfaces compared to the morphology of the heater surfaces.
- Particle diameter showed the strongest influence on the heat transfer coefficient compared to the rest of the morphological parameters.
- The BHTC is strongly influenced by the surface inclination angle of the heater surface.
- By dropping the surface inclination angle from the modeled parameters, R^2 is reduced to 0.967.
- The proposed methodology can be applied to a wider range of data in order to determine the highly influential surface and liquid parameters for boiling heat transfer assessment.

Author Contributions: Conceptualization, U.S.; data curation, U.S., I.H. and K.R.; formal analysis, U.S., I.H. and K.R.; funding acquisition, C.-C.W., K.R., S.M.S. and A.Y.E.; investigation, U.S., I.H., S.M., K.R. and E.E.H.; methodology, U.S. and I.H.; project administration, U.S., C.-C.W. and A.Y.E.; resources, S.M.S., A.Y.E. and E.E.H.; software, U.S., I.H. and S.M.; supervision, C.-C.W. and A.Y.E.; validation, I.H. and C.-C.W.; visualization, M.S., S.M. and E.E.H.; writing—original draft, U.S. and I.H.; writing—review and editing, M.S., K.R., S.M.S. and E.E.H. All authors have read and agreed to the published version of the manuscript.

Funding: This work was financially supported by the Ministry of Science and Technology, Taiwan under contracts 108-2221-E-009-058-MY3 and 109-2622-E-009-0015. This work was also financially supported by the Taif University Researchers Supporting Project No. TURSP-2020/32, Taif University, Taif, Saudi Arabia.

Institutional Review Board Statement: Not applicable.

Informed Consent Statement: Not applicable.

Data Availability Statement: Data are contained within the article.

Acknowledgments: This work was financially supported by the Taif University Researchers Supporting Project No. TURSP-2020/32, Taif University, Taif, Saudi Arabia. The authors are also indebted to the financial support from Ministry of science and technology, Taiwan under contracts 108-2221-E-009-058-MY3 and 109-2622-E-009-0015.

Conflicts of Interest: The authors declare no conflict of interest.

References

1. Sajjad, U.; Hussain, I.; Wang, C.-C. A high-fidelity approach to correlate the nucleate pool boiling data of roughened surfaces. *Int. J. Multiph. Flow* **2021**, *142*, 103719. [CrossRef]
2. Sajjad, U.; Sadeghianjahromi, A.; Ali, H.M.; Wang, C.-C. Enhanced pool boiling of dielectric and highly wetting liquids-A review on surface engineering. *Appl. Therm. Eng.* **2021**, *195*, 117074. [CrossRef]
3. Sajjad, U.; Sadeghianjahromi, A.; Ali, H.M.; Wang, C.-C. Enhanced pool boiling of dielectric and highly wetting liquids-a review on enhancement mechanisms. *Int. Commun. Heat Mass Transf.* **2020**, *119*, 104950. [CrossRef]
4. Sajjad, U.; Sadeghianjahromi, A.; Wang, C.-C. Enhancing Boiling Heat Transfer for Electronics Cooling by Embedding an Array of Microgrooves into Sandblasted Surfaces. *Heat Trans. Res.* **2021**, *52*, 71–89. [CrossRef]
5. Tran, N.; Sajjad, U.; Lin, R.; Wang, C.-C. Effects of surface inclination and type of surface roughness on the nucleate boiling heat transfer performance of HFE-7200 dielectric fluid. *Int. J. Heat Mass Transf.* **2020**, *147*, 119015. [CrossRef]
6. Manetti, L.L.; Ribatski, G.; de Souza, R.R.; Cardoso, E.M. Pool boiling heat transfer of HFE-7100 on metal foams. *Exp. Therm. Fluid Sci.* **2020**, *113*, 110025. [CrossRef]
7. Sajjad, U.; Wang, C.-C. Nucleate pool boiling of high flux sintered coated porous surfaces with dielectric liquid, HFE-7200. *J. Enhanc. Heat Transf.* **2020**, *27*, 767–784. [CrossRef]

8. Pastuszko, R.; Wójcik, T.M. Experimental investigations and a simplified model for pool boiling on micro-fins with sintered perforated foil. *Exp. Therm. Fluid Sci.* **2015**, *63*, 34–44. [CrossRef]
9. Halon, T.; Zajaczkowski, B.; Michaie, S.; Rulliere, R.; Bonjour, J. Enhanced tunneled surfaces for water pool boiling heat transfer under low pressure. *Int. J. Heat Mass Transf.* **2018**, *116*, 93–103. [CrossRef]
10. Jaikumar, A.; Kandlikar, S.G. Enhanced pool boiling heat transfer mechanisms for selectively sintered open microchannels. *Int. J. Heat Mass Transf.* **2015**, *88*, 652–661. [CrossRef]
11. Xu, Z.; Qu, Z.; Zhao, C.; Tao, W. Pool boiling heat transfer on open-celled metallic foam sintered surface under saturation condition. *Int. J. Heat Mass Transf.* **2011**, *54*, 3856–3867. [CrossRef]
12. Jun, S.; Kim, J.; Son, D.; Kim, H.Y.; You, S.M. Enhancement of pool boiling heat transfer in water using sintered copper microporous coatings. *Nucl. Eng. Technol.* **2016**, *48*, 932–940. [CrossRef]
13. Mo, D.-C.; Yang, S.; Luo, J.-L.; Wang, Y.-Q.; Lyu, S.-S. Enhanced pool boiling performance of a porous honeycomb copper surface with radial diameter gradient. *Int. J. Heat Mass Transf.* **2020**, *157*, 119867. [CrossRef]
14. Pastuszko, R.; Kaniowski, R.; Wójcik, T.M. Comparison of pool boiling performance for plain micro-fins and micro-fins with a porous layer. *Appl. Therm. Eng.* **2020**, *166*, 114658. [CrossRef]
15. Sajjad, U.; Hussain, I.; Hamid, K.; Bhat, S.A.; Ali, H.M.; Wang, C.-C. A deep learning method for estimating the boiling heat transfer coefficient of porous surfaces. *J. Ther. Anal. Calorim.* **2021**, *145*, 1911–1923. [CrossRef]
16. Ahmad, S.W.; Lewis, J.S.; McGlen, R.J.; Karayiannis, T.G. Pool boiling on modified surfaces using R-123. *Heat Transf. Eng.* **2014**, *35*, 1491–1503. [CrossRef]
17. Dąbek, L.; Kapjor, A.; Orman, Ł.J. Distilled water and ethyl alcohol boiling heat transfer on selected meshed surfaces. *Mech. Ind.* **2019**, *20*, 701. [CrossRef]
18. Deng, D.; Wan, W.; Feng, J.; Huang, Q.; Qin, Y.; Xie, Y. Comparative experimental study on pool boiling performance of porous coating and solid structures with reentrant channels. *Appl. Therm. Eng.* **2016**, *107*, 420–430. [CrossRef]
19. Dewangan, A.K.; Kumar, A.; Kumar, R. Experimental study of nucleate pool boiling of R-134a and R-410A on a porous surface. *Heat Transf. Eng.* **2019**, *40*, 1249–1258. [CrossRef]
20. Gupta, S.K.; Misra, R.D. Development of micro/nanostructured-Cu-TiO_2-nanocomposite surfaces to improve pool boiling heat transfer performance. *Heat Mass Transf.* **2020**, *56*, 2529–2544. [CrossRef]
21. Hu, Y.; Zhang, S.; Li, X.; Wang, S. Heat transfer enhancement of subcooled pool boiling with self-rewetting fluid. *Int. J. Heat Mass Transf.* **2015**, *83*, 64–68. [CrossRef]
22. Jun, S.; Kim, J.; You, S.M.; Kim, H.Y. Effect of subcooling on pool boiling of water from sintered copper microporous coating at different orientations. *Sci. Technol. Nucl. Install.* **2018**, *2018*, 8623985. [CrossRef]
23. Li, C.; Peterson, G. Evaporation/Boiling in Thin Capillary Wicks (II)—Effects of Volumetric Porosity and Mesh Size. *ASME J. Heat Transfer.* **2006**, *128*, 1320–1328. [CrossRef]
24. Li, C.; Peterson, G. Parametric Study of Pool Boiling on Horizontal Highly Conductive Microporous Coated Surfaces. *ASME J. Heat Transfer.* **2007**, *129*, 1465–1475. [CrossRef]
25. Liu, F. A Study of Sintered Copper Porous Surfaces for Pool Boiling Enhancement. Rochester Institute of Technology. ProQuest Dissertations Publishing, 2016; p. 10248320. Available online: https://www.proquest.com/openview/aa59d1b109e23f558dc5d16aeee8f927/1?pq-origsite=gscholar&cbl=18750 (accessed on 27 September 2021).
26. McHale, J.P.; Garimella, S.V.; Fisher, T.S.; Powell, G.A. Pool boiling performance comparison of smooth and sintered copper surfaces with and without carbon nanotubes. *Nanoscale Microscale Thermophys. Eng.* **2011**, *15*, 133–150. [CrossRef]
27. Mori, S.; Aznam, S.M.; Okuyama, K. Enhancement of the critical heat flux in saturated pool boiling of water by nanoparticle-coating and a honeycomb porous plate. *Int. J. Heat Mass Transf.* **2015**, *80*, 1–6. [CrossRef]
28. Nasersharifi, Y.; Kaviany, M.; Hwang, G. Pool-boiling enhancement using multilevel modulated wick. *Appl. Therm. Eng.* **2018**, *137*, 268–276. [CrossRef]
29. Rahimian, A.; Kazeminejad, H.; Khalafi, H.; Akhavan, A.; Mirvakili, S.M. Boiling Heat Transfer and Critical Heat Flux Enhancement Using Electrophoretic Deposition of SiO_2 Nanofluid. *Sci. Technol. Nucl. Install.* **2019**, *2019*, 1272156. [CrossRef]
30. Rioux, R.P.; Nolan, E.C.; Li, C.H. A systematic study of pool boiling heat transfer on structured porous surfaces: From nanoscale through microscale to macroscale. *AIP Adv.* **2014**, *4*, 117133. [CrossRef]
31. Sarangi, S.; Weibel, J.A.; Garimella, S.V. Effect of particle size on surface-coating enhancement of pool boiling heat transfer. *Int. J. Heat Mass Transf.* **2015**, *81*, 103–113. [CrossRef]
32. Wen, M.-Y.; Ho, C.-Y.; Jang, K.-J. An optimal parametric design to improve pool boiling heat transfer of sintered surfaces. *J. Eng. Technol. Res.* **2012**, *4*, 49–56.
33. Xu, H.; Dai, Y.; Cao, H.; Liu, J.; Zhang, L.; Xu, M.; Cao, J.; Xu, P.; Liu, J. Tubes with coated and sintered porous surface for highly efficient heat exchangers. *Front. Chem. Sci. Eng.* **2018**, *12*, 367–375. [CrossRef]
34. Zhang, K.; Bai, L.; Lin, G.; Jin, H.; Wen, D. Experimental study on pool boiling in a porous artery structure. *Appl. Therm. Eng.* **2019**, *149*, 377–384. [CrossRef]
35. Asfahan, H.M.; Sajjad, U.; Sultan, M.; Hussain, I.; Hamid, K.; Ali, M.; Wang, C.-C.; Shamshiri, R.R.; Khan, M.U. Artificial intelligence for the prediction of the thermal performance of evaporative cooling systems. *Energies* **2021**, *14*, 3946. [CrossRef]
36. Hamid, K.; Sajjad, U.; Yang, K.S.; Wu, S.-K.; Wang, C.-C. Assessment of an energy efficient closed loop heat pump dryer for high moisture contents materials: An experimental investigation and AI based modelling. *Energy* **2022**, *238*, 121819. [CrossRef]

37. Chang, J.Y.; You, S.M. Boiling heat transfer phenomena from microporous and porous surfaces in saturated FC-72. *Int. J. Heat Mass Transf.* **1997**, *40*, 4437–4447. [CrossRef]
38. Jakob, M. *Heat Transfer*; John Wiley & Sons: New York, NY, USA, 1949.
39. Pioro, I.L.; Rohsenow, W.; Doerffer, S.S. Nucleate pool-boiling heat transfer. I: Review of parametric effects of boiling surface. *Int. J. Heat Mass Transf.* **2004**, *47*, 5033–5044. [CrossRef]
40. Berenson, P.J. Experiments on pool-boiling heat transfer. *Int. J. Heat Mass Transf.* **1962**, *5*, 985–999. [CrossRef]

sustainability

Article

Development and Experiments on a Batch-Type Solar Roaster—An Innovative Decentralized System for Coffee Roasting

Faizan Majeed [1,2,*], Ali Raza [3], Anjum Munir [4] and Oliver Hensel [1]

[1] Department of Agricultural and Biosystems Engineering, University of Kassel, D-37213 Witzenhausen, Germany; ohensel@uni-kassel.de
[2] Department of Agricultural Engineering, Bahauddin Zakariya University, Multan 60800, Pakistan
[3] Department of Farm Machinery and Precision Engineering, Faculty of Agricultural Engineering and Technology, Pir Mehr Ali Shah-Arid Agriculture University, Rawalpindi 46300, Pakistan; draliraza@uaar.edu.pk
[4] Department of Energy Systems Engineering, University of Agriculture Faisalabad, Faisalabad 38000, Pakistan; anjum.munir@uaf.edu.pk
* Correspondence: faizanmajeed@bzu.edu.pk

Abstract: About 70% of the harvested coffee is exported to the industrialized nations for value addition due to lack of processing and logistic facilities in developing coffee producer countries, thus leaving behind a marginal economic return for the growers. This research was conducted to investigate the roasting capacity of an innovatively developed batch-type directly solar radiated roasting system for the decentralized processing of coffee using solar energy. Central composite rotatable design (CCRD) was employed to design the experiments to optimize the coffee roasting process. Experimental results revealed that with an average solar direct normal irradiance (DNI) of 800 W/m^2, the roaster was capable of roasting a batch of 2 kg coffee beans in 20, 23, and 25 min subjected to light roasts, medium roasts, and dark roasts, respectively at a drum speed of two revolutions per minute (rpm). The batch-type solar roaster has the capacity to roast 28.8–36 kg of coffee beans depending on dark to light roasting conditions on a clear sunny day with DNI ranging from 650 to 850 W/m^2. The system thermal efficiency during coffee roasting was determined to be 62.2%, whereas the roasting efficiency at a corresponding light roast, medium roast, and dark roast was found to be 97.5%, 95.2%, and 91.3%, respectively. The payback period of the solar roaster unit was estimated to be 1038 working sunshine hours, making it viable for commercialization.

Keywords: renewable energy; Scheffler concentrator reflector; batch-type solar roaster; response surface methodology; coffee roasting

Citation: Majeed, F.; Raza, A.; Munir, A.; Hensel, O. Development and Experiments on a Batch-Type Solar Roaster—An Innovative Decentralized System for Coffee Roasting. *Sustainability* **2022**, *14*, 2217. https://doi.org/10.3390/su14042217

Academic Editor: Manosh C. Paul

Received: 24 January 2022
Accepted: 12 February 2022
Published: 15 February 2022

1. Introduction

Coffee (*Coffea* spp.) is widely cultivated throughout the tropical regions comprising more than 70 species, all of them originating from Africa. Among them, Arabica (*Coffea arabica*, 64% of world production) and Robusta (*Coffea canephora*, var. *Robusta*, 35%) are economically important varieties that are being grown worldwide on an estimated area of 10.3 million hectares and represent the sole economic income for more than 25 million families of the developing world [1,2]. In 2021, about 11.08 million US tons (MT) of coffee were being produced, but a majority of this produce (7.75 MT) was imported by the industrialized countries in perishable bean form for value additions due to the lack of processing and logistic facilities in developing coffee producer countries who were able to export only 1.45 MT processed coffee (roasted and ground, and soluble) form, offering a very marginal economic benefit to them [3]. The escalating fossil fuel prices further burden up the cost of processing for existing facilities in producer countries, especially in those rural and far-flung farm areas where centralized grid connections are not available and even if available, only the populated areas are connected to the transmission lines for meeting only the domestic needs, while the majority of the agricultural processing operations are carried out at farmlands away from residential areas of the villages [4]. The individual villages are small socioeconomic units but often underappreciated in centralized energy planning models due to the higher cost of infrastructure and transmission lines for scattered populations, which ultimately uplift the overall consumption of carbon-based fuels and lead to environmental degradation. This whole scenario overburdens the deprived farmers and indirectly enforces them to sell their perishable agricultural produce in the local markets without adding any remarkable value, thus returning them a meagre profit [4,5]. Therefore, there is a dire need to devise sustainable solutions for the coffee growers that simultaneously can address their energy-deficient scenario and upsurges decentralized coffee processing facilities in a scientific manner using their local renewable resources, i.e., solar.

With the development of state-of-the-art technologies, immensely available solar power can be utilized for adding value to green coffee beans [6]. In adding value to the coffee, roasting is an energy-intensive unit operation in which green coffee beans are exposed to heat treatment at high temperatures over 200 °C for a specific time to attain color, aroma, and taste [7]. The roasting process is comprised of three main phases viz., drying, roasting, and cooling during which heat and mass transfer occur inside the coffee beans by means of convection and conduction mechanisms. Consequently, with the increase in temperature, various physical and chemical changes take place inside the beans including exothermic reactions and water evaporation that results in the development of color, aroma, and taste characteristics of the coffee [8,9]. The heat energy is used to evaporate the water in the early stages of the roasting process [10]. The rate of evaporation is high in the beginning and then slows down gradually toward the end of the roasting process. Several investigations have been made to determine the heat transfer properties in coffee during roasting. The specific heat of Green Arabica's was determined and reported as 1.85 Jg^{-1} $°C^{-1}$ in beans with 7.5% humidity, which is slightly higher than the specific heat of Robusta's (1.46 Jg^{-1} $°C^{-1}$) containing 4.5% humidity whereas, in the roasted coffee bean, it was also 1.46 Jg^{-1} $°C^{-1}$ at 2.5% humidity [11,12]. The degrees of roasting are controlled by roasting time and temperature, which are necessary for the required chemical reactions without burning the beans and compromising the flavor of the beverage [13] and were qualitatively assessed for color and classified as a light, medium, or dark roast [14,15]. Achieving an ideal roast is a goal that is complicated, since coffee beans behave differently and produce distinct results in physical properties, chemical composition, and biological activities when roasted under different conditions [16].

Traditional methods for roasting coffee beans use large iron pans that were exposed to fire by burning coals. The slender spoon was used for the mixing of coffee beans during roasting [17]. These conventional coffee roasting techniques are uncontrolled, energy-wasting, inefficient, and time consuming. Technological efforts have been increasingly made in recent years for developing sophisticated technologies for coffee roasting to produce coffee of the same quality as coffee using the manual roasting method [18]. Considering coffee roasting methods, generally, the batch or continuous roasting systems have employed that transfer the heat to the beans by direct contact with hot metal surfaces (conduction) or through hot air streaming media (convection), or by radiation [19]. Drum-type roasters comprising a rotating cylinder design are widely used roasting systems in which heat is provided through hot air to the coffee beans inside the cylinder through the perforated wall or from the center of the cylinder to ensure homogeneity of the roast. One of the recent developments in coffee roasting technology is fluidized bed roasting [20] in which high-velocity hot gas is directed toward the beans, usually from the bottom of the roasting machine, so that the gases heat and move the floating beans simultaneously. Industrial coffee roasters either use a gas-fired revolving oven, in batch type or continuous roasters for controlling the temperature, feeding rate, and time during the roasting process to achieve uniformly roasted coffee. However, all these large-scale roasters are not economically affordable by small and medium-level coffee farmers, rising costs for fossil fuel further increases roasting costs that turned the mindset to think out of the box for exploring the sustainable ideas of energy production.

Although the potential of solar thermal energy is vast in tropical regions that provide the most favorable conditions to address the high-temperature post-harvest processing facilities, unfortunately, the utilization of this extensive potential is mainly limited to low-temperature processing operations such as the drying of agricultural products [21,22]. Beyond the low-temperature applications, the roasting of different agricultural products is a promising field in the developing countries of tropical regions. The daily average DNI ranging from 5.5 to 7.5 kWh/m^2 with yearly sunshine of more than 300 days [23] provides an excellent opportunity for the development of innovative technologies to meet this challenging and need-based assignment. Up to now, there has been very little work on the value addition of post-harvest using solar thermal energy moving toward medium to the high-temperature range for the processing of perishable agricultural products [24] and roasting of nuts and beans [6,25,26]. This massive solar potential can be utilized through decentralized applications of solar thermal energy especially in rural areas through the development of innovative solar technologies for decentralized applications in coffee processing, which can scale up the farm-gate processing facilities for coffee growers of tropics.

In recent years, many innovative solar heat tapping devices such as parabolic trough (line focus), linear Fresnel (line focus), parabolic dish (point focus), and heliostat field (point focus) were introduced, but they are rarely applied for industrial purposes. There are also fixed focus and tracking problems present in them [27]. Among the available solar concentrating technologies, the Scheffler fixed-focus concentrator is the best suitable option for generating heat energy in medium to the high-temperature range with a variety of reflector sizes ranging from 2 to 60 m^2 [28–30]. The versatility of the Scheffler reflector is the fixed focus at the targeted position by automatically tracking the sun, which provides a uniform temperature distribution on the focus point throughout the day. At the focus point, a temperature up to 700 °C is achievable depending on the size of the Scheffler reflector [21,31]. The development and coupling of innovative solar concentrating devices in the medium to high-temperature range had opened new landmarks for decentralized processing of agricultural products such as roasting coffee.

Acknowledging all the above-cited literature, this research initiative has been taken to develop an on-farm coffee roasting system that is capable of roasting coffee of 1–3 kg batch capacity by using a Scheffler solar concentrator (8 m^2) that focuses on direct normal irradiance (DNI) throughout the day by a photovoltaic-driven gear motor-tracking mechanism to achieve maximum available temperature on the roasting drum. The extensively used tool for the optimization of food processes is Response Surface Methodology (RSM), which has been incorporated in research methodology for investigating roaster optimum operational parameters i.e., roasting time, drum rotational speed, and feeding capacity for achieving a light, medium, and dark-roasted coffee in comparison with industrial roasters coffee quality. The developed roaster had probed into the possibility of shifting the roasting process to solar thermal energy, thereby providing a decentralized cost-effective solution by negating the environmental impact. Alternative resources of non-carbon-based energies are the only futuristic, viable, and long-lasting alternatives. Coffee roasting using an unlimited vast source of solar energy would be the most simplistic tool for the efficacious on-farm roasting, which is easily operable by the unschooled coffee farmers for greater socio-economic benefit.

2. Materials and Methods

2.1. Experimental Setup and Data Acquisition

The solar batch-type roasting machine was equipped with a pyranometer and thermocouples (K-types) via a data logger attached to the computer. The data were recorded in 10 s time intervals during roasting experiments. Thermocouples were connected with the roaster drum to measure the focus temperature, drum air temperature, and coffee beans temperature throughout the roasting experiments. The pyranometer was fixed at the Scheffler reflector for continuously facing the sun through a black pipe of length 100 mm fixed on it for only allowing to record the direct normal irradiance (DNI). The Scheffler concentrator was mounted with a daily tracking mechanism working on an automatic photovoltaic (PV) sun tracking system comprised of a PV-powered direct current (DC) gear motor and for the seasonal adjustments due to the sun declination angle throughout a year, the manually adjusted telescopic clamps were installed to precisely focus the DNI onto the roasting drum. Moisture content (MC) was determined at various stages of roasting by evaluating samples (100 g each) collected in an airtight container to prevent moisture and foreign contamination. The process for determining the MC of roasted and green beans was in accordance with the method (925.09) specified in the Official Methods by Analytical Chemists (AOAC) [32]. To determine the color parameters of green, light roast, medium roast, and dark roast coffee beans (*Coffea arabica* L.), a colorimeter Konica Minolta CR400 (Osaka, Japan) was used. The colorimeter has been calibrated before every single measurement by standard tile (i.e., white). The measured values per coffee beans sample were averaged to describe the color parameters of coffee beans within the CIELab color space. The coordinates of the experimental site for roasting were 51°20′45.76″ N (latitude), 9°51′52.08″ E (longitude) from a mean sea level elevation of 137 m. Figure 1 illustrates a solar coffee roasting system.

Figure 1. Solar roasting system. (Legends: A: 8 m^2 Scheffler concentrator; A1: Sun tracking sensor; A2: Scheffler concentrator reflector; B: Roasting unit; B1: Feeding hopper; B2: Discharge chute handle; B3: Wooden handle sampler; B4: Gear motor for rotating drum; B5: Glass window; B6: Roasted product coolant tray; B7: Control panel).

2.2. Description and Working Mechanism of Solar Batch-Type Roaster

The development work of the batch-type solar roasting machine consists of a technical drawing for designing a roaster, development of various roaster components, material selection for various components of the roaster, and experimental test on the newly constructed roaster. The whole roasting unit was fabricated using stainless steel material. The size (diameter × length) of the roasting cylinder was set at 400 mm × 400 mm. A gear motor was installed for rotating the drum during the process. The rotating drum brings the conduction process quite high during roasting. The solar roasting machine uses the most efficient heat transfer conduction process for coffee beans roasting by the drum rotating technique. In this type, the coffee beans are fetched into the cylindrical-shaped chamber called a roasting drum. As the drum rotates, it mechanically fluidizes the coffee beans mass by turning horizontally, and the blades were fixed inside the rotating drum to axially mix the coffee beans during roasting. The drum blades inside drag the coffee beans forward in the drum, and the inclined drum surface brings the coffee beans back. So, in this way, a single coffee bean passed through every point of the drum during rotation absorbs uniform thermal energy from the drum surface by conduction. The solar roaster's main components are the rotating drum, electric motor, cooling tray with fan blower, and the heat source i.e., solar concentrator.

The basic criterion for the selection of solar concentrators was to perform roasting experiments. From the various studies on different solar reflectors [33], it has been extracted that the Scheffler solar concentrator would be the most appropriate for roasting experiments. In traditional parabolic concentrators design, the challenge was to perform continuous tracking on two axes with a fixed receiver as an integral part of the reflector on its focal point. Moreover, the focus position was in the path of incoming direct normal irradiance. The high temperature could be achieved by these types of concentrators, but for conducting roasting experiments, a concentrator with frequently changing focus was not suitable due to inadequate handling during roasting. However, this issue is resolved by using a fixed focus Scheffler solar concentrator that enables precise automatic tracking and keeps the path of incoming direct normal irradiance away from the focus. Furthermore, its fabrication work could be carried out in a simple workshop with minimum tools, hence providing a cheaper solution that is easily adoptable for small-scale applications in the food and agro-industry. Therefore, the Scheffler concentrator reflector with a surface area of 8 m^2 was used as a solar thermal energy source to conduct roasting experiments. The main components of the Scheffler reflector are an elliptical reflector frame, rotating support, tracking channel, reflector stand, and tracking devices for both daily and seasonal variations in the sun angle. The crossbars are designed precisely to form the required paraboloid section in the elliptical frame of the Scheffler reflector. The rotating support is fabricated (steel pipe material) as a reflector integral part to provide an axis of rotation and a tracking channel. The operating principle of the daily tracking system is to counterbalance the earth rotation effect with an angular velocity of one revolution per day by tracking the sun along an axis parallel to the polar axis of the earth. The tracking system is comprised of a photovoltaic panel, a solar sensor, and a geared motor. For the seasonal adjustments, manually operated telescopic clamps were used to acquire the required paraboloid reflector shape for precisely targeted fixed focus throughout the year. The Scheffler solar concentrator reflects the incoming direct normal irradiance from its 8 m^2 surface area onto a solar roaster drum having a diameter of 400 mm. The Scheffler reflector design of the paraboloid lateral part that is inclined at an angle of $(43.23 \pm \alpha/2)$, so the reflector's actual area of aperture (A_c) is calculated as $A_c \times \cos(43.23 \pm \alpha/2)$. For measuring the total available energy (Q_a), the direct normal irradiance (E_{DNI}) is multiplied by the fraction of the actual aperture area as equated below [34],

$$Q_a = E_{DNI} A_c \cos\left(43.23 \pm \frac{\alpha}{2}\right), \quad (1)$$

where α = angle of solar declination that can be equated as under [35]:

$$\alpha = (180/\pi)[0.006918 - 0.399912\cos(n-2)2\pi/365 + 0.070257\sin(n-1)2\pi/365 - 0.006758\cos 2(n-1)2\pi/365 + 0.000907\sin 2(n-1)2\pi/365 - 0.002679\cos 3(n-1)2\pi/365 + 0.00148\sin 3(n-1)2\pi/365], \quad (2)$$

where n = day of the year.

Furthermore, the available energy is divided in terms of absorbed (Q_{ap}) and reflected (Q_{pr}) radiations. The absorbed energy radiant (Q_{ap}) is estimated that is depending on the material reflectance (reflective aluminum > 88%). The energy available after the concentrator (Q_{pr}) is given in Equation (3) [34]:

$$Q_{pr} = R_p Q_a \quad (3)$$

where R_p = reflectance of the reflector surface material.

From the total radiations striking at the concentrator surface, some of the radiations were reflected out of focus. This happens due to imperfection in concentrator profiles, dust particles on the surface of the concentrator, and inadequate tracking of daily and seasonal variations. The part of the available quotient reaching targeted focus (F_f) was assumed to be 0.88 in the overall calculation. The energy received at the solar roaster (Q_{rcv}) is calculated as [24]:

$$Q_{rcv} = Q_{pr} F_f. \quad (4)$$

To measure the energy available at the roaster drum (Q_{rcv}), the roaster cylindrical drum was fabricated with a stainless-steel (S.S) food-grade material and insulated by polyurethane with 60 mm insulation thickness. The roaster drum absorbs thermal energy and desorbs through conduction to beans in contact with the drum surface. After the coffee beans were inserted inside the roasting drum through the feeding hopper, the tilted position of the roasting drum slides the coffee beans toward the discharge chute, allowing heat transfer through conduction from every point of the drum surface to the beans effectively. The energy absorbed by the coffee beans during the roasting process can be measured as [36]:

$$Q_u = \frac{mc\Delta T}{t}, \quad (5)$$

where Q_u = energy absorbed by coffee beans, m = mass of beans, c = specific heat capacity of the coffee beans, ΔT = change in temperature of coffee beans, and t = roasting time.

2.3. Efficiency of Solar Roaster

The roaster thermal efficiency (η_{th}) is calculated by dividing the energy absorbed by the coffee beans with the total energy available, as given in Equation (6) [36]:

$$\eta_{th} = \frac{Q_u}{Q_a} \times 100. \quad (6)$$

The roasting efficiency is calculated by dividing the mass of undamaged roasted coffee beans by the total mass of roasted coffee beans, as expressed in Equation (7) [36]:

$$\eta_{pr} = \frac{\mathrm{m_r} - \mathrm{m_{dr}}}{\mathrm{m_r}} \times 100, \quad (7)$$

where η_{pr} = roasting efficiency of the solar roaster, $\mathrm{m_r}$ = total mass of roasted coffee beans, and $\mathrm{m_{dr}}$ = mass of damaged (broken or burnt) coffee beans after roasting.

The roasting performance was also determined at varying rotational speeds (2, 3, and 4 rpm) of the roasting drum.

2.4. Optimal Operating Parameters for Roasting Coffee in Solar Batch-Type Roaster

The operating parameters (roasting time, drum rotation, and beans quantity) were selected for controlling the roasting temperature, feeding rate, and degree of roasting. These parameters have a substantial effect on the coffee roasting process. The central composite rotatable design (CCRD) technique by using Design-Expert software was employed to design the experiments, which was initially established by Box and Hunter [37] and improved by Box and Wilson [38]. The experiments were conducted by setting operating parameters (roasting time, drum rotation, and beans quantity) ranges at 15–25 min, 2–4 rpm, and 1.5–2.5 kg, respectively, to predict their optimal values using Design-Expert software. The operating parameters impact on the moisture content and colorimeter value of lightness factor (L^* = 0 represent black and L^* = 100 represents diffuse white) were recorded after two replications of the predicted values, and the average value for each response was examined during roasting experiments. The roasting experimental findings were assessed by applying the CCRD technique and second-order quadratic equation fitted for moisture content (MC) and colorimeter value of lightness factor (L^*) by incorporating second-order multiple regression analysis. For each predicted response, the generalized model is given as:

$$Y = \beta_o + \sum_{i-1}^{3} \beta_i x_i + \sum_{i-1}^{3} \beta_{ii} x_i + \sum_{i<j-1}^{3} \beta_{ij} x_i x_j + e \quad (8)$$

where Y = response variable, β_o, β_i, β_{ii} and β_{ij} = regression coefficients for the model intercept, linear, quadratic, and interaction terms, x_i, x_j = independent variables, and e = random error [39].

The results from roasting experiments were compared with the model predicted values, and the sufficiency of the designed model was validated for each response by applying the analysis of variance (ANOVA). If the p-value will be less than 0.05 at the level of significance of a 95% confidence interval (C.I), then the model is considered acceptable, and for the coefficients of variation (CV) less than 10%, the lack of fit for the model is non-significant.

From the ANOVA results, a probability to consider the observed Fisher's F statistic value and a p-value less than 0.05 at 95% C.I verifies a significant impact of the parameters on the responses.

2.5. Optimization of Roasting Parameters Using Desirability Function Technique

The roasting parameters optimization procedure was carried out in the Design-Expert software using a desirability function. In this approach, a multivariable problem through applying mathematical methods is converted to a single response problem [40]. The aim for optimization was to achieve maximum roasting capacity by employing the maximum available solar thermal energy to roast coffee beans. Therefore, the optimal drum rotation was set to examine the optimal roasting time to reach standardized color lightness values of L^* 41.5 + 1, 39.2 + 1, and 37.5 + 1 at corresponding light roast, medium roast, and dark roast coffee beans. The optimization process of each operating parameter and for responses with their set goals are given in Table 1.

Table 1. Optimization criteria of operating parameters.

	Parameter	Goal
Independent variables	Roasting time (min)	In range
	Drum rotational speed (rpm)	Minimum
	Coffee beans Quantity (kg)	Maximize
Responses	MC (%)	Minimize
	L^* (light)	Target → 41.5
	L^* (medium)	Target → 39.2
	L^* (dark)	Target → 37.5

L^*: Lightness factor, MC: Moisture content.

3. Results and Discussion

3.1. Roaster Thermal Energy Distribution

The DNI was recorded from 09:00 to 17:00 on 14 June 2021, and the corresponding temperature at the focus and drum air was recorded. The data were recorded every 10 min using a solar pyranometer and k-type thermocouples, respectively. The results of the experiment recorded are graphically illustrated in Figure 2.

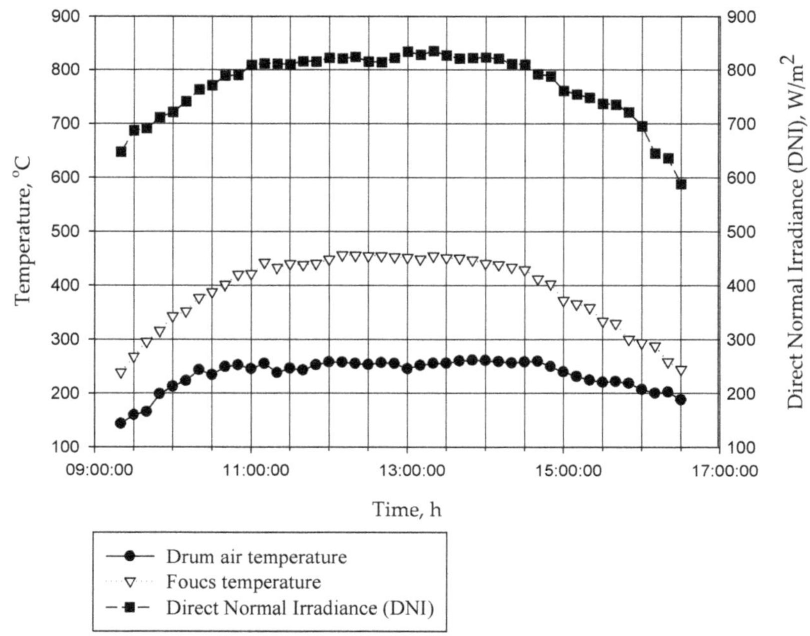

Figure 2. DNI, focus temperature, and drum air temperature.

It can be noticed from the roaster drum air temperature line in Figure 2 that the roasting cylinder air temperature of 200–250 °C was attained from 10 am to 4 pm, which is the most appropriate time for performing the roasting of coffee beans. This full-day experiment aims to record a suitable time to achieve the required temperature for roasting coffee beans. Based on the data collected, the available temperature (≥200 °C) that was desired for roasting could be achieved for 6 sunshine hours on a clear sunny day at a corresponding DNI value ranging from 650 to 850 W/m^2. The maximum and minimum temperature at the focus was recorded 456–173 °C, with DNI at site ranging from 835 to 455 W/m^2, respectively. The time required to roast 2 kg batch size coffee beans at the light, medium, and dark roasting conditions was 20 ± 0.1, 23 ± 0.1, and 25 ± 0.1 min, respectively. From the results of the roasting experiment, it can be derived that a solar roaster has a roasting capacity of 36, 31.3, and 28.8 kg of coffee beans per day subjected to light roasts, medium roasts, and dark roasts, respectively with average DNI ranging from 800 to 850 W/m^2.

Thermal Efficiency of Solar Roasting System

The roasting experiment was performed on clear sunny weather conditions with an average DNI value of 800 W/m^2 recorded at a concentrator surface area of 8 m^2, and the total energy available was calculated to be 6400 W. The optical losses of the Scheffler reflector were estimated by considering the actual useable aperture area (4.6 m^2) of the reflector; the reflectance (aluminum > 88%) from the reflective surface and out of focus radiations were estimated to be 10%, which could be due to the inadequate geometric accuracy of reflector profiles or imprecise tracking. The thermal losses from the roaster unit were estimated by considering the losses from the aluminum receiver (10%) and the calculations of thermal losses due to convection, conduction, and radiation from the roaster drum were made.

The energy distribution from the reflector surface to the roasting drum containing coffee beans was estimated to optimize the roasting system and to calculate how much energy in terms of power is available for the system under study. The power available at the Scheffler reflector, the power available after the reflector, the power available at the solar receiver, and finally the power available for roasting coffee were calculated using the respective equations as described before. Figure 3 illustrates that the power of 5400 W was available at 8 m^2 area; out of this, 3680 W of power was available at the Scheffler actual aperture area of 4 6 m^2, out of which 2870 W of power was available at the receiver. This power is further transferred to the drum for the roasting process; the power available inside the roasting drum was 2583 W and the final power available for roasting coffee beans was 2291 W. Furthermore, to achieve a uniform roasted product, roasting drum rotation is the main factor. It is obvious from both simulation and experimental results that temperature is quite uniformly distributed inside the roasting chamber. The power difference between the Scheffler reflector and the receiver is due to energy losses from different components of the roasting units. These losses include the reflectivity of the aluminum surface, due to incomplete absorbance and heat losses from different parts of a solar roaster, i.e., through conduction, convection, and radiations. Therefore, 2291 W power was utilized for coffee beans roasting out of the total 3680 W power available at Scheffler; from Equation (6), the thermal efficacy of the solar roasting system was 62.2%.

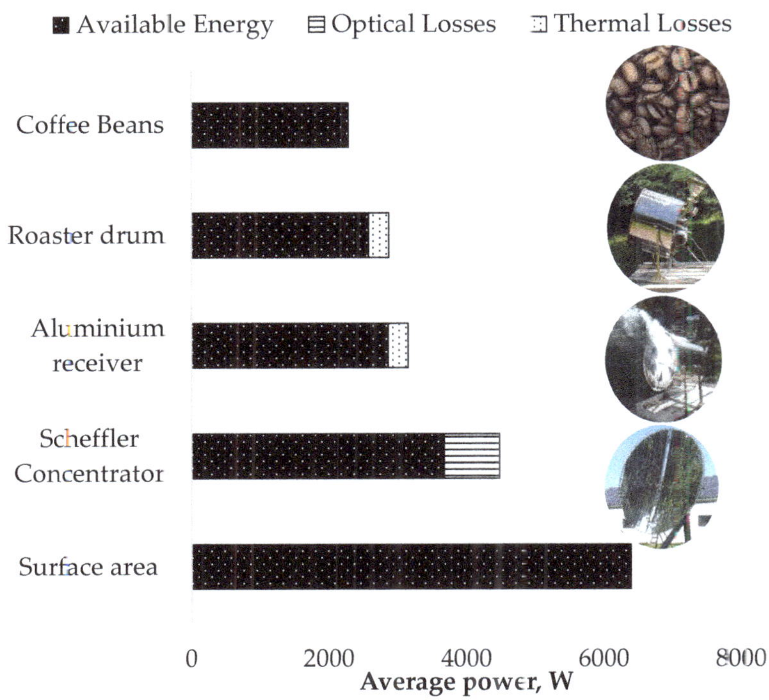

Figure 3. Power distribution at solar roasting system.

3.2. Roasting Efficiency of a Solar Roaster

The experiments were conducted for roaster performance evaluation by measuring the weight of roasted and damaged (broken or burnt) coffee beans, and the results at different degrees of roasting are illustrated in Figure 4. The results of solar roasting machine performance were 97.5 ± 1.0%, 95.2 ± 1.0%, and 91.3 ± 1.0% at light roasts, medium roasts, and dark roasts at lower rotational speed (2 rpm) of the roasting drum. Moreover, at higher rotational speed (4 rpm), the roasting efficiency of the solar roaster was 96.5 ± 1.0%, 92.5 ± 1.0%, and 87.8 ± 1.0% for light roasts, medium roasts, and dark roasts, respectively. It can be depicted from the findings that the highest roasting efficiency was achieved during light roast as compared to dark roast, which could be explained by the fact that at higher temperature, dark roasted coffee beans are damaged more readily because of their brittle nature as compared to light roasted coffee beans, which were less prone to damages. Moreover, the roasting efficiency was slightly higher at a lower rotational speed of the roasting as compared to a higher drum speed because the beans have the potential to break apart. However, the speed of the drum should not be very much lower, because the beans could scorch. The findings were in good relation to the various studies conducted on roasting efficiencies [41–43].

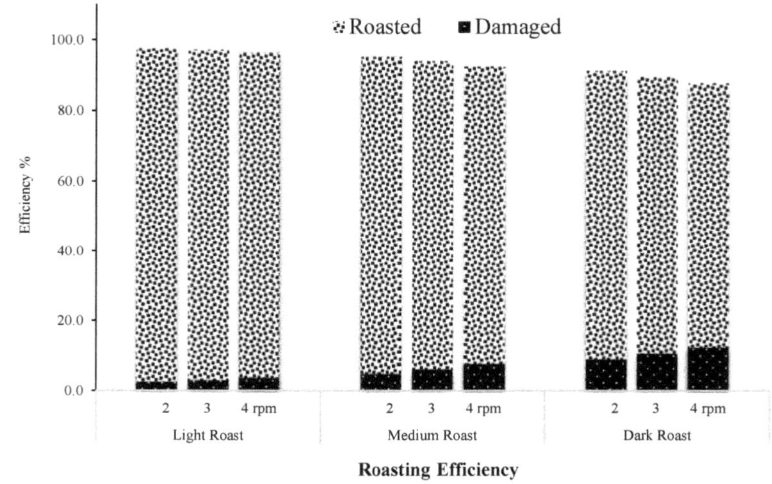

Figure 4. Roasting performance of a solar roaster.

3.3. Operating Parameters Impact on Responses

The results from both predicted design values and the findings of roasting experiments for each response (MC and L^*) are presented in Table 2. From experimental findings, the ranges for MC and L^* values were observed as 0.76 to 2.95% and 25.68 to 53.90, respectively. Equations for the MC and L^* in terms of coded factors are given below:

$$MC = 2.38 - 0.2A + 0.29B + 0.55C - 0.16\,AB + 0.14\,AC - 0.18\,BC - 0.11A^2 - 0.21B^2 - 0.068C^2, \quad (9)$$

$$L^* = 47.03 - 2.99A + 3.86B + 7.05C - 2.04\,AB + 1.62\,AC - 2.38\,BC - 1.71A^2 - 2.51B^2 - 1.53C^2, \quad (10)$$

where coded values A, B, and C represent the roasting time, drum rotation, and coffee beans quantity, respectively, whereas the statistical significance of the above equations is given in ANOVA in Tables 2 and 3, respectively.

The model F-value = 95.71, as illustrated in Table 3, indicates that the model was significant. There is only a 0.01% chance that noise can cause these large F-values. The p-value less than 0.05 implies significant model terms. The significant model terms are A, B, C, AB, AC, BC, A^2, B^2, and C^2. The values above 0.1000 imply that the model terms are not significant. The Lack of Fit F-value of 2.44 means that relative to pure error, the Lack of Fit was non-significant. There is a 17.47% chance that this high Lack of Fit F-value could be due to noise. The Predicted R^2 agrees reasonably with the adjusted R^2 and validates the model.

Table 2. Central composite response design (CCRD) of an experiment for quadratic response surface analysis on moisture content and colorimetric lightness value (L^*).

	Actual Level of Factor			Predicted Responses		Observed Responses	
Design Points	**Roasting Time (min)**	**Drum Rotational Speed (rpm)**	**Batch Capacity (kg)**	**Moisture Contents (%)**	***L****	**Moisture Contents (%)**	***L****
1	20	3	2	2.38	47.03	2.32	46.78
2	25	4	1.5	1.41	33.42	1.47	33.76
3	20	4.68	2	2.26	46.43	2.14	45.93
4	11.59	3	2	2.41	47.23	2.3	46.69
5	20	3	2	2.38	47.03	2.31	46.76
6	20	3	2	2.38	47.03	2.49	47.99
7	15	4	1.5	2.42	47.11	2.51	47.59
8	20	3	2	2.38	47.03	2.43	46.99
9	20	3	2	2.38	47.03	2.4	46.94
10	15	4	2.5	2.86	53.23	2.95	53.52
11	28.41	3	2	1.74	37.17	1.76	36.93
12	20	3	2	2.38	47.03	2.35	46.85
13	20	1.32	2	1.3	33.44	1.34	33.16
14	25	4	2.5	2.43	46.42	2.44	47.00
15	20	3	2.84	3.11	54.56	3.1	53.9
16	15	2	1.5	1.15	30.57	1.2	30.54
17	25	2	2.5	2.54	47.51	2.51	47.59
18	20	3	1.16	1.26	30.84	1.19	30.73
19	25	2	1.5	0.79	25.42	0.76	25.68
20	15	2	2.5	2.34	46.18	2.34	46.8

Table 3. Analysis of variance for quadratic response surface model on moisture content

Source	Sum of Squares	df	Mean Square	F-Value	*p*-Value	
Model	7.21	9	0.80	95.71	<0.0001	significant
A	0.54	1	0.54	65.07	<0.0001	
B	1.12	1	1.12	133.34	<0.0001	
C	4.13	1	4.13	493.34	<0.0001	
AB	0.20	1	0.20	24.45	0.0006	
AC	0.16	1	0.16	19.39	0.0013	
BC	0.27	1	0.27	32.69	0.0002	
A^2	0.17	1	0.17	20.41	0.0011	
B^2	0.64	1	0.64	76.93	<0.0001	
C^2	0.067	1	0.067	8.02	0.0178	
Residual	0.084	10	0.0083			
Lack of Fit	0.059	5	0.012	2.44	0.1747	Non-significant
Pure Error	0.024	5	0.0049			
Cor Total	7.30	19				

R^2 = 0.9885, Adjusted R^2 = 0.9782, Predicted R^2 = 0.9333, C.V. = 4.33%, df: degree of freedom.

The model F-value = 414.67, as illustrated in Table 4, indicates that the model was significant. There is only a 0.01% chance that noise can cause these large F-values. The *p*-value less than 0.05 implies significant model terms. The significant terms of the model are A, B, C, AB, AC, BC, A^2, B^2, and C^2. The values above 0.1000 imply that the model terms are not significant. The Lack of Fit F-value of 2.04 means that relative to pure error, the Lack of Fit was non-significant. There is a 22.58% chance that this high Lack of Fit

F-value could be due to noise. The predicted R^2 agrees reasonably with the adjusted R^2 and validates the model.

Table 4. Analysis of variance for quadratic response surface model on color lightness value *L**.

Source	Sum of Squares	df	Mean Square	F-Value	*p*-Value	
Model	1245.12	9	138.35	414.67	<0.0001	Significant
A	122.10	1	122.10	365.96	<0.0001	
B	203.64	1	203.64	610.38	<0.0001	
C	679.15	1	679.15	2035.63	<0.0001	
AB	33.13	1	33.13	99.30	<0.0001	
AC	21.00	1	21.00	62.93	<0.0001	
BC	45.13	1	45.13	135.25	<0.0001	
A^2	42.02	1	42.02	125.95	<0.0001	
B^2	90.67	1	90.67	271.78	<0.0001	
C^2	33.69	1	33.69	100.99	<0.0001	
Residual	3.34	10	0.33			
Lack of Fit	2.24	5	0.45	2.04	0.2258	Non-significant
Pure Error	1.10	5	0.22			
Cor Total	1248.46	19				

R^2 = 0.9973, Adjusted R^2 = 0.9949, Predicted R^2 = 0.9847, C.V. = 1.34%.

Analysis of variance showed that the models related to MC and *L** have a minor difference of values in each system (predicted and observed). Hence, it signifies the correlation of both models between operating parameters and their responses. Furthermore, these findings were confirmed by comparative analysis of both the actual results of roasting experiments and the model predicted values, as plotted in Figure 5a,b.

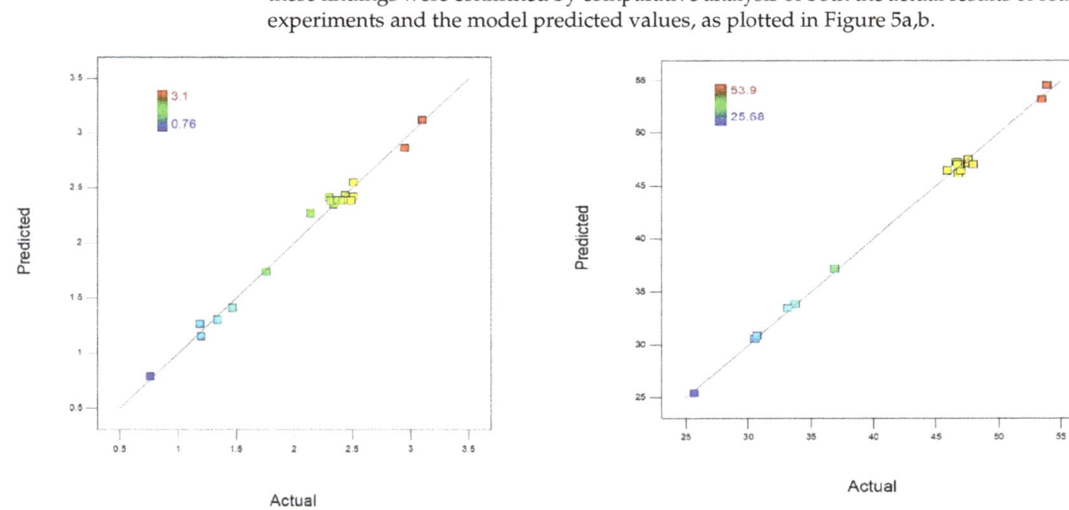

(a) (b)

Figure 5. Predicted vs. observed values of (**a**) MC; (**b**) *L**.

3.3.1. Impact of Operating Parameters on the Responses

The *p*-value is less than 0.05 for each single coded factor (A, B, and C) on both of the responses (MC and *L**), as illustrated in Tables 3 and 4. It confirmed the significant effect of the roasting time, drum rotational speed, and batch quantity on the responses. From Equations (9) and (10), the negative value of coefficient factor A represents a negative effect, and the positive value of corresponding coefficients B and C define the positive effects on both responses (MC and *L**). Hence, it can be derived that increasing the roasting time eventually decreases the moisture from the beans and vice versa. On the other hand, decreasing the roasting drum rotational speed rpm and lowering batch quantities will release more moisture from beans during roasting and vice versa [44].

It can be derived from the results presented in Table 3 that the highest F-value (493.34) of batch capacity has the most significant effect on the moisture content compared to drum rotation and roasting time. From these values, it can be narrated that by varying the values of operating parameters within the given experimental ranges, the most substantial changes in the bean moisture content were observed by varying batch capacity. The quantity of beans inside the drum absorbed the heat through conduction, as they are in direct contact with the drum surface, which further increases with fewer beans, hence giving every bean more opportunity to absorb more heat from the drum, and by increasing the quantity of beans, the moisture removal rate tends to be slower. Therefore, the lowest value of moisture content was noted with minimum batch quantities. Furthermore, the comparatively quiet moderate impact of the drum rotational speed was also observed on the moisture content, which can be explained. Rotating the drum at higher speed rpm reduces the contact time of the bean-roasting drum, resulting in the lower temperature of the beans and hence higher moisture levels as compared to lowering the drum speed, which enhances the contact time of the bean–drum surface to absorb more heat, resulting in a lower value of moisture content. The roasting time has a comparatively lower impact on the bean moisture content. Increasing the time for roasting decreases the moisture value, which was because the more time the beans spend in the roasting cylinder continuously raises the temperature of the beans, resulting in more moisture removal from the coffee beans [44].

For the response L^* (see Table 4), it can be depicted that the batch size has the most significant impact with the highest F-value (2035.63) as compared with the impacts of the drum rotational speed and roasting time. Roasting experiments at minimal quantities tend to quickly increase the bean temperature with significantly reducing moisture, which will restructure the beans accompanied by several chemical changes. The browning effect occurs due to the Maillard and caramelization process that tends to raise at higher temperatures of beans. At a lower drum rotational speed, it tends to darken the beans more readily, resulting in a lower L^* value by increasing the temperature of beans through increasing the conduction process between the drum surface and bean. The roasting time impacts the L^* by giving more time for chemical reactions to occur within the beans at higher temperatures by increasing the time span that beans spend in the high-temperature roasting chamber [45].

3.3.2. Interactive Impact of Parameters on MC

The roasting time and drum rotational speed have an interactive impact on the moisture content of coffee beans in a 2 kg batch size, which is presented in Figure 6a,b. It can be depicted from the observed impact on MC value from both the drum rotational speed and roasting time that the drum rotational speed has a more significant impact on moisture contact as compared to roasting time because it regulates the contact time between the bean and drum surface, which eventually decide the conductance time between the drum and beans It can be described from a 3D surface plot that at higher drum speed, the higher the percentage of MC, and the lower drum speed resulted in a lower percentage of moisture content. The reason is at lower drum rotation, the beans' temperature rises quickly due to more time in contact with the drum surface and vice versa [46].

Figure 6c,d illustrate the 3D surface and contour plots of the interactive impact of bean quantity and roasting time on MC by maintaining the drum rotational speed at 3 rpm. From the graph, it can be seen clearly that batch capacity has the most significant effect on the moisture content of beans during roasting. The lowest batch capacity tends to increase the temperature more rapidly, resulting in a lower percentage of moisture content of beans during roasting. The increase in roasting time removes more moisture from beans as it tends to further increase the temperature of beans and lower the coffee beans' moisture level during roasting [47].

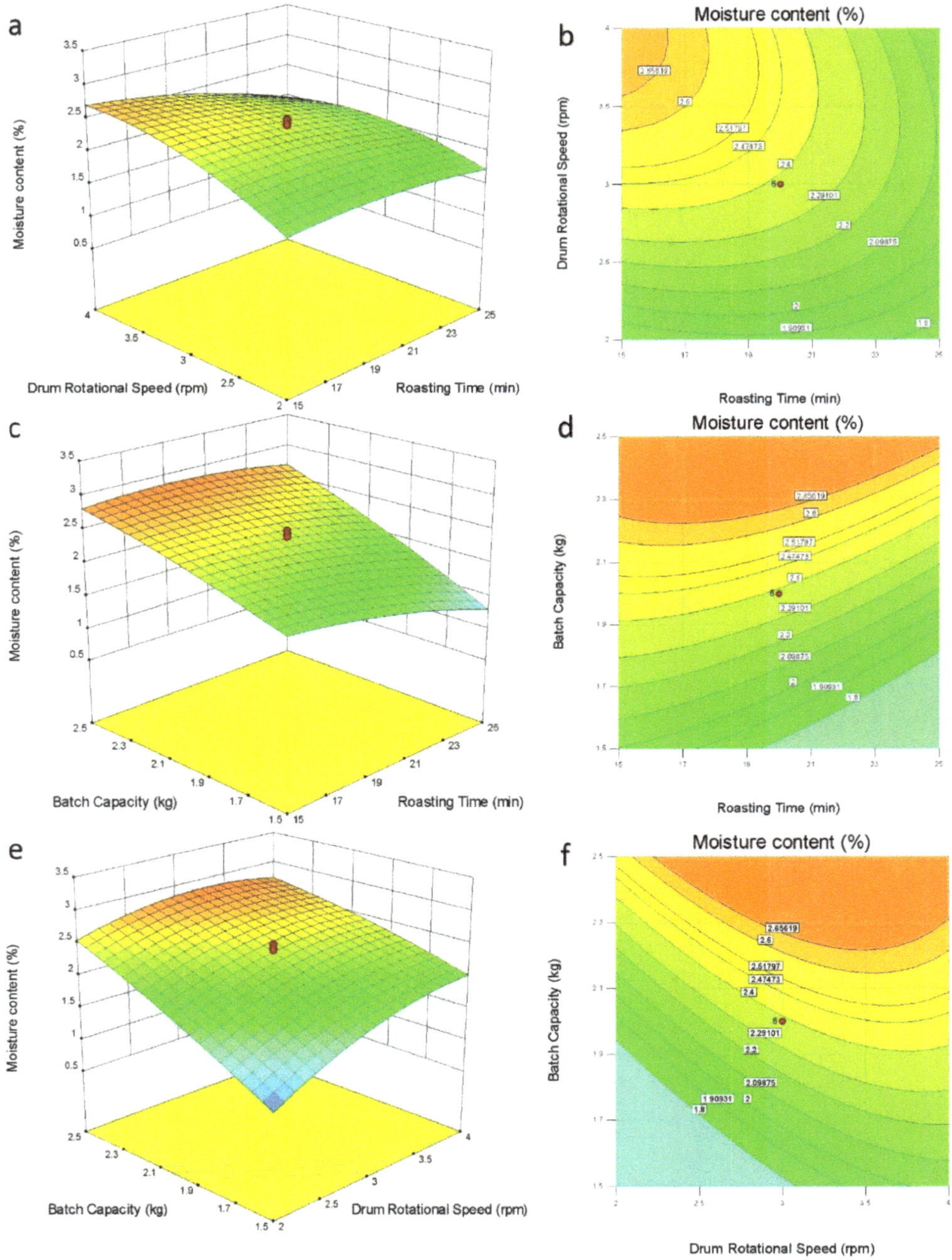

Figure 6. The 3D response surface and 2D contour plots depict the interactive impact of the roasting drum rotational speed, roasting time, and beans quantity on MC. (**a**,**b**) Roasting drum rotational speed interaction with roasting time; (**c**,**d**) Beans quantity interaction with roasting time; (**e**,**f**) Beans quantity interaction with roasting drum rotational speed.

Figure 6e,f illustrate the 3D surface and contour plots of the interactive impact of bean quantity and roasting drum rotational speed on MC. A positive impact on the moisture content has been recorded by the impact of both batch capacity and drum rotational speed. However, batch capacity has a more significant effect on moisture content, as it tends to raise the temperature of beans more rapidly by allowing beans to have more surface to absorb heat from the drum as compared to higher beans volume that reduces the temperature of beans. At a lower drum rotational speed, it allows a greater rise in temperature that will add more heat energy to beans for moisture removal, resulting in a lower percentage of moisture contents [46].

3.3.3. Interactive Impact of Parameters on Colorimeter Value (L^*)

The roasting time and drum rotational speed interactive impact on the color L^* of coffee beans having a batch size of 2 kg is illustrated in Figure 7a,b. It can be seen from the observed effect on L^* by both drum rotational speed and roasting time that the drum rotational speed has a more significant impact on L^* because it regulates the contact time between the bean and drum surface that eventually determine the conductance time between the drum and beans. It can be seen from the 3D surface plot below that at a higher drum rotational speed, the higher value of color L^* was observed due to less heat transfer through the drum surface and giving less browning color to the coffee beans and vice versa. The reason for the lower value for L^* at lower drum rotation was due to the temperature of the coffee beans, which abruptly rises by allowing more time to contact with the hot drum surface, resulting in higher conduction and vice versa [45].

Figure 7c,d illustrate the 3D surface and contour plots of the interactive impact of beans quantity and roasting time on L^* by maintaining the drum rotational speed at 3 rpm. From the graph, it can be seen clearly that batch capacity has the most significant effect on the L^* of beans during roasting. The lowest batch capacity tends to increase the temperature more rapidly, resulting in a lower L^* of beans during roasting. The increase in roasting time removes more moisture from beans as it tends to raise the temperature of beans and lower the value L^* during roasting [45].

Figure 7e,f illustrate the 3D surface and contour plots of the interactive impact of beans quantity and roasting drum rotational speed on L^* by maintaining the roasting time at 20 min. A positive impact on the L^* has been recorded by the impact of both batch capacity and drum rotational speed. However, batch capacity has a more significant effect on L^*, as it tends to raise the temperature of beans more rapidly by allowing beans to have more surface to absorb heat from the drum as compared to higher beans volume, which reduces the temperature of the beans. At a lower rotational speed of a rotating drum, it allows a greater rise in temperature that will add more heat energy to beans, resulting in a lower value of L^* [47].

3.4. Optimal Operating Parameters of Solar Roaster

Taking the optimization criteria into account, several solutions for the color response parameters L^* at the light, medium, and dark roasting degree were available at their corresponding moisture content values. Among the different available solutions, only the most desirable solutions were considered. The optimal feeding capacity of 2 kg was suggested by the quadratic response surface model at a drum rotational speed of 2 rpm for all three responses, while the optimal roasting times were predicted to be 20, 23, and 25 min for a light roast, medium roast, and dark roast degree, respectively. The optimal moisture contents were predicted to be 1.88, 1.82, and 1.74% at corresponding L^* values of 40.66, 39.47, and 37.99 compared to the selected solutions.

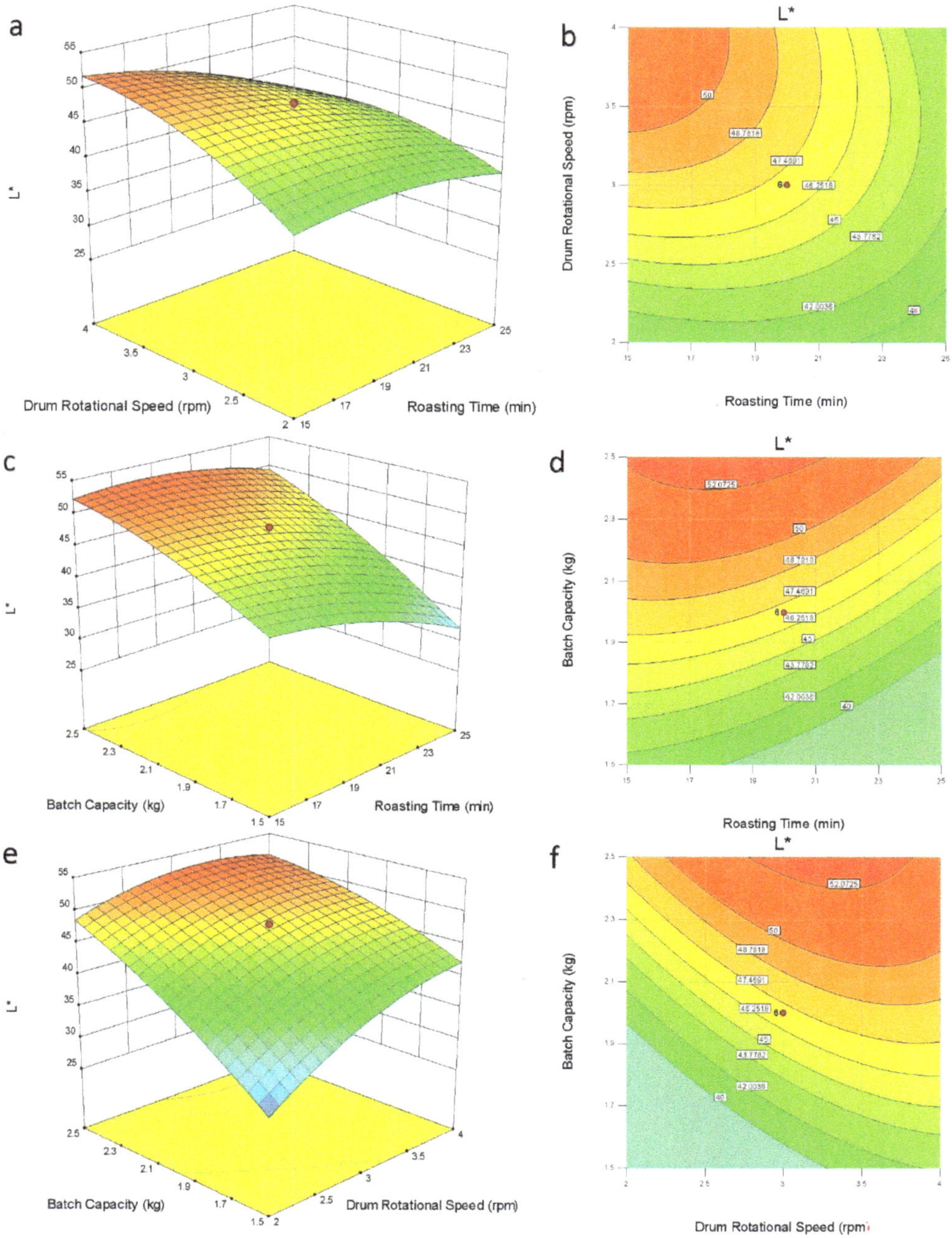

Figure 7. The 3D response surface and 2D contour plots depict the interactive impact of the roasting drum rotational speed, roasting time, and beans quantity on L^*. (**a**,**b**) Roasting drum rotational speed interaction with roasting time; (**c**,**d**) Beans quantity interaction with roasting time; (**e**,**f**) Beans quantity interaction with roasting drum rotational speed.

Coffee Roasting

The model's predictability was verified by coffee roasting experiments under optimal conditions. The experiments were repeated three times, and the average value of MC and L^* were measured for each experiment. The average MC values are 1.89, 1.83, and 1.75%, resulting in L^* values of 40.75, 39.64, and 38.21 for light roasts, medium roasts, and dark roasts, respectively. The values for MC and L^* both from experiments and predicted have indicated a very slight difference among them, as given in Table 5. For each response, the relative error of less than 1% is recorded; hence, these slight differences from the prediction results validate the model. The recorded time using a stopwatch during light roasts, medium roasts, and dark roasts was observed to be 20, 23, and 25 min, respectively. From the roasting experiments optimal time, it can be revealed that the roasting capacity of a solar roaster was 6, 5.2, and 4.8 kg of coffee beans per hour at corresponding light roasts, medium roasts, and dark roasts conditions, respectively.

Table 5. Predicted and observed responses.

Roasting Degree	Response	Predicted Value	Roasting Condition			Average Value	Error (%)
			Run 1	Run 2	Run 3		
Light	MC (%)	1.88	1.89	1.86	1.92	1.89	0.53
	L^*	40.66	40.72	40.63	40.89	40.75	0.21
Medium	MC (%)	1.82	1.83	1.81	1.84	1.83	0.35
	L^*	39.47	39.65	39.41	39.86	39.64	0.43
Dark	MC (%)	1.74	1.77	1.73	1.74	1.75	0.38
	L^*	37.99	38.77	37.86	38.01	38.21	0.58

3.5. Economic Analysis

The cost analysis study has been performed on an annual basis on a solar-based roasting machine to finally find out the cost-effectiveness of the system for its commercialization purpose. The useful working hour of the solar coffee roasting machine is subjected to sunshine hours for conducting coffee roasting. The long-term average sunshine hours are peak hours of sunshine. The total investment cost for the material and fabrication of a complete roasting system (Scheffler reflector, Roaster drum, Coolant tray) was 3000 USD. The expenses and revenue of the roasting unit were analyzed to calculate the break-even point as follows:

$$\mathrm{TR} = \mathrm{TC},$$

$$\mathrm{P} \times \mathrm{X} = \mathrm{TFC} + \mathrm{V} \times \mathrm{X},$$

$$\mathrm{X} = \frac{\mathrm{TFC}}{\mathrm{P} - \mathrm{V}}$$

$$\mathrm{X} = \frac{384}{25 - 24.63}$$

$$\mathrm{X} = 1037.8 \text{ h or } 173 \text{ days},$$

where TR = total revenue, TC = total cost of the roasting unit, P = revenue per hour, V = variable cost that includes routine maintenance and labor cost, TFC = total fixed cost that comprises the labor benefits and interest on financing, and X = operating time in hours. In this study, a batch size of 2 kg green coffee beans (*Coffea arabica*) was roasted in 20 min. On a typical summer day, the system was capable of roasting 36–28.8 kg of light–dark roasted coffee. For cost analysis, an estimated price of 1 kg green coffee beans is taken as \$3.32, and the roasted coffee is worth \$5.00 [48]. From this estimation, the average revenue of about \$25 can be generated per useful working sunshine hour. The number of sunny days in most of the coffee growing countries is more than 200 days per year [23]. From the break-even analysis, it is assessed that the cost of the complete system will be recovered after 173 days; i.e., around 6 months. Assuming a roaster life of 10 years, a great revenue can be generated by coffee farmers using solar roasters. It is worth mentioning that the cost will be even more quickly recovered in the countries receiving more sunshine hours to operate the roaster.

In summary, it can be depicted from the Figure 8 break-even analysis that the payback period in terms of cost is estimated to be half a year, and the revenue obtained will exceed its total cost. Furthermore, a solar roaster state-of-art design requires no gas or electricity connection, hence saving an additional amount as compared to other traditional roasters that require an additional cost of energy supplies transmission for roasting.

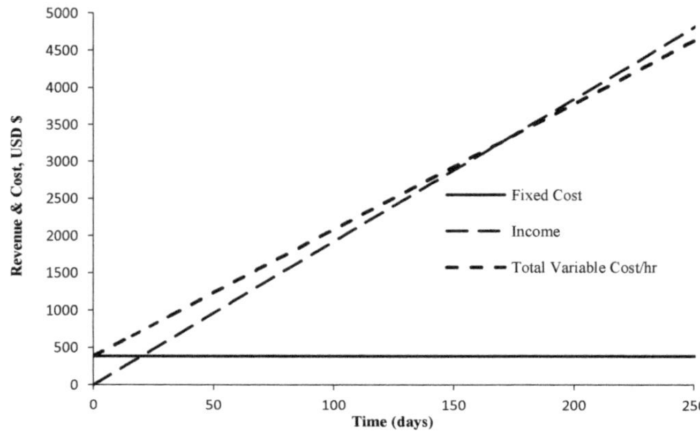

Figure 8. Break-even analysis (1 day = 6 h).

4. Conclusions

This study was conducted to investigate a solar roaster capacity as a decentralized coffee roasting approach. A standing 8 m^2 Scheffler reflector was used to concentrate the incoming DNI to the roaster drum focus. The system was completely independent of grid connections, and both thermal as well as electrical power was generated using solar energy. Experimental data show that the optimal times for roasting light, medium, and dark coffee at a drum temperature of 250 °C was 20 ± 0.1, 23 ± 0.1, and 25 ± 0.1 min, respectively. On a sunny day with a DNI of 650–850 W/m^2, the solar roaster was able to roast 28.8 kg, 31.3 kg, and 36 kg coffee beans with roasting efficiency of 97.5%, 95.2%, and 91.3% at the corresponding light roast, medium roast, and dark roast, respectively. The roasted coffee beans' final moisture content was 1.89, 1.83, and 1.75% at corresponding L^* values for a light roast, medium roast, and dark roast of 40.75, 39.64, and 38.21, respectively. The power distribution shows that out of the 3680 watts of total available energy at the Scheffler reflector, approximately 2291 watts were ultimately consumed by the coffee beans during roasting with a total thermal efficiency of 62.2%. Total investments on solar roasters were expected to return after 1038 useful working sunshine hours. Thus, the enormous potential of solar thermal energy can be used to meet globally faced rising energy demands in processing, particularly at the farm-gate as a decentralized approach for coffee beans roasting.

Author Contributions: Conceptualization, F.M., A.R., A.M. and O.H.; methodology, F.M., and A.R.; software, F.M. and A.R.; validation, F.M. and O.H.; formal analysis, F.M. and A.R.; investigation, F.M. and A.R.; resources, O.H.; data curation, F.M. and A.R.; writing—original draft preparation, F.M.; writing—review and editing, F.M., A.M. and O.H.; visualization, F.M. and A.R.; supervision, A.M. and O.H.; project administration, A.M. and O.H.; funding acquisition, F.M. and O.H. All authors have read and agreed to the published version of the manuscript.

Funding: This research was funded by the German Federal Ministry of Education and Research [grant number 031A247A] within the framework of the GlobE initiative through the Reduction of Post-Harvest Losses and Value Addition in East African Food Value Chains (RELOAD) project. The Article Processing Charges (APC) was funded by the University of Kassel, Germany.

Institutional Review Board Statement: Not applicable.

Informed Consent Statement: Not applicable.

Data Availability Statement: Not applicable.

Acknowledgments: This study is a part of Faizan Majeed's PhD research work at the Department of Agricultural and Biosystems Engineering, Faculty of Organic Agriculture, University of Kassel, Germany. Special thanks for the contribution of the Department of Post-Harvest Management, Jimma University, Jimma, Ethiopia. The great thanks and acknowledgement of the Higher Education Commission (HEC) Pakistan for financial support through a development project entitled "Strengthening of Bahauddin Zakariya University (BZU) Multan, Pakistan".

Conflicts of Interest: The authors declare no conflict of interest.

References

1. International Coffee Organization—Daily Coffee Prices. Available online: https://www.ico.org/coffee_prices.asp?section=Statistics (accessed on 21 June 2021).
2. Krishnan, S. Sustainable Coffee Production. *Oxf. Res. Encycl. Environ. Sci.* **2017**, 1–33. [CrossRef]
3. United States Department of Agriculture (USDA). Coffee: World Markets and Trade. 2021. Available online: https://apps.fas.usda.gov/psdonline/app/index.html#/app/downloads (accessed on 11 January 2022).
4. REN21 (Renewable Energy Policy Network for the 21st century). *Renewables 2020 Global Status Report*; REN21 Secretariat: Paris, France. 2020; ISBN 978-3-948393-00-7.
5. Taghizadeh-Hesary, F.; Rasoulinezhad, E.; Yoshino, N. Energy and Food Security: Linkages through Price Volatility. *Energy Policy* **2019**, *128*, 796–806. [CrossRef]
6. Hadzich, M.; Veynandt, F.; Delcol, J.; Hadzich, L.M.; Pérez, J.P.; Vergara, S. Design of a Solar Coffee Roaster for Rural Areas. *Energy Procedia* **2014**, *57*, 3215–3224. [CrossRef]
7. Laukaleja, I.; Kruma, Z. Influence of the Roasting Process on Bioactive Compounds and Aroma Profile in Speciality Coffee: A Review. In Proceedings of the Baltic Conference on Food Science and Technology and North and East European Congress on Food, Jelgava, Latvia, 2–3 May 2019. Available online: http://llufb.llu.lv/conference/foodbalt/2019/Laukaleja_Kruma_N080_FoodBalt2019.pdf (accessed on 13 January 2022).
8. Buffo, R.A.; Cardelli-Freire, C. Coffee flavour: An overview. *Flavour Fragr. J.* **2004**, *19*, 99–104. [CrossRef]
9. Franca, A.S.; Oliveira, L.S.; Oliveira, R.C.S.; Agresti, P.C.M.; Augusti, R. A preliminary evaluation of the effect of processing temperature on coffee roasting degree assessment. *J. Food Eng.* **2009**, *92*, 345–352. [CrossRef]
10. Palma, F.D.; Iacono, F.; Toffanin, C.; Ziccardi, A.; Magni, L. Scalable model for industrial coffee roasting chamber. *Procedia Comput. Sci.* **2021**, *180*, 122–131. [CrossRef]
11. Nagaraju, V.D.; Murthy, C.T.; Ramalakshmi, K.; Srinivasa Rao, P.N. Studies on roasting of coffee beans in a spouted bed. *J. Food Eng.* **1997**, *31*, 263–270. [CrossRef]
12. Raemy, A.; Lambelet, P. A calorimetric study of self-heating in coffee and chicory. *Int. J. Food Sci. Tech.* **1982**, *17*, 451–460. [CrossRef]
13. Mendes, L.C.; de Menezes, H.C.; Aparecida, M.; da Silva, A.P. Optimization of the roasting of Robusta coffee (*C. canephora conillon*) using acceptability tests and RSM. *Food Qual. Prefer.* **2001**, *12*, 153–162. [CrossRef]
14. de Carvalho, M.A.; de Rezende, J.C.; Rezende, T.T.; Ferreira, A.D.; Rezende, R.M.; Mendes, A.N.G.; Carvalho, G.R. Relationship between the sensory attributes and the quality of coffee in different environments. *Afr. J. Agric. Res.* **2016**, *11*, 3607–3614. [CrossRef]
15. Bauer, D.; Abreu, J.; Jordão, N.; Rosa, J.S.; da Freitas-Silva, O.; Teodoro, A. Effect of Roasting Levels and Drying Process of *Coffea canephora* on the Quality of Bioactive Compounds and Cytotoxicity. *Int. J. Mol. Sci.* **2018**, *19*, 3407. [CrossRef]
16. Rosa, J.S.D.; Freitas-Silva, O.; Rouws, J.R.C.; Moreira, I.G.D.S.; Novaes, F.J.M.; Azevedo, D.A.; Schwab, N.; Godoy, R.L.D.O.; Eberlin, M.N.; Rezende, C.M. Mass spectrometry screening of Arabica coffee roasting: A non-target and non-volatile approach by EASI-MS and ESI-MS. *Food Res. Int.* **2016**, *89*, 967–975. [CrossRef]
17. Ukers, W.H.; William, H. *All about Coffee*; The Tea and Coffee Trade Journal Company: New York, NY, USA, 1922; p. 615. Available online: http://archive.org/details/ (accessed on 14 January 2022).
18. Syafriandi; Fachruddin, F.; Lubis, A.; Maulina, H.; Nazura, P. Testing coffee roasting machine with electric heater as energy source. *IOP Conf. Ser. Earth Environ. Sci.* **2021**, *922*, 012073. [CrossRef]
19. Wang, X.; Lim, L.-T. A Kinetics and Modeling Study of Coffee Roasting Under Isothermal Conditions. *Food Bioprocess Technol.* **2014**, *7*, 621–632. [CrossRef]

20. Diallo—2019—A Heated Comparison Ikawa vs Probat.pdf. Available online: https://caravela.coffee/wp-content/uploads/2019/04/A-Heated-Comparison-Ikawa-vs-Probat-FINAL.pdf (accessed on 18 January 2022).
21. Dafle, V.R.; Shinde, N.N. Design, Development & Performance Evaluation of Concentrating Monoaxial Scheffler Technology for Water Heating and Low-Temperature Industrial Steam Application. *Int. J. Eng. Res. Appl.* **2012**, *2*, 848–852.
22. Nwakuba, N.R.; Asoegwu, S.N.; Nwaigwe, K.N. Energy requirements for drying of sliced agricultural products: A review. *Agric. Eng. Int. CIGR J.* **2016**, *18*, 144–155.
23. Solargis. World map of DNI. 2021. Available online: http://solargis.info/doc/_pics/freemaps/1000px/dni/SolarGIS-Solar-map-DNI-World-map-en.png (accessed on 17 January 2022).
24. Munir, A.; Hensel, O.; Scheffler, W. Design principle and calculations of a Scheffler fixed focus concentrator for medium temperature applications. *Sol. Energy* **2010**, *84*, 1490–1502. [CrossRef]
25. Raza, A.; Munir, A.; Ghafoor, A.; Ali, S. Soybean roasting using continuous solar roaster–A decentralized approach for agricultural processing. *J. Food Process Eng.* **2019**, *42*, e13184. [CrossRef]
26. Raza, A.; Majeed, F.; Munir, A.; Hensel, O. Development and Experimental Results of a Thermal Oil Based Roasting System for Decentralized Processing of Groundnuts. *Appl. Sci.* **2019**, *9*, 4342. [CrossRef]
27. EESI. *Solar Thermal Energy for Industrial Uses*; Issue Brief; Environmental and Energy Study Institute (ESSI): Washington, DC, USA, 2011.
28. Scheffler, W.; Bruecke, S.; von Werdenbergstr, G. Development of a solar crematorium. In Proceedings of the 6th International Conference on Solar Cooker, Granada, Spain, 15 July 2006.
29. Patil, R.; Awari, G.; Singh, M. Experimental analysis of Scheffler reflector water heater. *Therm. Sci.* **2011**, *15*, 599–604. [CrossRef]
30. Panchal, H.; Patel, J.; Parmar, K.; Patel, M. Different applications of Scheffler reflector for renewable energy: A comprehensive review. *Int. J. Ambient. Energy* **2020**, *41*, 716–728. [CrossRef]
31. Munir, A.; Hensel, O.; Scheffler, W.; Hoedt, H.; Amjad, W.; Ghafoor, A. Design, development and experimental results of a solar distillery for the essential oils extraction from medicinal and aromatic plants. *Sol. Energy* **2014**, *108*, 548–559. [CrossRef]
32. AOAC. *Official Methods of Analysis of AOAC International*, 17th ed.; Association of Analytical Communities: Gaithersburg, MD, USA, 2000.
33. Wang, L.; Yuan, Z.; Zhao, Y.; Guo, Z. Review on Development of Small Point-Focusing Solar Concentrators. *J. Therm. Sci.* **2019**, *28*, 929–947. [CrossRef]
34. Ayub, I.; Munir, A.; Ghafoor, A.; Amjad, W.; Nasir, M.S. Solar Thermal Application for Decentralized Food Baking Using Scheffler Reflector Technology. *J. Sol. Energy Eng.* **2018**, *140*, 061005. [CrossRef]
35. Duffie, J.A.; Beckman, W.A. *Solar Engineering of Thermal Processes*; John Wiley & Sons, Inc.: Hoboken, NJ, USA, 2006.
36. Olaniyan, A.M.; Karim, O.R.; Eromoselec, E.O. Design and Development of a Small-Scale Peanut Roaster. 2017. Available online: https://www.semanticscholar.org/paper/Design-and-Development-a-Small-scale-Peanut-Roaster-Olaniyan-Karim/a277049d55e266be60eb2c5d60f81887330c957b (accessed on 3 January 2022).
37. Box, G.E.P.; Hunter, J.S. Multi-Factor Experimental Designs for Exploring Response Surfaces. *Ann. Math. Stat.* **1957**, *28*, 195–241. [CrossRef]
38. Box, G.E.P.; Wilson, K.B. On the Experimental Attainment of Optimum Conditions. In *Breakthroughs in Statistics: Methodology and Distribution*; Kotz, S., Johnson, N.L., Eds.; Springer: Berlin/Heidelberg, Germany, 1992; pp. 270–310. [CrossRef]
39. Montgomery, D.C. *Design and Analysis of Experiments*, 9th ed.; John Wiley & Sons: Hoboken, NJ, USA, 2017.
40. Derringer, G.; Suich, R. Simultaneous Optimization of Several Response Variables. *J. Qual. Technol.* **2017**, *12*, 214–219. [CrossRef]
41. Sutarsi Soekarno, S.; Widyotomo, S. Performance Evaluation of Rotating Cylinder Type Coffee Bean Roaster. *J. Keteknikan Pertan.* **2010**, *24*, 106931.
42. Gallagher, J. Understanding Roaster Drum Speed & Its Affect on Your Coffee. Perfect Daily Grind. 2019. Available online: https://perfectdailygrind.com/2019/06/understanding-roaster-drum-speed-its-affect-on-your-coffee/ (accessed on 3 January 2022).
43. Getaneh, E.; Fanta, S.W.; Satheesh, N. Effect of Broken Coffee Beans Particle Size, Roasting Temperature, and Roasting Time on Quality of Coffee Beverage. *J. Food Qual.* **2020**, *2020*, e8871577. [CrossRef]
44. Jokanovic, M.; Dzinic, N.; Cvetkovic, B.; Grujic, S.; Odzakovic, B. Changes of physical properties of coffee beans during roasting. *Acta Period. Technol.* **2012**, *43*, 21–31. [CrossRef]
45. Bicho, N.C.; Leitão, A.E.; Ramalho, J.C.; Lidon, F.C. Use of colour parameters for roasted coffee assessment. *Food Sci. Technol.* **2012**, *32*, 436–442. [CrossRef]
46. Pittia, P.; Nicoli, M.C.; Sacchetti, G. Effect of moisture and water activity on textural properties of raw and roasted coffee beans. *J. Texture Stud.* **2007**, *38*, 116–134. [CrossRef]
47. Edzuan, A.M.F.; Aliah, A.M.N.; Bong, H.L. Physical and Chemical Property Changes of Coffee Beans during Roasting. *Am. J. Chem.* **2015**, *5*, 56–60.
48. Coffee: World Markets and Trade. USDA Foreign Agricultural Service. 2021. Available online: https://www.fas.usda.gov/data/coffee-world-markets-and-trade (accessed on 18 September 2021).

sustainability

MDPI

Article

Development and Experimental Study of Smart Solar Assisted Yogurt Processing Unit for Decentralized Dairy Value Chain

Syed Nabeel Husnain [1,2,*], Waseem Amjad [2], Anjum Munir [2] and Oliver Hensel [1]

1 Department of Agricultural and Biosystems Engineering, University of Kassel, 37213 Witzenhausen, Germany; agrartechnik@uni-kassel.de
2 Department of Energy Systems Engineering, University of Agriculture, Faisalabad 38000, Pakistan; waseem_amjad@uaf.edu.pk (W.A.); anjum.munir@uaf.edu.pk (A.M.)
* Correspondence: nabeel.husnain@uaf.edu.pk

Abstract: Yogurt production at the farm level is important for adding value to milk. In this study, a solar-assisted yogurt processing unit capable of performing the three processes of heating, fermentation, and cooling in a single unit was developed. It consisted of a circular chamber surrounded by a coil for heating by a solar vacuum tube collector and a pillow plate for cooling by a solar PV-powered chiller unit. Experiments were performed using 50, 40 and 30 L of raw milk under a constant water circulation rate of 50 L per minute for heating followed by a cooling process under 36, 18 and 6 rpm of stirrer speeds. The heat absorption rates of the milk were 5.48–0.31, 4.75–0.16 and 4.14–0.24 kW, and the heat removal rates from water ranged from 6.28–0.49, 5.58–0.49 and 4.88–0.69 kW for 50, 40 and 30 L of milk volume, respectively. The overall heat transfer efficiency was above 80% during the heating process. A stirring speed of 18 rpm was found to be optimal in terms of cooling speed and consistency of the yogurt. The total energy consumed was calculated to be 6.732, 5.559 and 4.207 kWh for a 50, 40 and 30 L batch capacity, respectively. The study offers a sustainable energy solution for the decentralized processing of raw milk, particularly in remote areas of the developing countries where access to electricity is limited.

Keywords: yogurt processing; solar energy; solar-based heating and cooling; thermal analysis

Citation: Husnain, S.N.; Amjad, W.; Munir, A.; Hensel, O. Development and Experimental Study of Smart Solar Assisted Yogurt Processing Unit for Decentralized Dairy Value Chain. *Sustainability* **2022**, *14*, 4285. https://doi.org/10.3390/su14074285

Academic Editor: Andrea Pezzuolo

Received: 15 March 2022
Accepted: 2 April 2022
Published: 4 April 2022

1. Introduction

Milk and its products are considered to be a good medium for the infectious growth of bacteria and other pathogens which grow faster at ambient temperature. Raw milk and yogurt are spoiled due to an increase in temperature, and these losses are significant in developing countries due to the non-availability of processing facilities at the farm level. Therefore, producers have to sell their products at low prices. Pakistan is ranked as the fourth largest milk-producing country in the world after the USA, China and India [1] by producing about 42 billion liters of milk annually [2], while the majority of producers are small-scale farmers (>80%). Unfortunately, only 5% of this milk is processed while the remainder is handled by milkmen which are mostly unhygienic and pose high health risks. About 15–19% of the total milk produced in the country is wasted due to a lack of processing facilities while the rest is handled improperly [3]. Yogurt is one of the popular dairy products in the Indo-Pak subcontinent. In Pakistan, the yogurt share is about 70% of total fermented dairy products [4], but diminutive attention is given to the fermentation of milk to increase its shelf life, aroma and nutritional value.

About 70% of dairy farms have limited access to the market, forcing milk producers to sell raw milk at a low price to middlemen and depriving them of a reasonable profit. Moreover, lack of handling and processing facilities at the farm level, poor financial support to farmers, and importantly, frequent interruption of the power used for farm processing are major hindrances to processing raw milk. There is a need not only to handle the raw milk (pasteurization), but also to convert it into a highly demandable and value-increased byproduct, i.e., yogurt and milk powder [5]. Although few commercial brands of yogurt are available in Pakistan, their prices are almost double those at the local market. Due to an increase in inflation, the majority of consumers are price conscious and buy yogurt from local shops which are exposed to dust and flies and possess no refrigeration facility, thus compromising on shelf life and the quality of the yogurt.

The yogurt-making process (heating and cooling) from raw milk is an energy-intensive process, which is why dairy is considered the fifth most energy-consuming industry [6]. Major energy demand in the dairy processing sector is still being met using fossil fuels [7] which not only increases operational cost, but also leads to environmental pollution [8]. Therefore, on-farm yogurt processing using renewable energy (solar energy with a potential average value of 5.3 $kWh/m^2/d$) can help to overcome such issues [9]. For solar thermal heating at low–medium temperature range, Michael et al. (2016) [10] highlighted the significance of using flat plate and evacuated tube collectors and reported their contribution as about 30%, while parabolic troughs, dishes and Fresnel collectors contribute 22% of industrial segments. Amjad et al. (2021) [11] reported the use of solar evacuated tube collectors for the decentralized processing (drying) of agricultural produce. A thirty-tube evacuated-tube solar collector was attached with a water–air heat exchanger to warm up the drying medium. It was reported that proper integration of the evacuated tube collector performs well in terms of meeting low temperature demand (up to 100 °C). Ismail et al. (2021) [12] also reported the potential of solar thermal applications in the food industry to meet low temperature demand (up to 100 °C). In this study, the use of flat plate collectors in the food industry has been estimated to be 38% for pre-heating, pasteurization and cleaning processes. Similarly, in the dairy sector, providing decentralized, energy-efficient and low-cost milk handling and yogurt processing units can play a vital role in uplifting the rural economy through value addition and income generation.

Various studies reported the use of solar energy in the dairy industry. Khawer et al. (2020) [13] developed a solar-based milk chilling system comprising a one-ton vapor compression refrigeration unit and a 2 kW_p PV system to process 200 L of milk using less than 1 kW power. Desai et al. (2013) [14] conducted a study to highlight the need for solar energy to assist vapor absorption cooling systems in maintaining cold storage conditions for milk handling in India and concluded its high scope in the dairy industry. Mekhilef et al. (2011) [15] comprehensively reviewed the use of solar thermal and PV systems for industrial applications. It is concluded that the greatest efficiency can be achieved through proper systems integration and suitable selection of solar collectors for water heating, solar refrigeration and steam generation. Anderson and Duke (2007) [16] reported the potential of solar thermal applications employing flat plate and evacuated tubes collectors for heating and cooling in the dairy industry. Zahira et al., 2009 [17] investigated the potential of solar energy to pasteurize raw milk at a temperature range of 65 °C to 75 °C while ambient air temperature was 40 °C. It was found that a pasteurizer unit fabricated from shipping cardboard easily attained the required temperature. Atia, 2011 [18] reported the working of a milk pasteurizer connected with a solar flat plate collector. It was found that 73.9 L of raw milk was successfully pasteurized at 63 °C, but fluctuation in solar radiation showed direct impact on the performance of the flat plate collector. Similarly, Wayua et al., 2013 [19] developed a milk pasteurizer made of a stainless steel cylindrical vessel having a jacket for the circulation of water being heated with a flat plate collector Although the capacity of the container was 80 L, it was found that 40 L milk can be pasteurized optimally. In all these reported studies, it can be assessed that use of a flat plate collector cannot provide consistent energy and possess low energy efficiency. For this to get better energy efficiency, Yaseen et al. 2019 [20] reported the use of a vacuum tube collector to pasteurize 200 L milk using steam. Milk was pasteurized at 63 °C followed by cooling up to 30 °C with tap water which further cooled down to 4 °C in a PV-powered rotary compressor chiller. This study reported the use of a vacuum tube collector only for milk pasteurization. A single-glazed flat plate solar collector and water in a glass evacuated tube solar water heater were primarily used [21]. However, the use of solar collectors, especially evacuated tube collectors (ETC), is limited to water heating only [22,23] and none of the studies reported its use for yogurt fermentation. Moreover, at the industrial scale, for cooling and heating processes involved in yogurt production, the raw milk is transferred in separate containers requiring more infrastructure and clean-in-place (CIP) cost, which could be viable at large scale, but would not be good practice for handling a small quantity of milk (less than 100 L). There is no study about a system capable of performing all the processes required for yogurt making in a single unit/system using solar energy.

Keeping in view the aforementioned facts, in this study a sustainable energy solution for the decentralized handling and processing of raw milk, especially in remote areas of the developing countries where access to electricity is reduced, has been presented. A solar-assisted three-in-one (heating, fermentation and cooling) yogurt processing unit for the value addition of raw milk has been developed. Integration of solar evacuated tube collectors (for heating) and a solar PV system (for cooling) with the yogurt processing unit was evaluated not only to pasteurize raw milk but also to make yogurt in a single unit. The salient feature of the system is the design of a single container capable of performing both heating and cooling processes which not only reduce capital cost but also make it user friendly.

2. Material and Methods

The complete unit for yogurt processing was developed and fabricated in the Department of Energy Systems Engineering Workshop, Faculty of Agricultural Engineering and Technology at the University of Agriculture Faisalabad (UAF) Pakistan in collaboration with the International Center for Development and Decent Work (ICDD), University of Kassel, Germany and Dairy Industries, Okara-Pakistan).

2.1. System Description

The design and selection of the yogurt processing unit was largely based on some basic parameters such as energy efficiency and maintenance, and especially on the product life cycle and environmental sustainability. Figure 1 shows a solar-assisted yogurt processing unit designed to process raw milk and its fermentation into yogurt in a timely manner at the production site. It consists of a cylindrically shape fermentation chamber (560 mm diameter and 230 mm depth) made of stainless steel (food grade SS 304) having a capacity of 50 L which is surrounded by a heating coil (3.5 m long, 40 mm wide and 12.5 mm high). The walls and bottom of the fermentation chamber were insulated by 100 mm thick PU (Polyurethane) material, so that heat loss through conduction and convection can be reduced. A variable frequency drive (VFD) electric motor was installed on the top of the chamber to rotate a stirrer for maintaining uniform temperature in the processed product. For cooling purposes, the bottom surface of the chamber was fabricated by a pillow plate which itself acts as an evaporator. The use of a pillow plate heat exchanger not only reduces the size and cost of the unit but also provides a higher heat transfer coefficient in comparison to the conventional coil heat exchanger. For cooling of yogurt, one ton of rotary compressor compatible with R-410A (environmentally friendly) gas was installed employing an inverter kit to reduce torque load to run on 2 kW_p PV modules.

For the heating of raw milk, the yogurt processing unit was connected with a hot water storage tank (100 L capacity) which receives heat from a solar evacuated tube collector (2.46 m^2) having a connection through polyvinyl chloride (PVC) pipe fittings as shown in Figure 2. The outer and inner diameters of the ETC tubes were Ø58 mm ± 0.7 mm and Ø47 mm ± 0.7 mm, respectively. The glass tube length was 1800 mm ± 5 mm and the vacuum was $P < 5 \times 10^{-3}$ Pa. The thermal energy absorbed by the collector is transferred to 100 L of water in the storage tank to raise the water temperature to 90 °C. A centrifugal pump (Wilo-SP106) was installed between the hot water storage tank and the evacuated tube collector for the circulation of propylene glycol solution (50% by volume). The pump can be operated at three variable speeds (600 L/h, 900 L/h and 1100 L/h), and it required 80 W power at maximum speed. The current research was conducted at a flow rate of

600 L/h. Heated glycol solution entered the storage tank and transferred its heat to water while passing through the helix-type heat exchanger present in the storage tank. In order to transfer heat from the storage tank to the yogurt processing unit, another water circulation pump (stainless steel centrifugal, WB50/025D, 50 L/min) was installed between the outlet of the hot water storage tank and the inlet of the yogurt processing unit to circulate the hot water through the square spiral coil heat exchanger to increase the temperature of milk up to 80 °C.

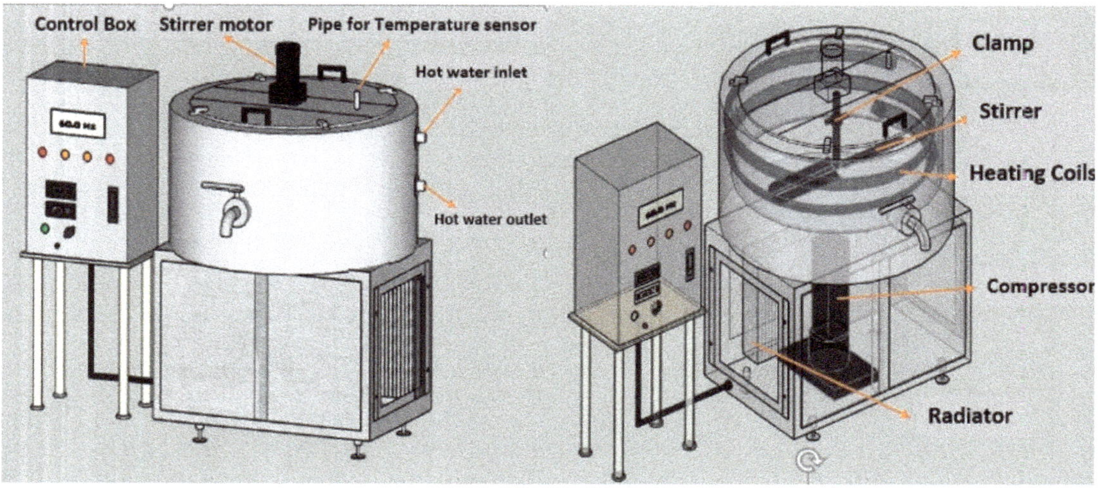

Figure 1. Three-dimensional (3D) layout of solar-assisted yogurt processing unit.

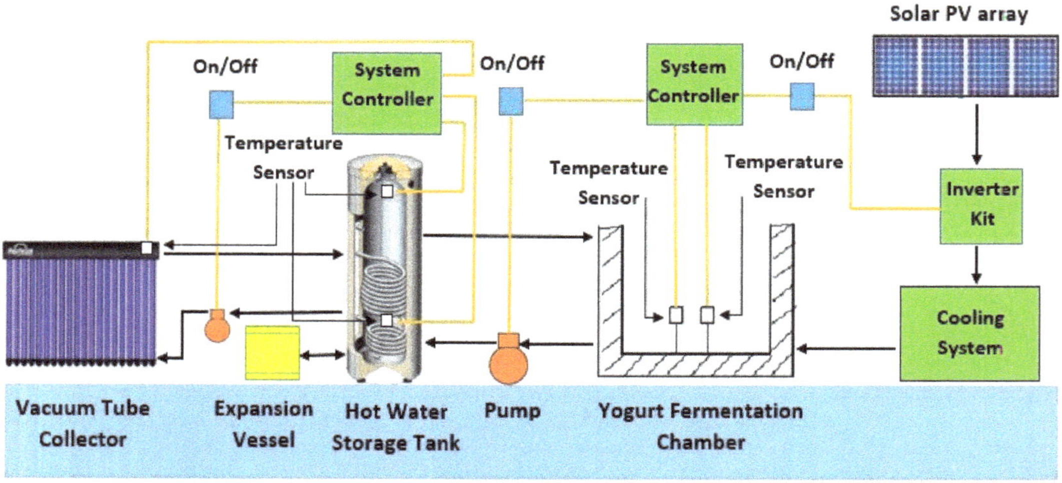

Figure 2. Layout of the experimental setup of the solar-assisted yogurt processing unit.

Being a closed cycle, an expansion vessel (12 Liter) was also provided to avoid high-pressure build-up in the system. The controller turns on the circulation pump (Wilo-SP106) when the temperature differential between the glycol solution leaving the collector and the water in the lower portion of the storage tank exceeds 5 °C and turns the pump off when the differential is below 5 °C or when the water temperature of the storage tank exceeds to 90 °C.

2.2. Sizing of Fermentation Chamber

The size of the fermentation chamber depends on the quantity of the product to be processed. Considering a milk capacity of 50 L, the chamber was designed by keeping the diameter to depth ratio in such a way that it could not only get enough exposure to the pillow plate fabricated at the bottom of the chamber but also to increase the surface area of surrounding heating coils. Moreover, the calculation was made for 56.6 L to provide space for the stirrer, stirrer shaft and spacing required during the shaking phenomenon. Normally, semi-circle type chambers are used which are suitable only for the cooling of milk. To perform both heating and cooling in the same chamber, a cylindrical-type chamber was designed having the provision of installing a heating coil around the chamber.

$$\text{Height} = \frac{V}{(\pi \mathrm{d}^2/4)} \quad (1)$$

where V is the volume of the chamber (m^3), and d is the diameter of the fermentation chamber (m).

The height and diameter of the chamber were calculated to be 230 mm and 560 mm, respectively. A 13.2% provision was given in the total volume of the chamber for air and stirrer.

2.3. Sizing of Refrigeration System

The size of a refrigeration system depends on the cooling load of the fermentation chamber, which further depends on the mass of milk or yogurt and the time required to reach the chilling temperature (4 to 8 °C).

Heat transfer through the walls and bottom can be calculated using the following equation [24]:

$$\mathrm{Q_t} = \frac{\mathrm{UA}(\mathrm{T_h} - \mathrm{T_c}) \times 24}{1000} \quad (2)$$

where Q_t stands for the total heat transfer through the walls and bottom (kWh/d), U is the overall heat transfer coefficient (W/m^2 °C), A is the heat transfer area (m^2), T_h is the hot face temperature of the fermentation chamber (°C) and T_c is the cold face temperature of fermentation chamber (°C).

Similarly, using the same equation, heat transfer through the walls (Q_{t1}) and heat transfer through the top of the chamber (Q_{t2}) were also calculated to be 0.162 kWh/d and 0.2179 kWh/d using U values of 0.28 and 0.95, respectively.

The cooling load of the product can be calculated by using Equation (3)

$$\mathrm{Q_R} = \frac{\mathrm{mC_p\Delta T}}{3600} \quad (3)$$

where Q_R stands for the cooling load of the product (2.10 kWh/d), m is the mass of the product (kg) and C_p is the specific heat of the milk (3.89 kJ/kg K). Adding the outcomes from Equations 2 and 3, the total heat load was calculated to be 2.4799 kWh/d.

The required refrigeration capacity was calculated by dividing the total heat load by the specified time of cooling (2 h in this case), and it was found to be 0.709 TR. Considering the 20% factor of safety, the required refrigeration capacity was calculated to be 0.851 TR. Keeping in view other losses and the availability of standard size, one ton of refrigeration system was used for yogurt processing.

2.4. Photovoltaic System Design

The size of the solar system was selected as per the load requirements of the developed yogurt processing unit, and it can be estimated for any size of the yogurt processing unit. The size of the photovoltaic (PV) system based on the peak power (P_p) in kWp required to operate the compressor, stirrer motor and water pumps can be estimated by using the following equation [25].

$$P_p = \frac{L_e I_b}{H_{avg}\eta_{inv}\ \eta_{bat}\ T_{CF}} \quad (4)$$

where P_p is the peak power of the solar system (kWp); L_e is the electric load (kWh/d) which is the product of the power required to run appliances and the time of operation (hours per day); I_b is the solar irradiance (kW/m^2) and 1 is taken as its peak value for calculation; H_{avg} is the average global horizontal irradiance (kWh/(m^2 d)) and its value lies between 5 and 6 kWh/(m^2 d) for Faisalabad, Pakistan; and minimum value taken for calculation, η_{inv} is the efficiency of the inverter (95–98%); η_{bat} is the efficiency of battery (85–95%); and T_{CF} is the temperature correction factor obtained by subtracting the product of the loss factor (0.4% per °C) and the change in the PV temperature from unity, and for the current study it was 0.92. This means that the power will reduce by 0.4% per degree rise in temperature from its optimum value, i.e., 25 °C at standard testing conditions. Solving Equation (4), the peak power was calculated to be almost 2 kW$_p$. So, eight PV panels (polycrystalline, each 250 W$_p$) were installed with a 3 kW inverter. Here the surge factor has been taken as 1 due to the presence of inverter technology which stats the compressor with zero torque load.

In order to maintain the required load for the cooling system, a battery bank was used to charge and discharge with the varying solar intensity throughout the day. The size of the battery bank was calculated by the following equation [25]

$$C_{Bat} = \frac{N_{ccd} L_e}{D_d \eta_{bat}\ V_{bat}} \quad (5)$$

where C_{Bat} is the capacity of the battery bank (Ah), N_{ccd} is the autonomy and taken as the number of continuous cloudy days, D_d is the depth of discharge in fraction, and V_{bat} is the nominal voltage of the battery. All of the specifications of the solar system used in the current study have been tabulated in Table 1.

Table 1. Datasheet and specification of PV modules, inverter and battery bank.

Component of Solar PV System	Attributes	Rating	Unit
PV Module Model (TSPP-6U (60)-250W)	Power Output Tolerance	±3.00	%
	Temperature Coefficient	−0.0047	%/°C
	Open Circuit Voltage (V_{oc}) at STC	37.30	V
	Short Circuit Current (I_{sc}) at STC	8.50	A
	Peak Power Watts (Wp) at STC	250.00	W
	Maximum Power Voltage (V_{mpp}) at STC	30.90	V
	Maximum Power Current (I_{mpp}) at STC	8.00	A
	Efficiency of Module	0.155	%
	Dimension of solar panel	1650 × 1000 × 40	mm^3
Inverter Model ((Premier-PW-24DC)	Maximum Input Current	50.00	A
	Maximum input DC Voltage to controller	60.00	V
	Operated Voltage Range	30–32	V
	Output Voltage to battery	24.00	V
	Max. Power can be connected	3.00	kW
Battery bank (Lithium-ion Battery)	Nominal Voltage	12.00	V
	Nominal capacity	150.00	Ah
	Depth of discharge	0.60	
	Charging current	100.00	A
	Continuous discharging current	100.00	A
	Discharging working temperature	−20~60 °C	°C
	Charging working temperature	0–45	°C
	Dimensions	479 × 482 × 132	mm^3
	Efficiency of battery	0.95	%
	Cell combination	36,130,170 8S2P + Smart BMS	

2.5. Sizing of Evacuated Tube Collector (ETC)

For the thermal application of the yogurt processing unit, an evacuated tube collector (ETC) was used. The ETC area was calculated by using the following formula [25]

$$A_c = \frac{mC_p\Delta T}{I_b\tau\alpha t} \tag{6}$$

where A_c is the area of the receiver tube exposed to radiation (m^2); m is the mass of water (kg); C_pC_p is the specific heat capacity of the receiver tube ($kJkg^{-1}\ K^{-1}$); ΔT is the change in temperature (K); I_b is the solar irradiance (W/m^2); τ is the transmission coefficient; α is the absorption coefficient and t is the time in seconds.

The amount of heat energy required to increase the temperature of milk (m = 50 kg, specific heat: 3.89 $kJkg^{-1}\ K^{-1}$) up to 50 °C (30 °C to 80 °C) in two hours time (t = 7200 s) using a vacuum tube collector (transmission coefficient $\tau\tau = 0.95$, absorption coefficient $\alpha = 0.95$) was calculated for a tropical region like Faisalabad (31.4303° N, 73.0672° E) lying in the solar belt having an average global horizontal irradiance (P) of 0.8 kWm^{-2}. Substituting these values into Equation (6), the area required is calculated to be 1.87 m^2. The absorber area of one tube is calculated by multiplying the circumference of the absorber plate and calculated to be 0.164 m^2, while the number of tubes is calculated by dividing the total absorber area by the absorber area of one tube and was calculated to be 11.4 ≈ 12 tubes. Therefore, an evacuated tube collector of 15 tubes was selected considering thermal losses in the system as shown in Figure 3.

Figure 3. Pictorial view of the installed solar-assisted yogurt processing unit.

2.6. Experiments

Trials were conducted to assess the heating and cooling performances using three different volumes of raw cow's milk (50, 40 and 30 L). Before conducting an experiment, the heating component of the system was turned on in order to raise the water temperature of the storage tank up to 90 °C, which normally took 2–3 h depending upon solar radiation. After that, raw milk is poured into the fermentation tank and the circulation of hot water from the storage tank to the fermentation chamber is turned on. The heating of raw milk up to 80 °C took about 140 min, and it continuously stirred at a speed of 36 rpm. To bring down the temperature of heated milk to 43 °C, which is recommended for fermentation of milk, tap water passed through the heating coil in an open loop controlled by manually operated valves. The inoculation of starter culture was carried out (2–3%) at this temperature, and it was maintained by a solenoid valve controlling the water circulation for a period of 5–6 h until the required pH (4.85–4.5) was attained. After that process, the cooling process of yogurt was started by turning on the refrigeration unit to bring down the temperature of yogurt below 8 °C to reduce the bacterial activities which normally took 48–103 min depending upon the quantity of the processed milk and stirrer speed. For this, three different speeds (36, 18 and 6 rpm) of the stirrer were selected using a variable speed motor. Each experiment was replicated thrice, and values were averaged for each volume of milk. For data acquisition, a controller with resistance temperature detector (RTD)-based temperature sensors was used to measure temperature at the inlet and outlet of the ETC, the top and bottom of the hot water storage tank, and inside the fermentation chamber. A portable pH meter (ML1010) was used to measure the pH of milk during the fermentation process. To access the performance of the installed PV system, a clamp meter (Fluke 345PQ), and pyranometer (METEON) were used.

Thermodynamic Analysis for Efficient Refrigeration System

While selecting a cooing machine, the coefficient of performance (COP) is the key factor to be considered for maximum cooling output with the lowest possible input energy requirement. Therefore, a comprehensive thermodynamic analysis is mandatory for evaluation and cooling system optimization for the best application. For this purpose, the system was connected with temperature sensors (thermocouples: K type, error: <0.1%) to record the temperature differentials of the circulating refrigerant in order to monitor the phase changes during the complete refrigeration cycle. The design and selection of the compressor was carried out with the aim that it must be capable of superheating the refrigerant to provide a reasonable degree of superheat to provide a high enthalpy drop during the expansion process (process (irreversible adiabatic process) with the help of a capillary tube for maximum cooling efficiency during the evaporation process to chill the milk/yogurt. Therefore, the rotary compressor was used to decrease the torque load and increase the pressure to achieve the required degree of superheat from the refrigerant.

R410a refrigerant was used in the compressor and the flow rate of the refrigerant was maintained as 1.42 L/min, having a specific volume of 0.0009 m^3 kg^{-1} ($m' = 1.58$ kg/min). The temperature after compression was recorded to be 347 K, while the temperature before compressing was noted as 261 K, and this line is represented on the T-S diagram by a vertical line showing an isentropic process owing to the high speed of the compressor. It has been noted that the temperature after the compression process is 347 K, and this temperature is well above the saturation temperature of the refrigerant (318 K at 2758 kPa), thus providing a degree of superheat as 29 K. The heat is removed at constant pressure in the condenser: the process first removes sensible heat from superheated vapor (347 K) to saturated vapor (318 K), i.e., a process from 2 to $2'$, then in the form of latent heat from dry saturated vapor to the saturated liquid line (process from $2'$ to $3'$) and then sensible heat from the saturated liquid line to 312 K as a sub-cooling process ($3'$ to 3). This high heat dissipation rate was due to the high heat transfer coefficient of copper coiling used and the efficient design of the condenser as well as the addition of two fans to dissipate heat at a faster rate. Therefore, the condensation process takes place partially in the superheated region and predominantly inside the saturation region to change the phase of the refrigerant from gas to liquid and then follows a sub-cooling line to decrease the refrigerant temperature. The expansion takes place through the capillary tube and undergoes an irreversible adiabatic (constant enthalpy) process as shown in process 3–4 in Figure 4. Process 4-1 is accomplished at constant pressure which evaporates the refrigerant once again at saturation temperature to complete the vapor compression refrigeration cycle using R410a refrigerant.

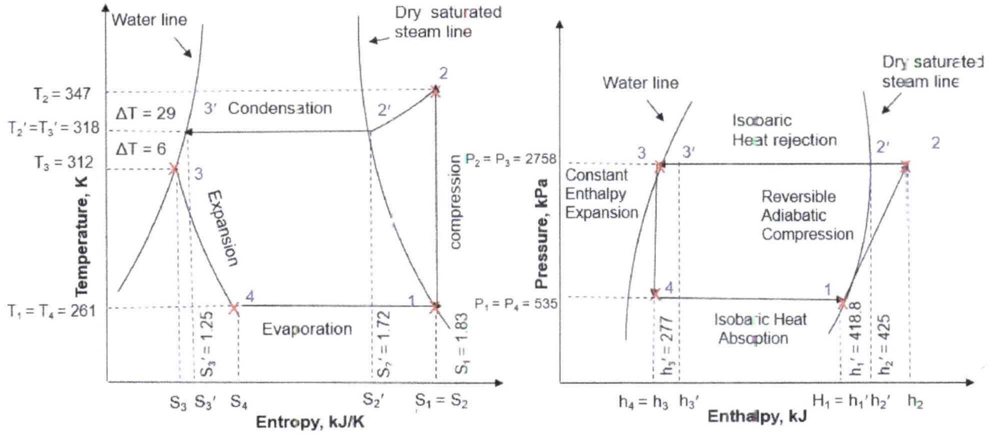

Figure 4. Temperature–entropy (T-s) and pressure–enthalpy (p-h) diagrams of the optimized compressor of the yogurt processing system.

The COP of the refrigerant is calculated using Equation (7)

$$\text{COP} = \frac{h_1 - h_{f3}}{h_2 - h_1} \quad (7)$$

where h_1, h_2 and h_{f3} are the enthalpies of the refrigerant used at the compressor inlet, compressor outlets and liquid enthalpy after the condenser outlet.

The values of dry saturated vapor before the compressor and after the compressor were found to be 418.8 and 425 kJ/kg at temperatures of 261 and 318 K, respectively (at 535 and 2758 kPa), while the value of the dry liquid line was found to be 277 kJ/kg at a temperature of 312 K and 2758 kPa pressure. The total enthalpy of actual refrigeration points after compression 'h_2' was calculated by using Equation (8):

$$h_2 = h_{2'} + C_{pg}(T_2 - T_{2'}) \quad (8)$$

where h_2 is the actual enthalpy of the refrigerant after the compression process, h_2' is the enthalpy of the refrigerant at the dry saturated steam line after the compression process and is taken from the enthalpy table, C_{pg} is the specific heat capacity of the refrigerant at constant pressure in gaseous state (0.84 kJ kg^{-1} K^{-1}), and T_2 and T_2' are the actual temperatures recorded of the refrigerant and saturation temperature at the compressor outlet, respectively.

The enthalpy of actual the refrigeration points after the condensation process 'h_3' was calculated by using Equation (9):

$$h_3 = h_{3'} + C_{pl}(T_{3'} - T_3) \quad (9)$$

where h_3 is the enthalpy of the refrigerant after the condensation process and is noted physically, $h_{3'}$ is the enthalpy of the refrigerant at the water line after condensation and is taken from the enthalpy table, C_{pl} is the specific heat capacity of the refrigerant at constant pressure in the liquid state (1.8 kJ kg^{-1} K^{-1}), T_3 is the actual value of the temperature of the refrigerant after condenser and T_3' is the saturation temperature at the liquid line after the condensation process, respectively.

By using Equations (8) and (9), the values of enthalpies h_2 and h_3 are calculated to be 454 kJ/kg and 266.2 kJ/kg, respectively. It is worth mentioning here that the value h_3 is the total enthalpy at point 3, and as there is no latent heat consideration as the refrigerant is in a liquid state after condensation in the condenser, this enthalpy h_3 is equal to h_{f3} (liquid enthalpy), as the refrigerant at the inlet of the compressor is in a dry saturation condition and enthalpy at point 1 is equal to 418.8 kJ/kg.

Using these values in Equation (7), the COP of the system was found to be 4.33. In fact, this is the thermodynamic value for the COP. However, the actual value of the refrigeration system for the cooling application is a little lower, as these values include the thermal losses and heat transfer losses through the pillow plate of the chiller unit where processing is carried out.

The capacity of refrigeration (TR), i.e., the heat extraction rate, is calculated using the following equation

$$TR = \frac{m'_r(h_1 - h_{f3})}{3.5} \tag{10}$$

where m'_r is the mass flow rate of the refrigerant (1.58 kg/min), h_1 is the enthalpy of the refrigerant used at the compressor inlet (kJ), and h_{f3} is the liquid enthalpy after the condenser outlet (kJ).

The capacity of refrigeration (TR) is calculated to be 1.15.

The power required to run the system is calculated by using Equation (11)

$$P = m'_r(h_2 - h_1) \tag{11}$$

where P is the power, m'_r is the mass flow rate of the refrigerant (1.58 kg/min), h_2 is the enthalpy of the refrigerant at the compressor outlet (kJ), and h_1 is the enthalpy of the refrigerant used at the compressor inlet (kJ)

The power required is calculated to be 0.93 kW; however, a 1 kW compressor motor was used considering electro-mechanical losses. The research depicts that through proper design and optimization of the refrigeration system, less than 1 kW of power is required to achieve one ton of cooling effect (3.5 kW/210 KJ/min or 12,000 BTU/h), which is in accordance with the findings investigated by [13,20,27,28] for high-performance cooling machines. The actual COP results were also conducted under field conditions using milk/yogurt by direct method, and the details of these trials are given in Section 3.4.

3. Results and Discussion

3.1. *Heating of Raw Milk*

Temperatures changes of hot water and milk during the heating process for three different product volumes (50, 40, 30 L) are shown in Figure 5. It can be observed that the initial temperatures of the milk and water storage tanks were 30 °C and 90 °C, respectively. During the first thirty minutes of the heating process, the temperature of the hot water dropped to 72 °C due to the high heat transfer rate, resulting in milk temperature increases of 62 °C, 66 °C and 70 °C for 50, 40 and 30 L of milk capacities, respectively. Thereafter, it was observed that the rate of temperature change was slower for all processed capacities. However, based on the quantity of the product to be processed, the time required to achieve the desired temperature (80 °C) was found to be 140, 110 and 80 min for 50, 40 and 30 L of milk capacities, respectively. In the case of a hot water storage tank, after thirty minutes, the temperature started to increase due to the continuous addition of thermal energy from ETC and a significantly lower rate of heat transfer to milk. This shows that the system can achieve the temperature (80 °C) required to denature and unfold most of the milk protein "lactoglobulin". This allows lactoglobulin to bind with some of the other proteins in milk, called caseins, thus forming a thick and well-structured yogurt. In addition to this, it can also be noted that the system can easily be used for the low-temperature pasteurization (63 °C) of up to 50 L of milk in approximately thirty minutes, which makes the system state of the art in terms of yogurt making and pasteurization.

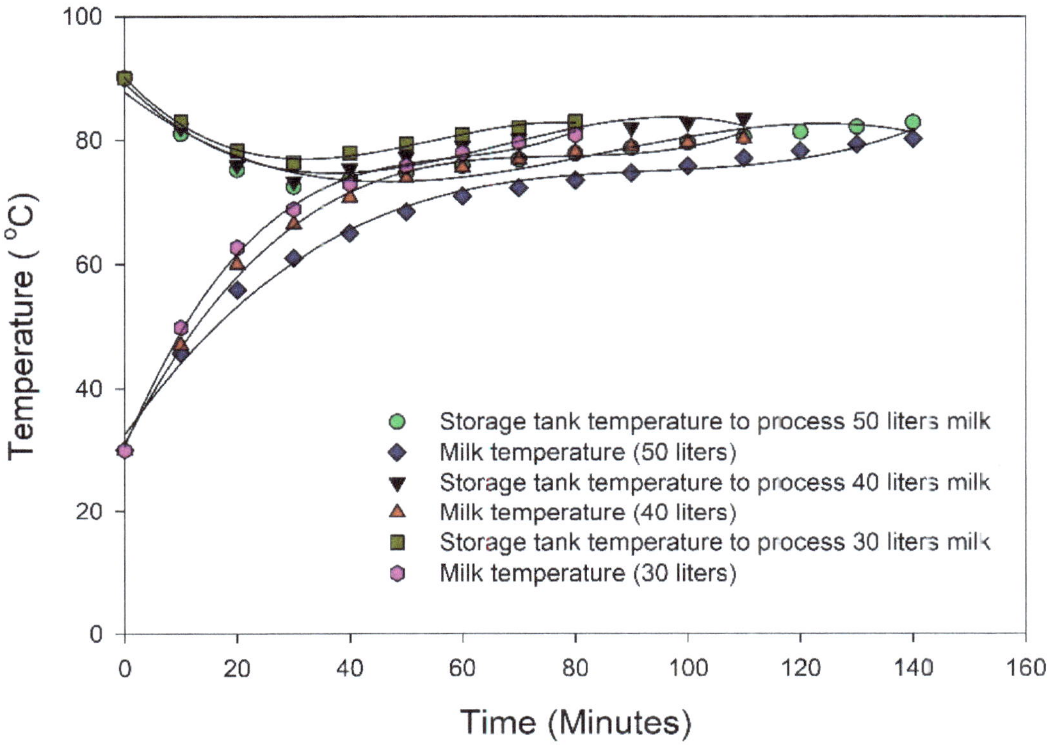

Figure 5. Relationship between change in the hot water of the storage tank and milk temperatures under different loading capacities.

The heating rates of milk in response to the heating rate of circulation water during the entire heating process under different product capacities have been shown in Figure 6. The solar irradiance and the total solar power available at ETC were also recorded and calculated during the experiments conducted on 2 July 2020, 3 July 2020, and 14 July 2020 for 50, 40 and 30 L of batch capacities, respectively by using a pyranometer (METEON, Accuracy ± 0.1%). These values were found in the ranges of 850–896 W/m^2, 840–891 W/m^2, 840–881 W/m^2 and 2.16–2.23 kW, 2.10–2.22 kW, 2.10–2.2 kW for 50, 40 and 30 L of milk capacities, respectively. At the early stages of the heating process, the rate of heat extraction from hot water was higher than that of heat addition, and it reduced as soon as milk temperature increased. It can be observed that the heating rate of milk is higher than the thermal energy provided by ETC due to the energy provided by the hot water of the storage tank at the beginning of the heating process. The heating rate of milk and the heat extraction rate from hot water was calculated at 10 min intervals throughout the heating process, and their values were found in ranges of 5.48–0.31 kW, 4.75–0.16 kW, 4.14–0,24 kW and 6.28–0.49 kW, 5.58–0.49 kW, 4.88–0.69 kW for 50, 40 and 30 L of milk capacities, respectively.

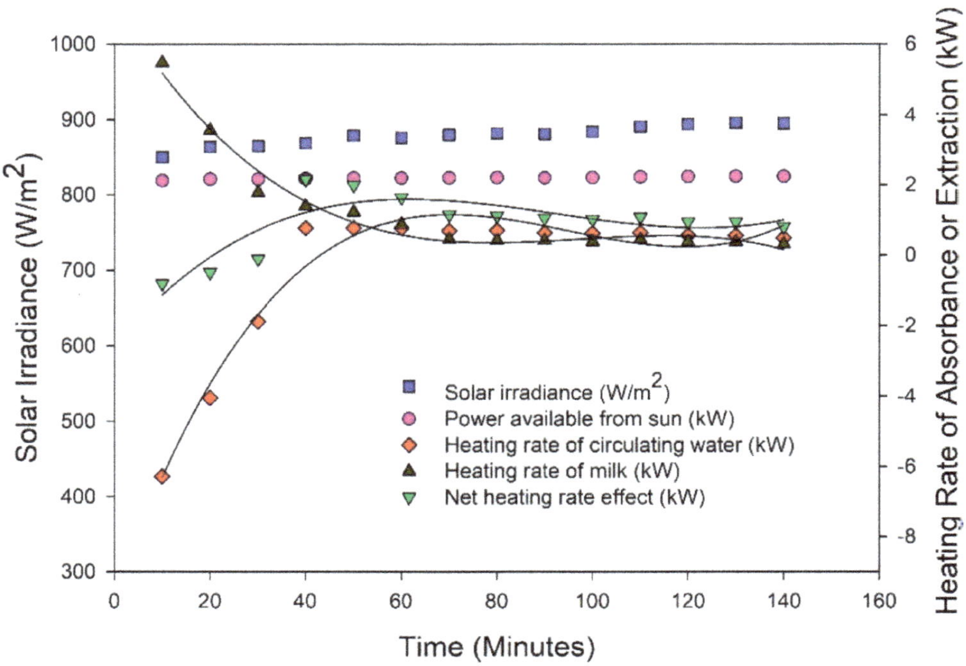

(a)

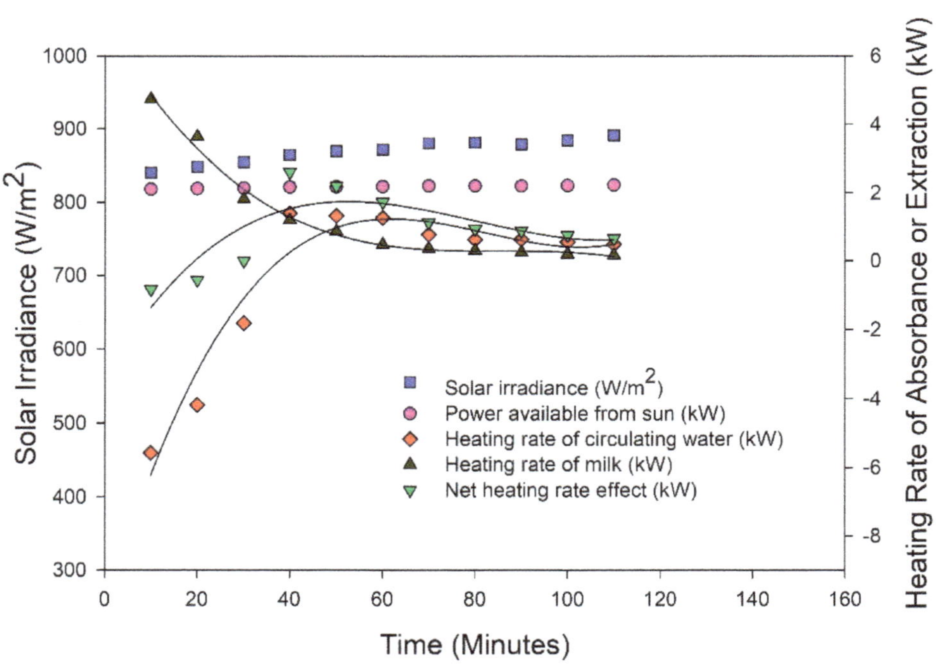

(b)

Figure 6. *Cont.*

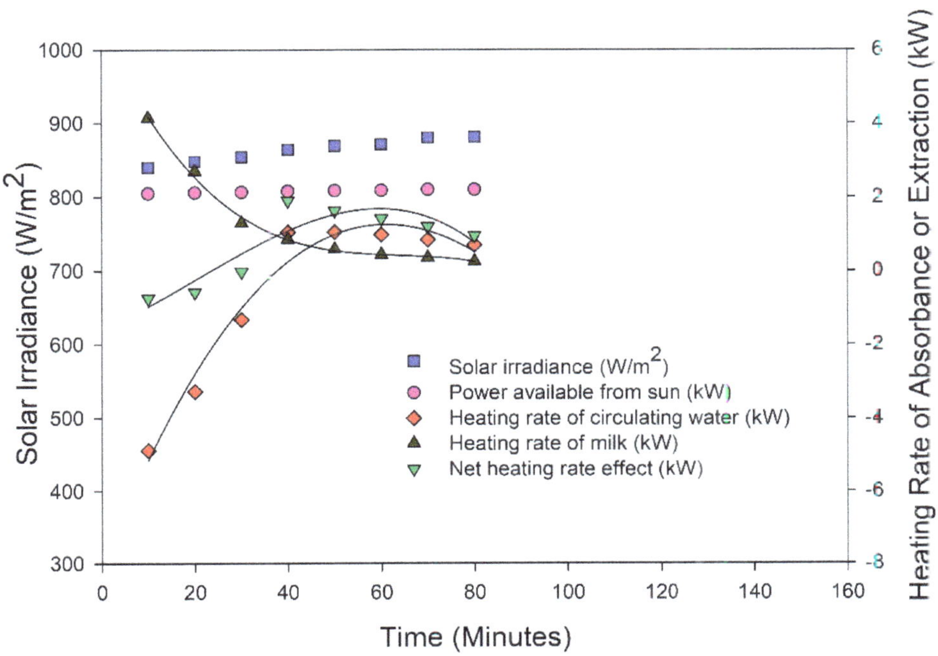

(**c**)

Figure 6. Effect of solar irradiance on the performance parameters of the system (heating rate of circulation water, milk and net heating rate) during the heating process of milk at 50 L (**a**), 40 L (**b**) and 30 L (**c**) of milk.

Figure 7 shows the cumulative heat energy released by hot circulating water, the heat energy absorbed by milk and the corresponding heat transfer efficiency. It was found that during the first 40 min of the heating process, hot water released 66%, 59%, and 74% of its total energy for 50, 40 and 30 L of milk capacities. The rest of the energy was transferred to milk in the next 100, 70 and 40 min for 50, 40 and 30 L of milk. Contrarily, the milk absorbed 69%, 81% and 84% of the total required energy to reach the desired temperature of 80 °C. Therefore, the quantity of milk plays an important role in energy consumption. The heat transfer efficiency was found to be above 80% during the whole heating process because milk received energy from the storage tank where any fluctuation was compensated continuously by ETC.

3.2. Cooling Process of Milk

The cooling rates of heated milk in response to the circulation of tap water through the coils under different product capacities have been shown in Figure 8. Tap water was passed through the coils of the fermentation chamber in the opposite direction of the stirrer in order to reduce the temperature rapidly. The temperature and flow rate of tap water was 29.8 °C and 32.25 L per minute, respectively. The time required to drop the temperature from 80 °C to 43 °C was found to be 19, 16 and 14 min for 50, 40 and 30 L of volume, respectively. Similar trend lines of heat extraction for all the milk volumes were observed with a slight difference in time taken to complete the process.

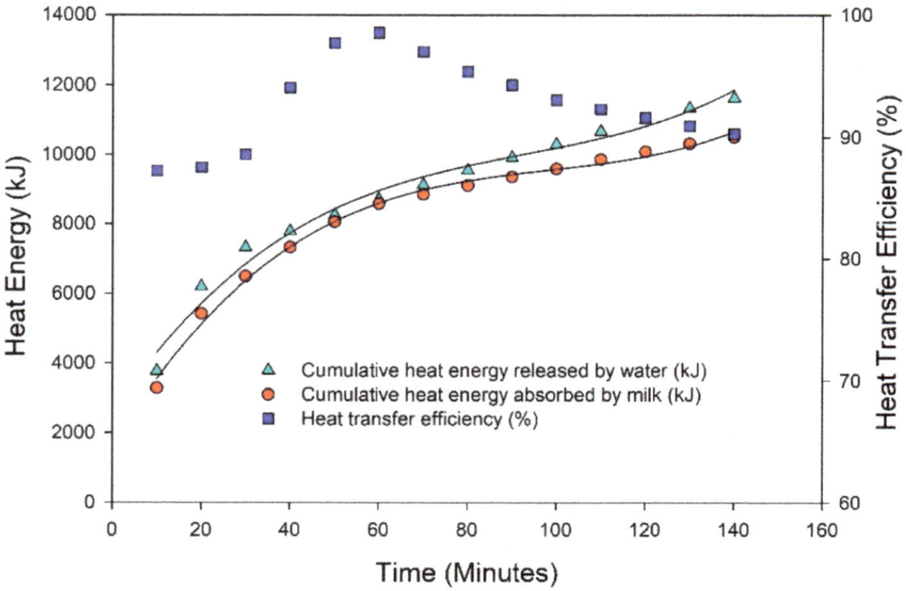

(a)

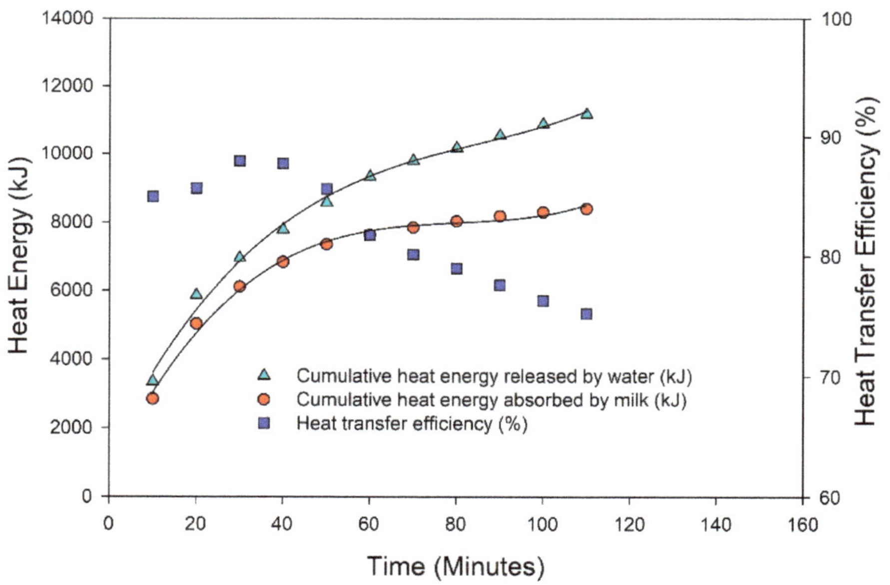

(b)

Figure 7. *Cont.*

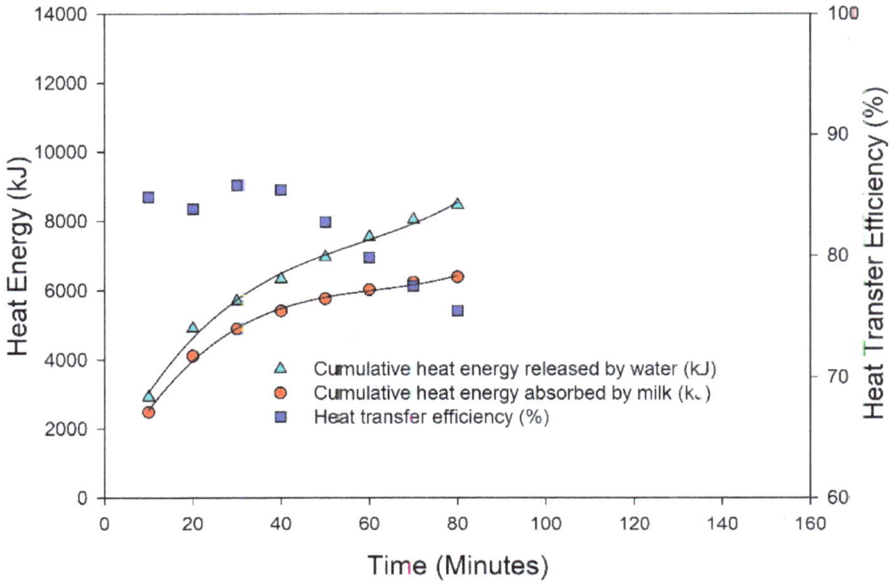

(**c**)

Figure 7. Total heat energy released by water and absorbed by milk during the processing of 50 L (**a**), 40 L (**b**) and 30 L (**c**) milk.

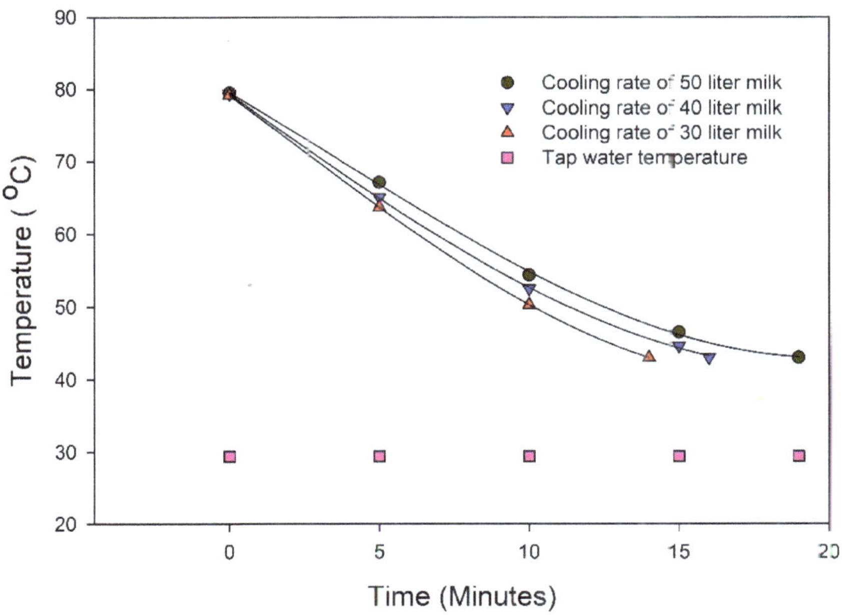

Figure 8. Time required to cool down the temperature of heated milk of different capacities up to 43 °C using tap water.

3.3. Fermentation Process

For the fermentation process, 2% of starter culture was added to milk to make yogurt and the stirrer was kept off during that period. It took five hours to complete the fermentation process. The pH value of milk was found to be 6.45, which changed to 4.35 after being made into yogurt. At this stage, the temperature was maintained between 43 °C to 44 °C, being controlled using a thermostatic valve and water circulation pump. The thermostatic valve opened automatically, letting the hot water flow through the coils when the temperature of the milk falls below than 43 °C and closing once it rises above 44 °C.

3.4. Cooling Process of Yogurt

A one-ton refrigeration system (using R410a refrigerant) energized by a solar hybrid photovoltaic system was used for the cooling of yogurt below 8 °C in order to store it without quality deterioration. In the current study, the temperature of yogurt was lowered to 4 °C by considering the effect of stirrer speed on the cooling rate and consistency (thickness) of yogurt. For this, three different stirrer speeds (36, 18 and 6 rpm) were selected using a variable-speed motor to cool down the product of three different capacities (50, 40 and 30 L), and the outcomes are shown in Figure 9. It can be observed that the time required to cool 50 L of yogurt from 43 °C to 4 °C was 64, 73 and 103 min with the stirrer speeds of 36, 18 and 6 rpm, respectively. Similarly, for 40 and 30 L, the time required for cooling was observed to be 59, 64, 97 min and 48, 54 and 65 min, respectively, at 36, 18 and 6 rpm stirrer speeds. It can be noted that the higher speed of the stirrer resulted in a higher cooling rate. Importantly, greater and lesser stirrer speeds caused rupturing of yogurt texture and non-uniform cooling, respectively. It was observed that at 36 and 18 rpm stirrer speeds, the cooling uniformity was satisfactory, but at 6 rpm, a non-uniform cooling behavior was observed for all volumes of product. Although stirring at 36 rpm gave a good cooling rate, the consistency of the yogurt was found to be on the higher side compared to stirring at 18 rpm.

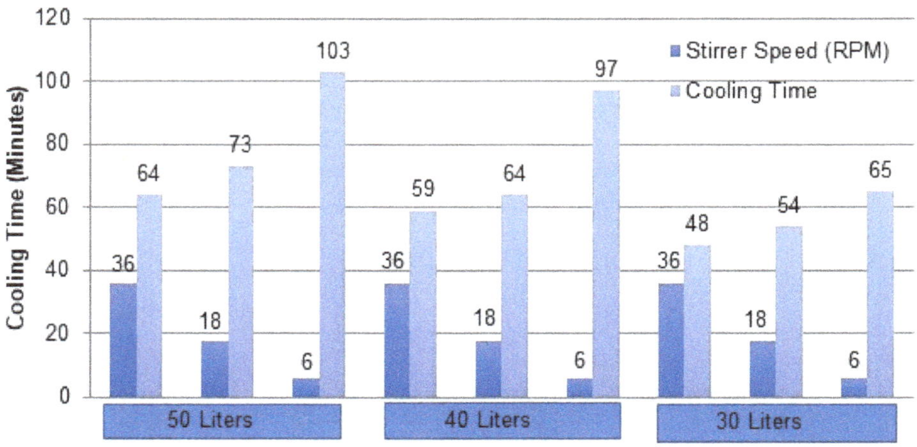

Figure 9. Effect of stirrer speeds on cooling rates for different volumes of yogurt.

To evaluate the cooling efficiency of the system, the coefficient of performance (COP) of the cooling unit was calculated after every 10 min throughout the process for 50, 40 and 30 L of yogurt. In this context, the output is the amount of heat removed from the yogurt. Under all operating conditions, it was found that the COP was higher at the beginning of the cooling process and lowered as the process proceeded to the end, which is in accordance with the second law of thermodynamics [29]. As discussed earlier, variation in stirrer speed affects the rate of cooling, so COP was also calculated using three different stirrer speeds

(36, 18, 6 rpm) as shown in Figure 10. Although the fluctuation of the COP followed an almost similar trend throughout the process under 36 and 18 rpm, the maximum value of COP (3.35) was achieved at 18 rpm for 50 L yogurt (Figure 10a). In the cases of 40 and 30 L, the maximum COP was calculated to be 2.41 (Figure 10b) and 2.75 (Figure 10c) under 36 and 18 rpm, respectively. Taking into account the cooling rate and consistency of yogurt, a stirrer speed of 18 rpm was found to be optimum.

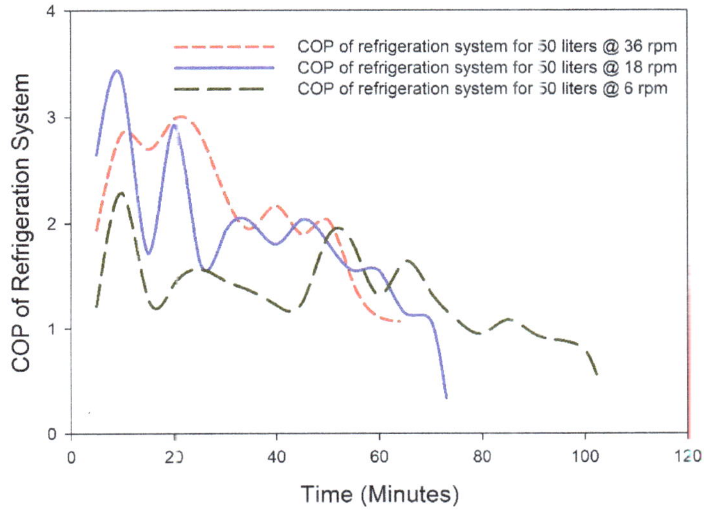

(**a**)

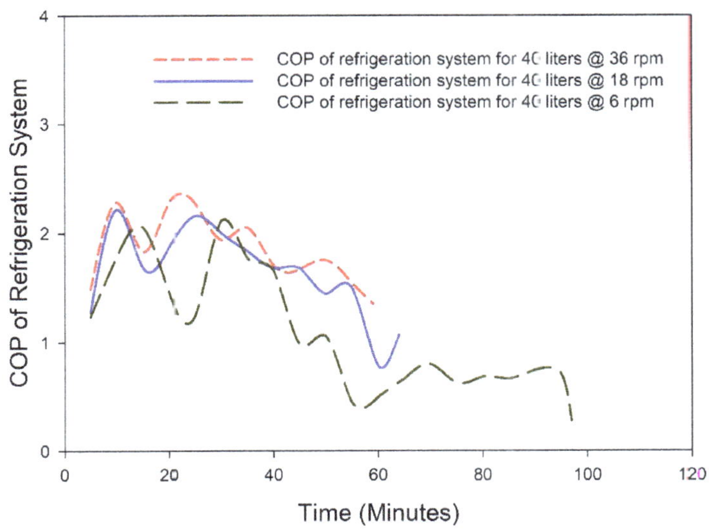

(**b**)

Figure 10. *Cont.*

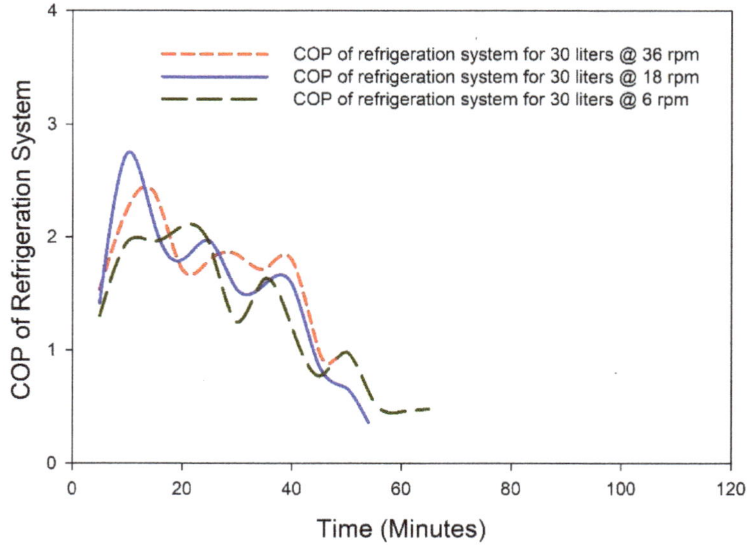

(c)

Figure 10. Effect of stirring speed on the COP of the refrigeration unit for 50 L (**a**), 40 L (**b**) and 30 L (**c**) yogurt volumes.

All the calculated data (temperature change in storage tank water, milk, heating rates of circulating water and milk, net heating rate effect, cumulative heat energy released by water and absorbed by milk and cooling rate of milk) with respect to process time were modeled using sigma plot-12, and it was found that polynomial cubic model fitted best to all the data obtained under the prevailing operating conditions. It helps to interpolate new, invisible data points. All the values of respective model constants for the parameters calculated are tabulated in Table 2.

Table 2. The parameters of fitted models (polynomial cubic) for the investigated parameters.

Parameter	Milk Quantity	Y_0	A	b	C	R^2
Storage tank temperature	50	87.7742	−0.7063	0.0104	−0.000040523	0.8903
	40	89.2370	−0.8957	0.0167	−0.000082776	0.9315
	30	90.1101	−0.9468	0.0208	−0.0001	0.9907
Milk temperature	50	32.5390	1.2812	−0.0133	0.000047356	0.9905
	40	31.0960	1.7573	−0.0225	0.000096685	0.9957
	30	30.6020	2.1874	−0.0358	0.0002	0.9978
Heating rate of circulating water (kW)	50	−9.7531	0.3859	−0.0044	0.000015318	0.9648
	40	−10.3078	0.4659	−0.0060	0.000024359	0.9377
	30	−8.1497	0.3307	−0.0033	0.0000059915	0.9672
Heating rate of milk (kW)	50	7.1966	−0.2242	0.0024	−0.0000081955	0.9751
	40	6.9267	−0.2331	0.0027	−0.000010638	0.9896
	30	6.5691	−0.2777	0.0042	−0.000021992	0.9956

Table 2. *Cont.*

Parameter	Milk Quantity	Y_0	A	b	C	R^2
Net heating rate effect	50	−2.5565	0.1617	−0.0020	0.0000071226	0.7301
	40	−3.3811	0.2328	−0.0033	0.000013721	0.7165
	30	−1.5806	0.0530	0.0010	−0.000015001	0.8302
Cumulative heat released by water (kJ)	50	2577.6611	189.5369	−1.7670	0.0063	0.9830
	40	1312.5611	254.9214	−2.6405	0.0104	0.9943
	30	388.2421	250.5796	−3.5952	0.0207	0.9936
Cumulative heat absorbed by milk	50	1624.8781	212.2518	−2.0389	0.0063	0.9950
	40	759.2427	257.3700	−3.1228	0.0129	0.9939
	30	534.7397	235.4172	−3.5727	0.0194	0.9967
Cooling rate of milk	50	79.5733	−2.4785	−0.0311	0.0032	0.9995
	40	79.4196	−3.0165	0.0122	0.0022	0.9998
	30	79.2000	−3.1207	−0.0148	0.0038	1.0000

3.5. Thermal Profile of the Entire System

The complete yogurt-making process consists of heating, cooling the milk, fermentation and cooling the yogurt, exchanging thermal energy. In order to verify that either system is capable of harvesting enough solar energy for all the processes to be completed effectively and economically, complete thermal profiles were developed for three different milk volumes with an optimized stirrer speed (18 rpm) as shown in Figure 11. It can be observed that the heating of 50 L of milk took 140 min to reach 80 °C, and after that its temperature fell down to 43 °C in 14 min using tap water (Figure 11a). This showed that the system was able to maintain the temperature of inoculated milk governed through a thermostat valve at 43 °C for five hours followed by the rapid cooling of yogurt in 73 min when the stirrer rpm was 18. Milk was converted to yogurt by lowering its pH value to 4.45 during those five hours. In order to estimate the amount of heat extracted and absorbed from the product during the entire process, heating and cooling rate trends were also developed. It was also noted that the product heating rate started from 5.09 kW and it reduced to 0.29 kW at the end of the heating process. After that, the cooling process of milk started which resulted in a maximum decline in heating rate of −8.68 kW followed by the fermentation process with an average heating rate of 0.129 kW. Negative values of the heating rate during the fermentation process were due to the heat loss when the thermostat valve was closed and hot water circulation was shut down. So, no transfer of heat energy would take place under this condition. Finally, the process of yogurt cooling occurred with a maximum heat extraction rate of −3.047 which reduced −0.1945 kW at the end of the cooling process of yogurt. The total process time was calculated to be 540 min (9 h).

Similar trends were found for the cases of 40 L and 30 L. However, due to having less volume than 50 L, it took 110 and 80 min to complete the heating process for 40 and 30 L, respectively. The heating rate was about 4.40 kW, 3.85 kW at the start and 0.15 kW, 0.23 kW at the end of the heating process for 40 and 30 L, respectively. During the milk cooling process, the heating rate was reduced to −6.48 kW and −4.94 kW, and at the end of the fermentation process, it was 0.104 kW and 0.077 kW for 40 and 30 L, respectively. The total process time was estimated to be 5.7% and 13.2% less than that of 50 L in the case of 40 and 30 L, respectively.

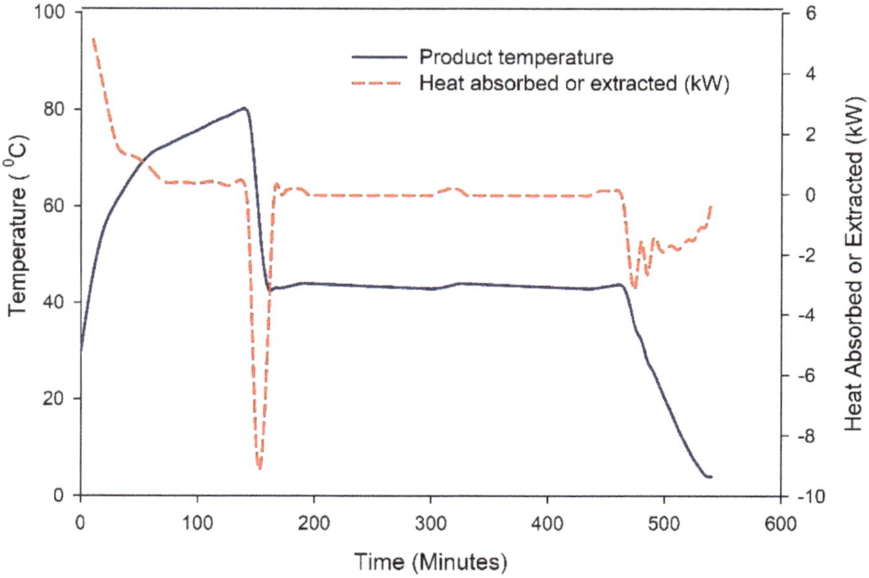

(a)

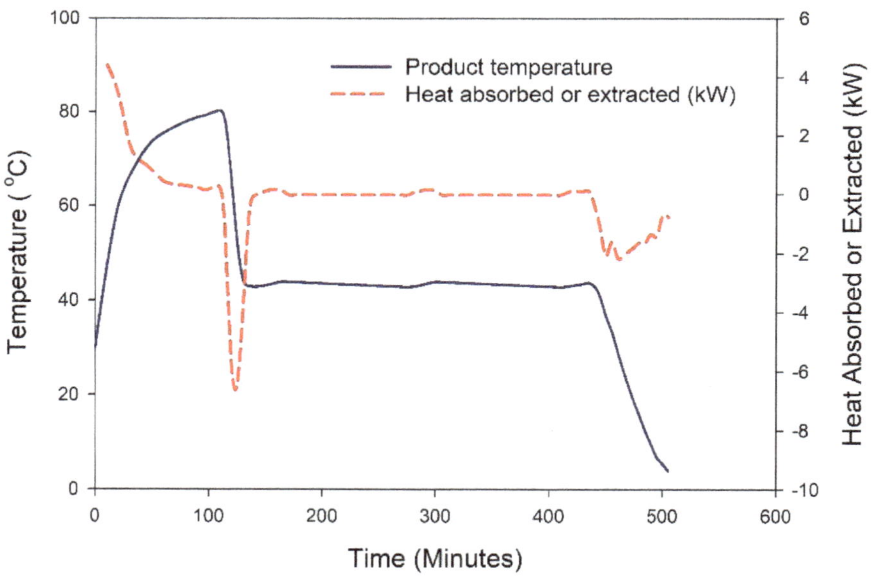

(b)

Figure 11. *Cont.*

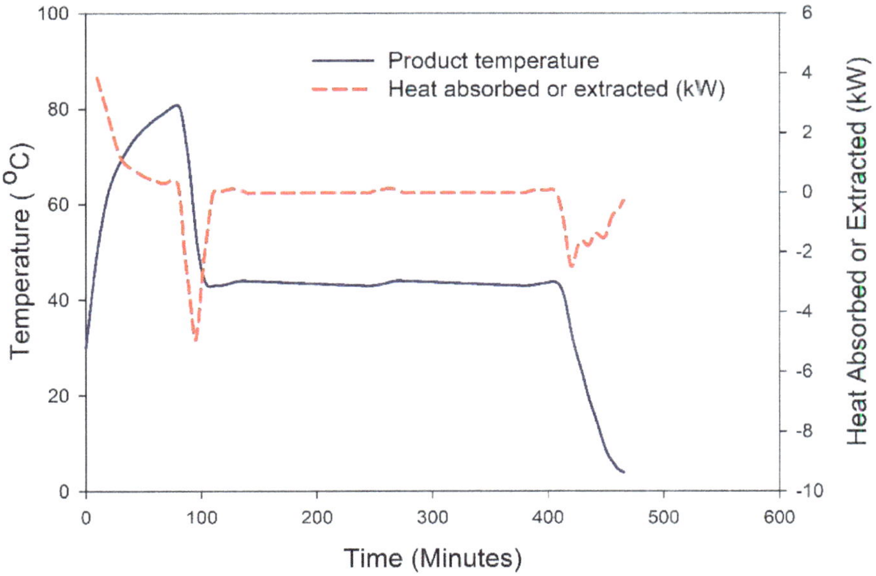

(**c**)

Figure 11. Thermal profile of the yogurt-making process for 50 L (**a**), 40 L (**b**) and 30 L (**c**) yogurt under a constant stirrer speed of 18 rpm.

The energy consumption pattern for the heating, fermentation and cooling processes is shown in Figure 12. The energy required to complete the yogurt-making process was provided by solar thermal and solar photovoltaic systems. The total energy consumed was calculated to be 6.732 kWh, 5.559 kWh and 4.207 kWh for 50, 40 and 30 L batches, respectively. It can be observed that, in all cases, a large part of the total energy was consumed during the heating process of the milk, followed by the cooling process of yogurt. The heating process was found to account for 40% of the total energy consumed during the 50 L process (Figure 12a), which was 39% of the respective total energies for the 40 L (Figure 12b) and 30 L cases. Similarly, the cooling of the yogurt process consumed 2.107 kWh (31% of total), 1.686 kWh (30% of the total) and 1.264 kWh (30% of the total) for the cases of 50, 40 and 30 L, respectively. Although a significant part of the energy is consumed during the cooling process of the milk, it is also important to note that this energy was neither provided by the solar photovoltaic nor by the solar thermal system or utility but through the thermal energy gained by circulating tap water. In addition, the specific product energy (SPE) was calculated to be 485 kJ/kg, 500 kJ/kg and 505 kJ/kg for 50, 40 and 30 L batches, respectively. In the case of 40 and 30 L milk processing, the contact area of the heating coil was not fully exposed to the product, resulting in higher heat loss, which is why the value of SPE is high. Although the process design, product quantity to be processed and operating conditions vary and effect the rates of energy usage, the process can be compared with other somehow similar work. It can be observed that the solar yogurt processing unit gave quite good results in terms of thermal analysis when compared with results reported by Yaseen et al. (2019) [20] who worked on milk pasteurization (100 L milk was heated up to 63 °C) and cooling (4 °C in the chiller within 90 min). In the current study, milk was pasteurized at 80 °C and it can also be noted that the system can easily be used for low-temperature pasteurization (63 °C) as well for 50 L of milk, making it feasible in terms of yogurt making and pasteurization. For cooling of the yogurt from 43 °C to 4 °C, the energy-utilization pattern showed similar rates as reported by Yaseen et al. (2019) [20] and Khan et al. (2020) [13] for milk cooling using rotary compressor.

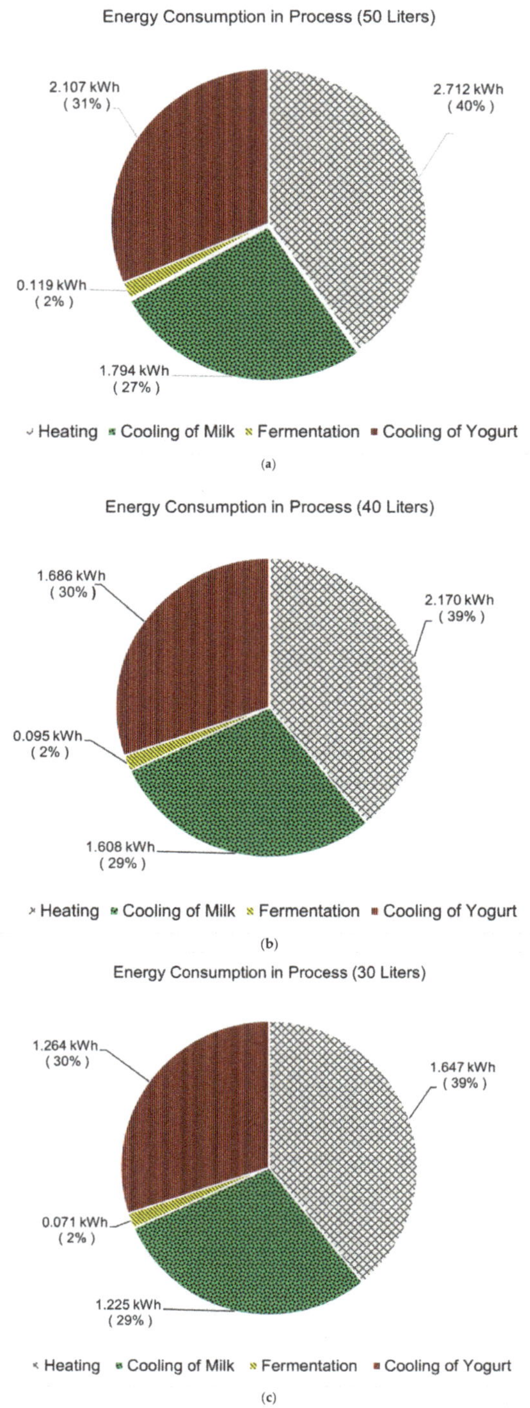

Figure 12. Energy consumption pattern during the yogurt-making process under 50 (**a**), 40 (**b**) and 30 (**c**) L.

4. Conclusions

In the current study, a solar-assisted decentralized solution for yogurt processing was presented, capable of performing all required processes (heating, fermentation and cooling) in a single chamber. The systems consisted of a cylindrically shape chamber surrounded by a squared helix coil heat exchanger for heating and a pillow plate at its bottom for cooling. Experiments were performed using raw milk from cows at different volumes and stirrer speeds. The total energy consumed was calculated to be 6.732 kWh, 5.559 kWh and 4.207 kWh for the cases of 50, 40 and 30 L, respectively. This shows that approximately 2020, 1668, and 1262 electrical units (kWh) can be saved annually to process 50, 40 and 30 L, respectively, by considering three hundred sunny days (one batch per day) throughout the year in Faisalabad, Pakistan. During milk heating and cooling, the optimized stirrer speed was found to be 36 rpm, while for the cooling of yogurt it was 18 rpm. It was found that a majority of the total energy was consumed by the heating processes followed by the cooling process. This supports the integration of solar evacuated tube collectors for heating purposes instead of converting electrical energy into thermal energy which can increase the operating and capital costs of the system. The use of solar thermal energy (for heating) and PV modules (for cooling) not only played a significant role in reducing the operational energy cost, but also facilitated its decentralized applications. The study provides a basis for design optimization to process any milk volume. Being a batch system, the integrated solar system (both PV and thermal) can be used for farm electrification and heating purposes in the event of non-operational periods.

5. Directions for Further Research

Once processing a large quantity of raw milk, the effect of the flow rates of the circulating hot water need to be investigated. Similarly, impact of heat transfer fluids, i.e., silicon fluids and paraffin oil, on the heating efficiency of the evacuated tube collector can also be considered for further research. Thirdly, use of the Scheffler fixed focus concentrator as a heat source can also be compared.

Author Contributions: Conceptualization, S.N.H. and O.H.; methodology, S.N.H. and W.A.; software, S.N.H. and W.A.; validation, S.N.H., W.A. and O.H.; formal analysis, S.N.H. and W.A.; investigation, S.N.H., W.A., A.M.; resources, O.H. and A.M.; data curation, S.N.H. and W.A.; writing—original draft preparation, S.N.H.; writing—review and editing, S.N.H., W.A., A.M. and O.H.; visualization, S.N.H. and W.A.; supervision, O.H. and A.M.; project administration, A.M. and O.H.; funding acquisition, A.M. and O.H. All authors have read and agreed to the published version of the manuscript.

Funding: The development cost of the system was provided by the International Center for Development and Decent Work (ICDD) Germany.

Institutional Review Board Statement: Not applicable.

Informed Consent Statement: Not applicable.

Data Availability Statement: Not applicable.

Acknowledgments: The author thanks the Higher Education Commission (HEC) of Pakistan and the German Academic Exchange Service (DAAD) for providing subsistence/research costs for the current study. The Author would also like to thank the University of Kassel, Germany for providing funds for publishing the work.

Conflicts of Interest: The authors declare no conflict of interest.

Nomenclature

Symbol

A_c	Surface area of evacuated tube collector (m^2)
A	Heat transfer area area (m^2)
C_{Bat}	Capacity of the battery bank (Ah)
C_p	Specific heat (kJ kg^{-1} k^{-1})

C_{pg}	Specific heat capacity of the refrigerant at constant pressure in gaseous state (kJ kg^{-1} k^{-1})
C_{pl}	Specific heat capacity of the refrigerant at constant pressure in the liquid state (kJ kg^{-1} k^{-1})
CIP	Clean-in-place
COP	Coefficient of performance
ETC	Evacuated tube collectors
d	Diameter (m)
D_d	Depth of discharge in fraction
GHI	Global Horizontal Irradiation) (W/m^2)
H_{avg}	Average global horizontal irradiance ($kWh/(m^2$ d))
h_1	Enthalpy of the refrigerant at the compressor inlet (kJ kg^{-1})
h2	Enthalpy of the refrigerant at the compressor inlet (kJ kg^{-1})
h_3	Enthalpy of the refrigerant after the condensation process outlets (kJ kg^{-1})
h_{f3}	Liquid enthalpy after condenser outlet (kJ kg^{-1})
h_2'	Enthalpy of the refrigerant at the dry saturated steam line after compression process
h_3'	Enthalpy of the refrigerant at the water line after condensation (kJ kg^{-1})
I	Current (A)
I_b	Solar irradiance (W/m^2)
Le	Electric load (kWh/d)
m	Mass (kg)
m′	Mass flow rate (kgs^{-1})
N_{ccd}	Autonomy and taken as the number of continuous cloudy days
P_p	Peak power of the solar system (kWp)
P	Power required to run the refrigeration system (kW)
PU	Polyurethane
PV	Photovoltaic
PVC	Polyvinyl chloride
Q_t	Total heat transfer through walls and bottom (kWh/d)
Q_R	Cooling load of the product (kWh/d)
RTD	Resistance Temperature Detector
SS	Stainless steel
SPE	Specific production energy (kJ/kg)
STC	Standard test conditions
T_c	Cold face temperature of fermentation chamber (°C).
T_h	Hot face temperature of the fermentation chamber (°C)
T_{CF}	Temperature correction factor (0.4% per °C)
T_2	Actual temperatures of the refrigerant at the compressor outlet (K)
T_2'	Saturation temperature at the compressor outlet (K)
T_3	Actual temperature of the refrigerant after condenser (K)
T_3'	Saturation temperature at the liquid line after condensation process (K)
TR	Capacity of refrigeration
ΔT	Change in temperature (K)
t	Time (sec)
U	Overall heat transfer coefficient (W/m^2 °C)
VFD	Variable frequency drive
V	Voltage
V	Volume (m^3)
Subscripts	
bat	Battery
inv	Inverter
r	Refrigerant
oc	Open circuit
mpp	Maximum power
Greek Symbols	
τ	Transmission coefficient,
α	Absorption coefficient
η	Efficiency (%)

References

1. Wang, L.; Bandyopadhyay, S.; Cosgrove-Davies, M.; Samad, H. *Quantifying Carbon and Distributional Benefits of Solar Home System Programs in Bangladesh*; Policy Research Working Paper 5545; The World Bank Environment Department: Washington, DC, USA, 2011. [CrossRef]
2. Shahid, H.; Shafique, O.; Shokat, A. Dairy Industry of Pakistan. *Eur. J. Bus. Manag.* **2012**, *4*, 1–4.
3. Charad, J.; Lanctuit, H. *Nestlé Dairy Factories in Pakistan: Losses Across the Value Chain—A Case Study*; The Food Loss & Waste Protocol: Washington, DC, USA, 2017.
4. Tamang, J.P. *Ethnic Fermented Foods and Alcoholic Beverages of Asia*; Chapter 5 Ethnic Fermented Foods in Pakistan; Springer: Delhi, India, 2016. [CrossRef]
5. Singh, G.; Chopra, K.; Tyagi, V.; Pandey, A.; Ma, Z.; Ren, H. A comprehensive energy, exergy and enviroeconomic (3-E) analysis with carbon mitigation for multistage evaporation assisted milk powder production unit. *Sustain. Energy Technol. Assess.* **2020**, *43*, 100925. [CrossRef]
6. Munir, M.T.; Yu, W.; Young, B.R. *Can Exergy Be a Useful Tool for the Dairy Industry? Computer Aided Chemical Engineering*; Elsevier: Amsterdam, The Netherlands, 2014; Volume 33. [CrossRef]
7. Brush, A.; Eric, M.; Ernst, W. Energy Efficiency Improvement and Cost Saving Opportunities, An ENERGY STAR® Guide for Energy and Plant Managers. In *Energy Analysis Department Environmental Energy Technologies Division Lawrence Berkeley National Laboratory*; Lawrence Berkeley National Laboratory: Berkeley, CA, USA, 2011.
8. Von Keyserlingk, M.A.G.; Martin, N.P.; Kebreab, E.; Knowlton, K.F.; Grant, R.J.; Stephenson, M.; Sniffen, C.J.; Harner, J.R., III; Wright, A.D.; Smith, S.I. Invited review: Sustainability of the US dairy industry. *J. Dairy Sci.* **2013**, *96*, 5405–5425. [CrossRef] [PubMed]
9. Tahir, Z.; Asim, M. Surface measured solar radiation data and solar energy resource assessment of Pakistan: A review. *Renew. Sustain. Energy Rev.* **2018**, *81*, 2839–2861. [CrossRef]
10. Michael, M.B.; Akinlabi, E.T.; Jen, T.C. A Review of Solar Thermal Systems Utilization for Industrial Process Heat Applications. *Lect. Notes Eng. Comput. Sci.* **2016**, *2226*, 859–863
11. Amjad, W.; Waseem, M.; Munir, A.; Ghafoor, A.; Asghar, F.; Gilani, G.A. Solar Assisted Dehydrator for Decentralized Controlled and Homogeneous Multi-Product Drying. *J. Sol. Energy Eng. Trans. ASME* **2021**, *143*, 11011. [CrossRef]
12. Ismail, M.I.; Yunus, N.A.; Hashim, H. Integration of Solar Heating Systems for Low-Temperature Heat Demand in Food Processing Industry—A Review. *Renew. Sustain. Energy Rev.* **2021**, *147*, 111192. [CrossRef]
13. Khan, K.S.; Amjad, W.; Munir, A.; Hensel, O. Improved Solar Milk Chilling System Using Variable Refrigerant Flow Technology (VRF). *Sol. Energy* **2020**, *197*, 317–325. [CrossRef]
14. Desai, D.D.; Raol, J.B.; Patel, S.; Chauhan, I. Application of Solar Energy for Sustainable Dairy Development. *Eur. J. Sustain. Dev.* **2013**, *2*, 131–140. [CrossRef]
15. Mekhilef, S.; Saidur, R.; Safari, A. A Review on Solar Energy Use in Industries. *Renew. Sustain. Energy Rev.* **2011**, *15*, 1777–1790. [CrossRef]
16. Anderson, T.N.; Duke, M. Solar Energy Use for Energy Savings in Dairy Processing Plants. *IPENZ Eng. TrNz* **2007** *2008*, 1–9. Available online: https://hdl.handle.net/10289/3204 (accessed on 20 September 2021).
17. Zahira, R.; Akif, H.; Amin, N.; Azam, M.; Haq, Z.U. Fabrication and Performance Study of a Solar Milk Pasteurizer. *Pak. J. Agric. Sci.* **2009**, *46*, 162–168.
18. Fathey Mohamed Atia, M.; Mostafa, M.M.; El-Nono, A.M.; Abdel-Salam, F.M. Solar Energy Utilization for Milk Pasteurization. *Misr J. Agric. Eng.* **2011**, *28*, 729–744.
19. Wayua, F.O.; Okoth, M.W.; Wangoh, J. Design and Performance Assessment of a Flat-Plate Solar Milk Pasteurizer for Arid Pastoral Areas of Kenya. *J. Food Process. Preserv.* **2013**, *37*, 120–125. [CrossRef]
20. Yaseen, M.U.; Ashraf, M.; Munir, A. Development and performance evaluation of solar assisted milk pasteurizer by using vacuum tube collector. *Pak. J. Agric. Sci.* **2019**, *56*, 693–700. [CrossRef]
21. Liu, Z.; Li, H.; Liu, K.; Yu, H.; Cheng, K. Design of high-performance water-in-glass evacuated tube solar water heaters by a high-throughput screening based on machine learning: A combined modeling and experimental study. *Sol. Energy* **2017**, *142*, 61–67. [CrossRef]
22. Ben Taher, M.; Benseddik, Z.; Afass, A.; Smouh, S.; Ahachad, M.; Mahdaoui, M. Energy life cycle cost analysis of various solar water heating systems under Middle East and North Africa region. *Case Stud. Therm. Eng.* **2021**, *27*, 101262. [CrossRef]
23. Mohammed, A.K.; Hamakhan, I.A. Analysis of energy savings for residential electrical and solar water heating systems. *Case Stud. Therm. Eng.* **2021**, *27*, 101347. [CrossRef]
24. Cengel, Y.A. *Heat Transference a Practical Approach*; MacGraw-Hill: New York, NY, USA, 2004; Volume 4, p. 874.
25. Ghafoor, A.; Munir, A. Design and Economics Analysis of an Off-Grid PV System for Household Electrification. *Renew. Sustain. Energy Rev.* **2015**, *42*, 496–502. [CrossRef]
26. Vendan, S.P.; Shunmuganathan, L.P.A.; Manojkumar, T.; Thanu, C.S. Study on Design of an Evacuated Tube Solar Collector for High Temperature Steam Generation. *Int. J. Emerg. Technol. Adv. Eng.* **2012**, *2*, 539–541.
27. Thornton, B.; Wagner, A. Prepared for the General Services Administration by Pacific Northwest National Laboratory Variable Refrigerant Flow Systems. 2012. Available online: https://www.gsa.gov/cdnstatic/GPG_VRFTechnologyReport_20121220.pdf (accessed on 22 August 2021).

28. Alsagri, A.S. Photovoltaic and Photovoltaic Thermal Technologies for Refrigeration Purposes: An Overview. *Arab. J. Sci. Eng.* **2022**, 1–34. [CrossRef] [PubMed]
29. Neri, I. Second law of thermodynamics at stopping times. *Phys. Rev. Lett.* **2020**, *124*, 40601. [CrossRef] [PubMed]

sustainability

MDPI

Article

TRNSYS Simulation and Experimental Validation of Internal Temperature and Heating Demand in a Glass Greenhouse

Misbaudeen Aderemi Adesanya [1], Wook-Ho Na [2], Anis Rabiu [1], Qazeem Opeyemi Ogunlowo [1,3], Timothy Denen Akpenpuun [4], Adnan Rasheed [2], Yong-Cheol Yoon [5] and Hyun-Woo Lee [1,2,*]

1 Department of Agricultural Civil Engineering, College of Agricultural and Life Sciences, Kyungpook National University, Daegu 41566, Korea; misbauadesanya@knu.ac.kr (M.A.A.); rabiuanis@knu.ac.kr (A.R.); ogunlowoqazeem@knu.ac.kr (Q.O.O.)
2 Smart Agriculture Innovation Centre, Kyungpook National University, Daegu 41566, Korea; wooks121@knu.ac.kr (W.-H.N.); adnanrasheed@knu.ac.kr (A.R.)
3 Department of Agricultural and Bioenvironmental Engineering, Federal College of Agriculture Ibadan, Ibadan PMB 5029, Nigeria
4 Department of Agricultural and Biosystems Engineering, University of Ilorin, Ilorin PMB 1515, Nigeria akpenpuun.td@unilorin.edu.ng
5 Department of Agricultural Engineering, Gyeongsang National University, Jinju 52828, Korea; ychyoon@gsnu.ac.kr
* Correspondence: whlee@knu.ac.kr; Tel.: +82-53-950-5736

Abstract: The energy demand in greenhouses is enormous, and high-performance covering materials and thermal screens with varying radiometric properties are used to optimise the energy demand in building energy simulations (BES). Transient System Simulation (TRNSYS) software is a common BES tool used to model the thermal performance of buildings. The calculation of the greenhouse internal temperature and heating demand in TRNSYS involves the solution of the transient heat transfer processes. This study modelled the temperature and heating demand of two multi-span glass greenhouses with concave (farm A) and convex (farm B) shapes. This study aims to investigate the influence of the different BES longwave radiation modes on greenhouse internal temperature in different zones and the heating demand of a conditioned zone. The standard hourly simulation results were compared with the experimental data. The results showed that the standard and detailed modes accurately predicted greenhouse internal temperature (the Nash–Sutcliffe efficiency coefficient (NSE) > 0.7 for all three zones separated by thermal screens) and heating demand (NSE > 0.8) for farms A and B. The monthly heating demand predicted by the simple and standard radiation modes for farm A matched the experimental measurements with deviations within 27.7% and 7.6%, respectively. The monthly heating demand predicted by the simple, standard, and detailed radiation modes for farm B were similar to the experimental measurements with deviations within 10.5%, 6.7%, and 2.9%, respectively. In the order of decreasing accuracy, the results showed that the preferred radiation modes for the heating demand were standard and simple for farm A, and detailed, standard, and simple for farm B.

Keywords: thermal screens; heating demand; TRNSYS; greenhouse internal temperature; building energy simulation; longwave radiation

Citation: Adesanya, M.A.; Na, W.-H.; Rabiu, A.; Ogunlowo, Q.O.; Akpenpuun, T.D.; Rasheed, A.; Yoon, Y.-C.; Lee, H.-W. TRNSYS Simulation and Experimental Validation of Internal Temperature and Heating Demand in a Glass Greenhouse. *Sustainability* **2022**, *14*, 8283. https://doi.org/10.3390/su14148283

Academic Editors: Muhammad Sultan, Yuguang Zhou, Uzair Sajjad and Walter Den

Received: 19 May 2022
Accepted: 3 July 2022
Published: 6 July 2022

Publisher's Note: MDPI stays neutral with regard to jurisdictional claims in published maps and institutional affiliations.

1. Introduction

Greenhouse farming aims to grow warm-season crops in winter and has been utilised to solve food shortages year-round. The control of indoor air, especially under adverse weather conditions, is crucial for crop production in greenhouses [1–3]. The energy consumption for the indoor air control contributes to 50% of the total cost of greenhouse production in many countries [4], making the greenhouse sector one of the most energy-intensive agricultural sectors, with significant economic and environmental impacts [5,6].

Passive energy-saving strategies reduce greenhouse energy consumption. This involves selecting high-performance greenhouse-covering materials and thermal screens to limit the amount of solar radiation the crop receives in summer and the rate of heat loss in winter [7]. Using thermal screens reduces the heat loss rate by 23–24% [8], and they are placed either internally or externally to ensure a favourable microclimate for the crop [9]. Most growers prefer internal thermal screens use owing to the increased heat retention [10] and reduced surface-to-volume ratio of the heating zone, leading to a reduction in the heating cost [11].

Several studies have used numerical modelling to assess the impact of different covering materials and thermal screens on greenhouse microclimate and energy consumption [12–14]. Zhang et al. [15] solved unsteady state equations using MATLAB to study the thermal performance of a greenhouse by altering the cover transmittance and absorbance for beam, diffuse, and ground-reflected radiations in a glass greenhouse. The surface radiative properties change with time and impact energy demand. Another study [16] investigated the effect of different plastic materials with spectral radiative properties ranging from 0.2 to 28 μm. The annual heating demand of the greenhouse was reduced by 6.3% using appropriate greenhouse materials in the developed model.

EnergyPlus and TRNSYS are building energy simulation (BES) tools for modelling the thermal performance of different types of buildings [17]. Within the simulation environment of these tools, there exist pre-existing components to construct a model [18]. The users have the flexibility of varying different parameters on each component to suit the application the model is intended for. The major setback of these tools for greenhouse modelling is their inability to account for the influence of crop and their environmental control systems [19,20]. Unlike EnergyPlus, TRNSYS permits users to develop new components and link with other simulation tools [7]. This makes TRNSYS, which was primarily developed for residential and commercial buildings, able to be continuously improved for greenhouse modelling. The Gembloux Dynamic Greenhouse Climate Model based on heat and mass balance of greenhouse layers was linked with TRNSYS as a sub-model for greenhouse air renewal rate and canopy resistance in a naturally ventilated greenhouse [21]. The model accurately predicted the greenhouse microclimate parameters, but there was an overestimation of the ventilation rate, mostly during night time. Ahamed et al. [22] also developed a TRNSYS model to predict the heating demand in a Chinese-style solar greenhouse (CSG). The results were compared with a heating simulation model (CSGHEAT) from MATLAB. The heating load from TRNSYS resulted in significant errors from assumptions such as constant infiltration rate, fixed scheduled for thermal screens and moisture gain.

A solar-radiation-based control for thermal screens as operated in the CSGHEAT model was implemented in TRNSYS using the equation editor in the simulation studio [23]. Based on the control, the multi-layer thermal screens reduced greenhouse energy consumption compared with double and single thermal screens investigated. The thermal screens also ensured a temperature difference of approximately 5 °C between the three zones in the multi-span greenhouse. However, with a unidirectional heat transfer model, there was no significant temperature difference within four zones of the greenhouse with moveable thermal screens on the north and south walls [24]. Asa'd et al. [25] performed a parametric analysis using TRNSYS to study the effect of the cooling and heating setpoint temperatures, cover materials, rock-bed air flow rate, and mechanical ventilation to improve the overall design of a greenhouse. The study concluded that the U-value of the materials and the rock-bed airflow rate have the most significant effect on the greenhouse internal temperature. Rabiu et al. [26] used TRNSYS and hotbox to evaluate the greenhouse energy-saving screen properties. The investigation also concluded that the screens' infiltration has the most significant impact on their U-values.

Scope of the Study

BES tools are used to investigate the effect of the different parameters on greenhouse microclimate and predict the most favourable operating conditions that minimise the energy consumption. With TRNSYS, several studies have reported the effect of the regime type, environmental control systems, covering materials, greenhouse shape, and orientation on the thermal performance of the greenhouse, with no consideration to the radiation modes [27–29]. The accurate modelling of the solar heat gains in the greenhouse with the different radiation modes in TRNSYS is also an important aspect that should be considered. To this end, a multi-span glass greenhouse area of 3942 m^2 was divided in two (farm A, 2160 m^2, and farm B, 1782 m^2) based on their geometry to study the effect of the radiation modes on the internal temperature and heating demand of the greenhouse. The resulting TRNSYS simulation results were validated with the experimental measurements of farm A (concave-shaped) and farm B (convex-shaped) greenhouses. The major findings from this research will assist greenhouse growers, researchers, and engineers to choose the best radiation mode to model the thermal performance of a greenhouse depending on the greenhouse shape.

2. Materials and Methods

2.1. Description of the Experimental Site

The experimental greenhouse (Purme social farm, Figure 1) is located in Ohak-dong, Yeoju-si, a small town in Gyeonggi-do, western South Korea. The experimental site was in a temperate climate with lengthy, hot, rainy, and partially cloudy summers, while winters are short, cold, snowy, and largely clear [30]. The farm covers a gross area of 4374 m^2 in three parts, farm A (2160 m^2), farm B (1782 m^2), and a packaging room (432 m^2), represented by A, B, and C, respectively, in Figure 2. Farms A and B are Venlo-roofed, multi-span, glass greenhouses with dimensions of 32 m × 67.5 m × 7.25 m and 36 m × 49.5 m × 7.25 m, respectively. The vertical view of one of the spans and sensor positions are shown in Figure 3. The total net conditioned volume of the greenhouse was 22,942.4 m^3, with a net conditioned area of 3942 m^2.

Figure 1. External greenhouse side view.

The greenhouse was fitted with passive, eco-friendly materials (transparent and opaque) for internal climate management. The sides and roof of the greenhouse were covered with horticultural glass (HG) and fluorine film. Retractable horizontal (Tempa and Luxous) and vertical (Obscura) thermal screens were installed (Figure 4). The white surface of Obscura helps to reflect light onto the crop to enhance crop growth. Tempa is a shading screen used during the summer to minimise excessive solar radiation during the day. It reflects and diffuses sunlight to moderate the solar heat load and cools the greenhouse. However, because most shading screens are used as thermal screens during winter, Tempa is also deployed at night to enhance heat retention. Luxous is a thermal screen that aids retaining heat at night to maintain a stable greenhouse temperature. The horizontal thermal screens divided the greenhouse into conditioned (zone 1) and unconditioned (zones 2 and 3) zones to reduce the greenhouse heating area. Sensors were fixed in each zone to monitor the effect of the screens on the macro-environment.

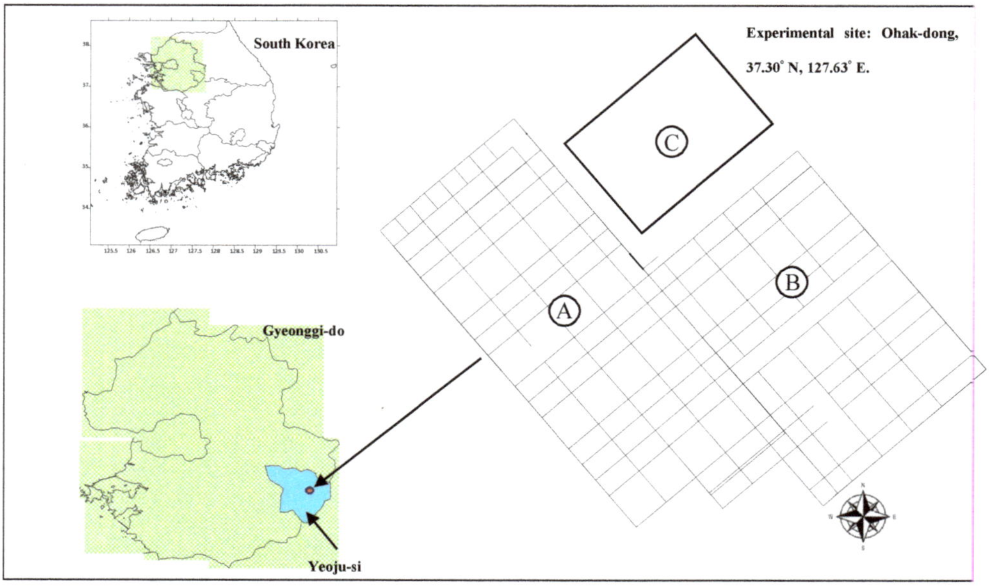

Figure 2. The geographic location of the experimental site.

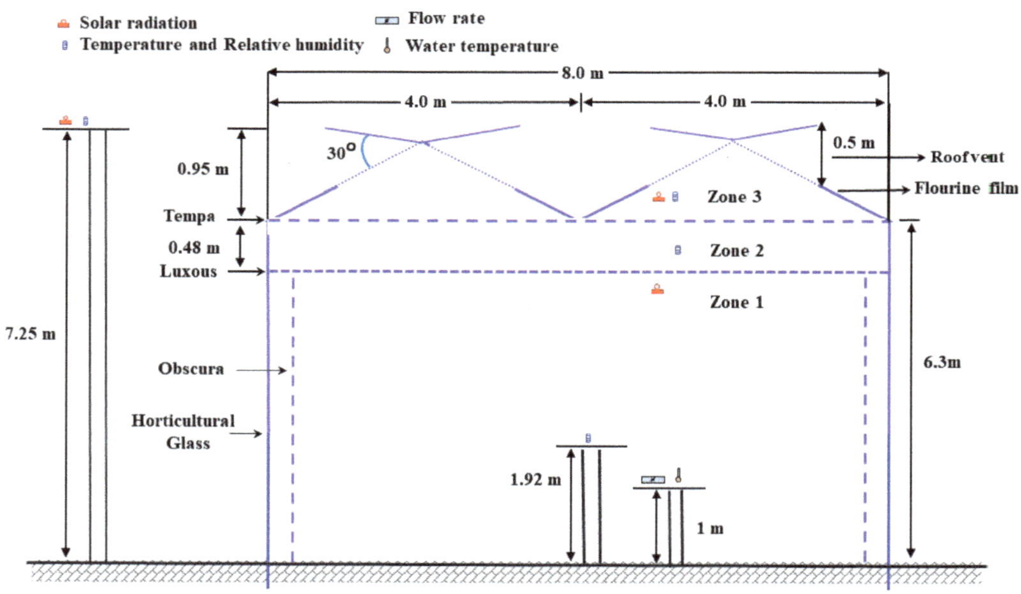

Figure 3. Dimension of a single span and sensor positions in the experimental greenhouse.

(a)

(b)

Figure 4. Thermal screens in the case study: (**a**) horizontal screens; (**b**) vertical screen.

Outside weather data were recorded during winter between December 2021 and March 2022. The recorded data included the solar radiation, temperature, and relative humidity. Other required outside weather data for the BES model, such as wind speed, wind direction, and ambient pressure, were downloaded from the Korean Meteorological Administration (KMA), Icheon (Station 203, 37.26° N 127.48° E) [31]. The details of the recorded data are listed in Table 1. The windspeed data were measured at a height that was different from the eave height of the greenhouse. The data were modified according to the power law in Equation (1):

$$V_h = V_0(\frac{h}{h_0})^{\propto}, \quad (1)$$

where V_h is the calculated wind speed (ms^{-1}) at height h (m) of the greenhouse, V_0 is the downloaded KMA station wind speed (ms^{-1}) at height h_0 (m), and $\propto$ is the power law exponent. The power law exponent is an empirically derived coefficient that increases in value with increase in terrain roughness. In an experiment conducted by Jung et al. [32] in a coastal region (Buan-gu) similar to the experimental site, the power law exponent was 0.28.

Table 1. Details of the recorded weather data.

Parameter	Measurement Method	Sensor	Precision of Sensor	Characteristics
Ambient temperature (°C)	9 places at 1.92 m above the ground (zone 1) and 1 place at the centre of zone 2 and 3	HOBO MX1102A	±0.5%	Field recorded
Relative humidity (%)	9 places at 1.92 m above the ground (zone 1) and 1 place at the centre of zone 2 and 3	HOBO MX1102A	±0.5%	Field recorded
Solar radiation (Wm^{-2})	7.25 m above the ground (outside the greenhouse)	CMP3 pyranometer	±2%	Field recorded
Water temperature (°C)	1 m above the ground (zone 1)	I-Sensor, PT 100	±0.3 °C	Field recorded
Flow rate (LPM)	1 m above the ground (zone 1)	KFCM-1000 K-2101083-2	±5%	Field recorded
Wind speed (ms^{-1})	10 m above the ground (at the weather station)	Clima sensor, US, Thies Clima	±5%	KMA
Wind direction (degree)	10 m above the ground (at the weather station)	Clima sensor, US, Thies Clima	±5%	KMA
Ambient pressure (hPa)	10 m above the ground (at the weather station)	PTB-220TS, VAISALA	±5 hPa	KMA

The greenhouse design combines different strategies to supply heating energy via a fan coil unit, growing tubes, and tube rails (Figure 5) to ensure adequate thermal comfort of crops under low ambient temperature conditions. The water temperature and flow rate were monitored to calculate the energy consumed using Equation (2). The locations and specifications of the water temperature and flow rate sensors are detailed in Figure 3 and Table 1, respectively.

$$Q = \dot{m} \cdot C_p \cdot \Delta t, \tag{2}$$

where Q is the amount of energy consumed ($kcalh^{-1}$), $\dot{m}$ is the mass flow rate (kgh^{-1}), C_p is the specific heat capacity of water ($kcalkg^{-1}K^{-1}$), and Δt is the difference between inlet and outlet water temperatures (K).

(**a**)

(**b**)

Figure 5. Heating energy supply units: (**a**) growing tubes and tube rails; (**b**) fan coil unit.

2.2. Greenhouse Material Properties

Some properties of the greenhouse-covering materials have been presented in the literature. Valera et al. [33], Rafiq et al. [34], and Rabiu et al. [26] detailed the properties of HG, Tempa, and Luxous, respectively (Table 2). Ground and steel are treated as opaque in TRNBuild (a TRNSYS building interface), with properties listed in Table 3. Fluorine film and Obscura are novel greenhouse materials whose properties were determined experimentally.

Table 2. Properties of greenhouse covering materials from the literature.

Cover Characteristics	Covering Material	Thermal Screens	
	HG	Tempa	Luxous
Thickness (mm)	4	0.31	0.30
Solar transmittance (front)	0.89	0.10	0.58
Solar transmittance (back)	0.89	0.12	0.57
Solar reflectance (front)	0.08	0.65	0.30
Solar reflectance (back)	0.08	0.51	0.25
Visible radiation transmittance (front)	0.91	0.10	0.58
Visible radiation transmittance (back)	0.91	0.12	0.57
Visible radiation reflection (front)	0.08	0.65	0.30
Visible radiation reflection (back)	0.08	0.51	0.25
Thermal radiation transmittance	0.1	0.05	0.38
Thermal radiation emission (front)	0.90	0.20	0.44
Thermal radiation emission (back)	0.90	0.33	0.44
Thermal conductivity ($Wm^{-1}K^{-1}$)	0.10	0.52	0.06
Infiltration ($m^3h^{-1}m^2$)	-	3.62	5.45

Table 3. Properties of the opaque materials.

Materials	Thickness (m)	Thermal Conductivity ($kJh^{-1}m^{-1}K^{-1}$)	Thermal Capacity ($kJkg^{-1}K^{-1}$)	Density (kgm^{-3})	Convective Heat Transfer Coefficient ($kJh^{-1}m^{-2}K^{-1}$)	
					Front	Back
Ground	0.1000	0.97	0.75	2900	11	0.001
Steel	0.05	54	1.8	7800	11	64

The thickness (mm) of the fluorine film and Obscura was measured using an electronic digital calliper (TED PELLA, Inc., Redding, CA, USA), and the thermal conductivity ($Wm^{-1}K^{-1}$) was measured using a QTM-500 (Kyoto Electronics MFG. Co., Ltd., Kyoto, Japan) thermal conductivity meter. The infiltration through the materials was measured using a procedure detailed in our previous study [26].

There are two main approaches for determining radiative properties: spectrophotometry and the radiation balance method (RBM) [35]. The suitability of fluorine film and Obscura in the BES model was confirmed using these two methods. The shortwave and longwave properties were determined in the laboratory using a haze meter (Murakami, HM-150) and Fourier transform infrared spectrophotometer. The RBM experimental procedure for longwave and shortwave radiation is shown in Figures 6 and 7. Figure 8 shows the experimental setup for the fluorine film (left) and Obscura (right). The experiments were conducted on the open roof of a building of height 15 m for unobstructed sky radiation.

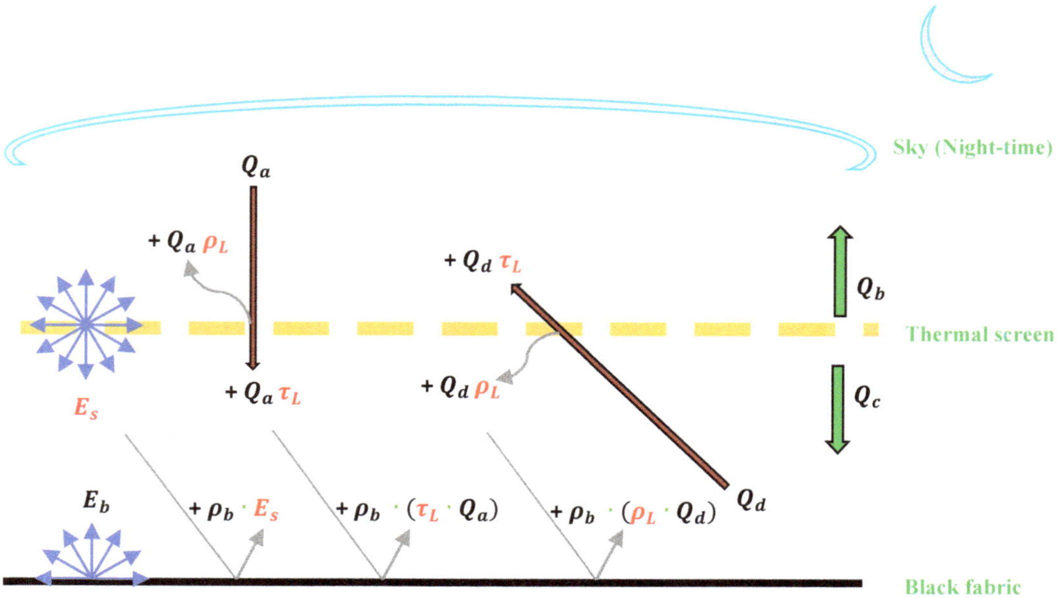

Figure 6. Schematic diagram of the incoming (Q_a and Q_d) and outgoing (Q_b and Q_c) longwave radiations for the greenhouse materials during the night [26].

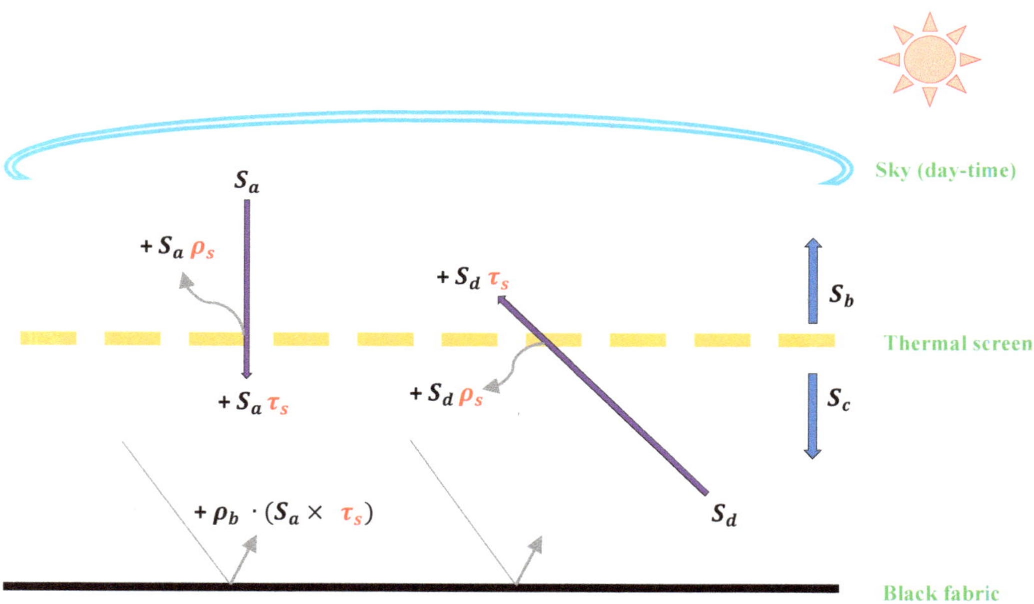

Figure 7. Schematic diagram of the inward (S_a and S_d) and outward (S_b and S_c) shortwave radiations for the greenhouse materials during the daytime [26].

Figure 8. Experimental setup for determining the radiometric properties of the greenhouse materials.

The longwave radiation is calculated using Equations (3)–(5):

$$Q_b = E_s + (\rho_L \cdot Q_a) + (Q_d \cdot \tau_L), \tag{3}$$

$$Q_c = E_s + (\tau_L \cdot Q_a) + (Q_d \cdot \rho_L), \tag{4}$$

$$Q_d = E_b + (\rho_b + \mu)E_s + (\rho_b - \mu)(\rho_L \cdot Q_d) + (\rho_b - \mu)(\tau_L \cdot Q_a), \tag{5}$$

where Q_b is the upward longwave radiation from the material to the sky (Wm^{-2}), Q_c is the upward longwave radiation from the material to the black fabric during the night (Wm^{-2}), Q_d is the inward longwave radiation from the black fabric toward the material (Wm^{-2}), Q_a is the inward sky radiation toward the material (Wm^{-2}), ρ_L is the longwave reflectance of the material, τ_L is the longwave transmittance of the material, and E_s is the emissive power of the material (Wm^{-2}) from the Stefan–Boltzmann's law. E_s is calculated using Equation (6):

$$E_s = \sigma {T_b}^4 \varepsilon_b, \tag{6}$$

where σ is the Stefan–Boltzmann constant, T_b and ε_b are the surface temperature and emissivity of the black fabric, respectively.

The shortwave radiation is calculated using Equations (7) and (8):

$$S_b = (\rho_s \cdot S_a) + (\tau_s \cdot S_d), \tag{7}$$

$$S_c = (\rho_s \cdot S_d) + (\tau_s \cdot S_a), \tag{8}$$

where S_b is the outward shortwave radiation from the material toward the sky, S_c is the outward shortwave radiation from the material toward the black material, ρ_s is the reflectance of the material, τ_s is the transmittance of the material, and S_a and S_d are the downward shortwave radiation from the sky and the radiation from the black fabric toward the material, respectively.

2.3. Greenhouse Simulation Modelling in TRNSYS 18

Dynamic simulation of the greenhouse environment in TRNSYS was conducted by linking several components from the TRNSYS library. After connecting the required components, the time steps were specified to run the simulation. Figure 9 shows the simulation studio interface, and a dynamic flowchart of the modelling and simulation procedure is shown in Figure 10. Prior to the simulation, a three-dimensional (3D) greenhouse model was designed using the add-in TRNSYS3D for Google SketchUp (Figure 11). In Google SketchUp, the greenhouse was divided into three zones separated by thermal screens, as shown in Figure 3.

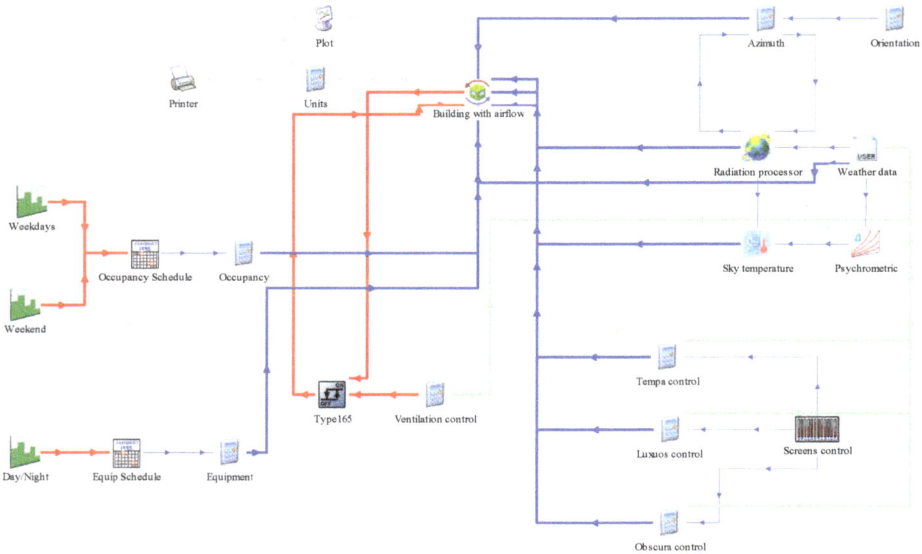

Figure 9. TRNSYS simulation studio interface.

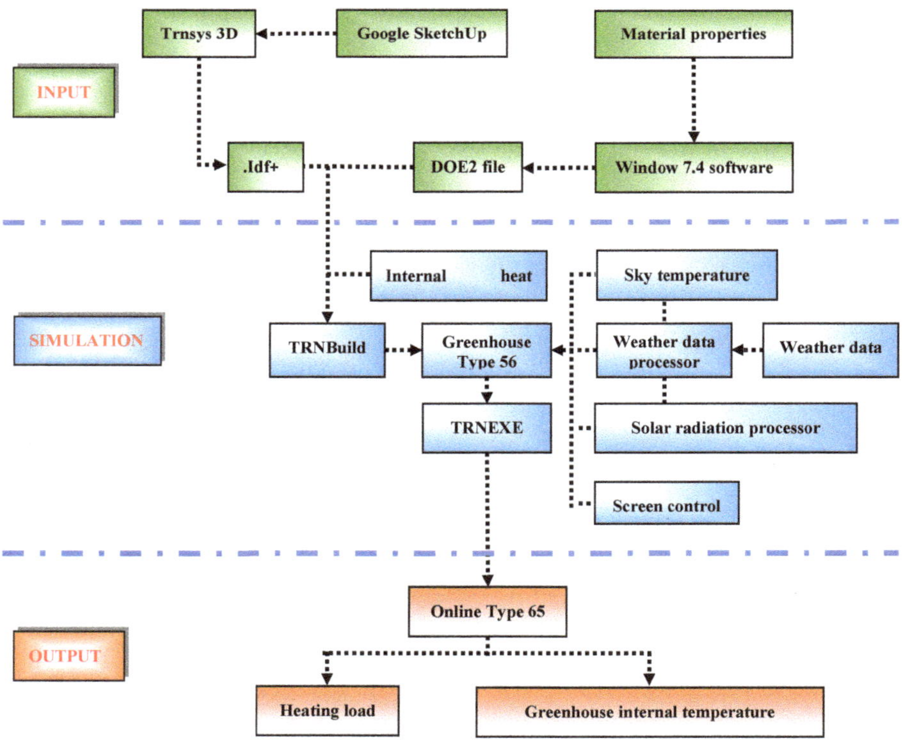

Figure 10. Dynamic flowchart of the TRNSYS model.

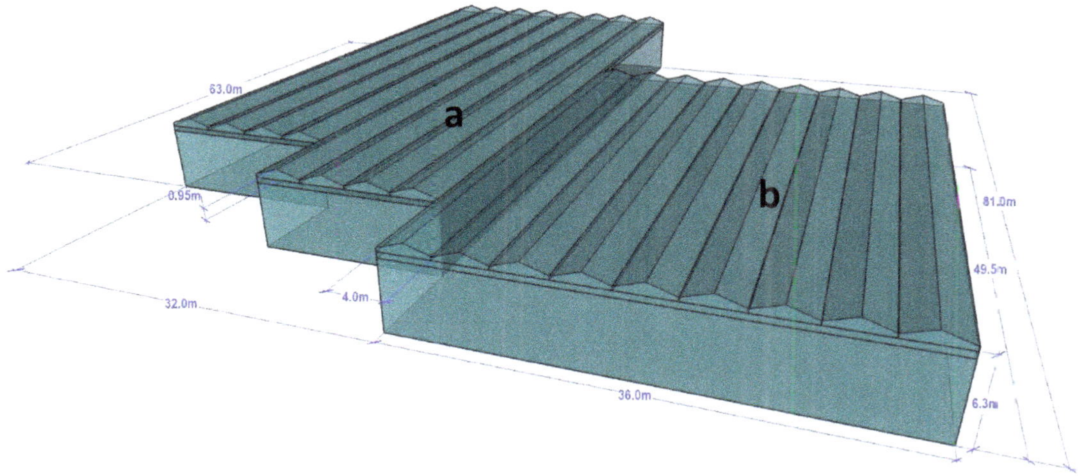

Figure 11. Three-dimensional model of the greenhouse: (**a**) Farm A; (**b**) Farm B.

The screens use the same shading device principle as that in a residential building to limit the conductive and radiative heat losses from the greenhouse. Similar to a residential building where the shading device is designed to open or close based on solar radiation, this study controls the thermal screens based on solar radiation using Equations (9) and (10). Equations (9) and (10) are for Tempa and Luxous control, respectively:

$$\text{If} \int_{T_i<18\ ^\circ\text{C}}^{SR<200\ \text{Wm}^{-2}} \text{then } (bool = 1) \text{ else } (bool = 0), \tag{9}$$

$$\int_{T_i<15\ ^\circ\text{C}}^{SR<150\ \text{Wm}^{-2}} \text{then } (bool = 1) \text{ else } (bool = 0). \tag{10}$$

The Boolean controller controls the closing (*bool* = 1) and opening (*bool* = 0) of the screens; SR is the solar radiation, and T_i is the greenhouse's internal temperature.

The 3D model was generated in the Intermediate Data Format using Google SketchUp and imported into TRNBuild. TRNBuild was used to define the building orientation and material properties. The TRNBuild database did not include the properties of the greenhouse covering materials and thermal screens used in this study. Therefore, the material properties detailed in Section 2.2 were defined using Berkeley Lab WINDOW 7.4 program, generating a DOE-2 file, and the file was imported into TRNBuild.

The building with airflow (Type 56) was selected in the TRNSYS 18 simulation studio interface to assess the internal greenhouse thermal comfort based on outside weather conditions. The recorded outside weather data were then input into the weather data component (Type 9). The solar radiation recorded was the total horizontal radiation on the horizontal surface. To calculate the direct, diffuse, and reflected solar radiation, Type 16c was linked with Type 9 to accurately account for the solar heat gain of the greenhouse. Type 16c uses different modes to calculate the tilted surface radiation. Rasheed et al. [36] calibrated a TRNSYS model by varying the Type 16c mode. They concluded that mode 2 was the most accurate for modelling the greenhouse internal temperature when the total horizontal radiation was input in Type 9.

Occupancy and equipment were implemented as internal heat gains in TRNBuild and were provided with a schedule using the Type 14 forcing function in the simulation studio. The ASHRAE standard 130 W gain type was chosen for the occupancy, and a radiative and convective power of 0.5 kJh^{-1} was used for the equipment gain [37]. However, internal gain due to plant evapotranspiration was not considered.

The sensible heat balance of the air within the thermal zones is described as follows:

$$\dot{Q}_{sens,i} = \dot{Q}_{surf,i} + \dot{Q}_{inf,i} + \dot{Q}_{ven,i} + \dot{Q}_{g,c,i} + \dot{Q}_{cplg,i} + \dot{Q}_{solar,i} + \dot{Q}_{ISHCC,i}, \quad (11)$$

where $\dot{Q}_{sens,i}$ is the sensible heat flux of the zone (kJh^{-1}), $\dot{Q}_{surf,i}$ is the convective gain from surfaces (kJh^{-1}), $\dot{Q}_{inf,i}$ is the infiltration gains (kJh^{-1}), $\dot{Q}_{ven,i}$ is the ventilation gains (kJh^{-1}), $\dot{Q}_{g,c,i}$ is the internal convective gains (kJh^{-1}), $\dot{Q}_{cplg,i}$ is the gains due to inter-connected air nodes (kJh^{-1}), $\dot{Q}_{solar,i}$ is the solar radiation entering an air node through external windows, transformed immediately into a convective gain to the internal air (kJh^{-1}), and $\dot{Q}_{ISHCC,i}$ is the solar radiation absorbed on all internal shading devices of the zone transformed immediately into a convective gain to the internal air (kJh^{-1}).

The latent heat flux of the air node is described as follows [37]:

$$\dot{Q}_{lat,i} = h_v \left[\dot{m}_{inf}(w_a - w_i) + \dot{m}_{vent}(w_{vent} - w_i) + W_g + \dot{m}_{ig}(w_j - w_i) - M_{eff}\left(\frac{w_i - \Delta T}{\Delta T}\right)\right], \quad (12)$$

where $\dot{Q}_{lat,i}$ is the latent energy flux of the zone (kJh^{-1}), h_v is the heat of vaporisation of water (kJkg^{-1}), $\dot{m}_{inf}$ is the mass flow rate of infiltration air (kgm^{-3}), w_a is the ambient humidity ratio (kg$_{water}$kg$_{air}$$^{-1}$), w_i is the air node humidity ratio (kg$_{water}$kg$_{air}$$^{-1}$), $\dot{m}_{vent}$ is the mass flow rate of ventilation air (kgm^{-3}), w_{vent} is the humidity ratio of ventilation air (kg$_{water}$kg$_{air}$$^{-1}$), W_g is the internal humidity gain (kg$_{water}$h^{-1}), $\dot{m}_{ig}$ is the mass flow rate due to couplings of two zones (kgm^{-3}), w_j is the adjacent air node humidity ratio (kg$_{water}$kg$_{air}$$^{-1}$), M_{eff} is the effective moisture capacitance (kg), and Δt is the change in time step.

The airflow due to natural ventilation was estimated using TRNFLOW. In TRNFLOW, fans, straight ducts, cracks, and large opening are the different ways of modelling the airflow network. Since the ventilation in the greenhouse involves the opening and closing of the screen, the large opening window was selected. The equation that governs the large opening airflow network is given as follows [38]:

$$\dot{m}_{1\to 2} = C_d \int_0^H \sqrt{2\rho(z) f_{1\to 2}(z)} \cdot \sqrt{\frac{(z \cdot \tan\alpha)^2 \cdot W^2}{(z \cdot \tan\alpha)^2 + W^2}} \cdot dz \quad (13)$$

where $\dot{m}$ is the mass flow rate of air (kgh^{-1}), $1 \to 2$ is the flow direction from one air node to another, C_d is the discharge coefficient, H is the total height of the window (m), $\rho(z)$ is the air density (kgm^{-3}) at height z, z is the height of the opening (m), α is the angle of opening (°), W is the width of rectangular opening (m), and $f(z)$ is the pressure difference at height z (pa).

The thermal zone energy balance is described as follows [39]:

$$C_m \frac{dT_m}{dt} = -\dot{Q}_{conv,in} - \dot{Q}_{lw,in} + \left[\dot{Q}_{solg,rad} + \dot{Q}_{ig,rad}\right] + \dot{Q}_{conv,out} + \dot{Q}_{solx} + \dot{Q}_{lw,out} + \dot{Q}_{cond}, \quad (14)$$

where T_m is the temperature of the masses (°C), C_m is the thermal capacitance of the zone masses (Kj °C^{-1}), $\dot{Q}_{conv,in}$ is the convective heat flux between the zone and the inner surface (kJh^{-1}), $\dot{Q}_{lw,in}$ is the longwave radiation exchange between two inner surfaces (kJh^{-1}), $\dot{Q}_{solg,rad}$ and $\dot{Q}_{ig,rad}$ are the radiative solar and internal gains, respectively (kJh^{-1}), $\dot{Q}_{conv,out}$ is the convective heat flux between the external surface and ambient (kJh^{-1}), Q_{solx} is the absorbed solar gain on the outside opaque surfaces (kJh^{-1}), $\dot{Q}_{lw,\,out}$ is the longwave radiation emitted by the outside surfaces to their surroundings (kJh^{-1}), and $\dot{Q}_{cond}$ is the heat conduction through the building envelope (kJh^{-1}).

Different radiation modes are available for computing the direct and diffuse shortwave and longwave radiations within a thermal zone in the TRNbuild. The radiation modes are explained in the following subsections.

2.3.1. Beam Radiation Distribution

Standard and detailed modes are available for the beam radiation distribution. The standard mode uses a user-defined factor called GEOSURF to distribute beam radiation within a zone. GEOSURF is the fraction of beam radiation that strikes the individual surfaces with a maximum value of 1 per zone. The detailed mode for the beam radiation distribution was used to distribute the solar radiation entering the zone through external windows. An external program, TRNSHD, is used by TRNBuild to generate the shading and insolation matrices required for the detailed mode. However, for beam radiation entering the zone through an adjacent window, GEOSURF was also applied.

2.3.2. Diffuse Radiation Distribution

Standard and detailed modes based on different principles are available for the diffuse radiation distribution. The standard mode is based on the absorption–transmission weighted area ratios for all surfaces, whereas the detailed mode uses the Gebhart factor for shortwave diffuse radiation. An auxiliary program, TRNVFM, based on the view factor matrix is used in the detailed mode.

2.3.3. Longwave Radiation Distribution

Simple, standard, and detailed modes are available for the longwave radiation within a zone. The simple mode is a one-node model with a combined transfer function, assuming a constant convective coefficients and longwave radiation resistance for the heat transfer process.

The standard mode calculates simultaneously the longwave radiation exchange between the surfaces in an air node and the convective heat flux from the inside surfaces to the air node using a star network. This approach uses an artificial temperature of the air node (T_{star}) to consider simultaneously the energy flow from a wall surface by convection to the air node and radiation to other surfaces. A mathematical description of the internal surface is given in Equations (15)–(17) [40]:

$$R_{star,i} = \frac{1}{Q_{surf,i}}(T_{star} - T_i), \tag{15}$$

$$\dot{q}_{comb,s,i} = \dot{q}_{c,s,i} + \dot{q}_{r,s,i}, \tag{16}$$

$$\dot{q}_{comb,s,i} = \frac{1}{R_{equiv,i} \cdot A_{s,i}}(T_{s,i} - T_{star}), \tag{17}$$

where $R_{star,i}$ and $R_{equiv,i}$ are the resistance of each surface and the equivalent resistance of all the surfaces (Ohms), respectively, $\dot{q}_{c,s,i}$ is the convective heat flux (kJh^{-1}), $\dot{q}_{r,s,i}$ is the radiative heat flux (kJh^{-1}), $\dot{q}_{comb,s,i}$ is the combined convective and radiative heat flux (kJh^{-1}), $A_{s,i}$ is the inside surface area (m^2), and $T_{s,i}$ and T_i are the surface and the equivalent air node temperatures (°C), respectively.

In addition, the total heat transfer at the outside surface is the sum of the convective and radiative heat fluxes calculated from Equations (18)–(21) [37]. The convective heat transfer at the outside surface is complicated due to wind direction, building size, local wind velocity, building surrounding, and building size [40].

$$\dot{q}_{comb,s,o} = \dot{q}_{c,s,o} + \dot{q}_{r,s,o}, \tag{18}$$

$$\dot{q}_{c,s,o} = h_{conv,s,o}(T_{a,s} - T_{s,o}), \tag{19}$$

$$q_{r,s,o}^{\cdot} = \sigma \cdot \varepsilon_{s,o}\left(T^4{}_{s,o} - T^4{}_{fsky}\right), \tag{20}$$

$$T_{fsky} = \left(1 - f_{s,sky}\right) \cdot T_{sgrd} - f_{s,sky} \cdot T_{sky}, \tag{21}$$

where $q_{c,s,o}^{\cdot}$ is the convective heat flux to the outside surface (kJh^{-1}), $q_{r,s,o}^{\cdot}$ is the radiative heat flux to the outside surface (kJh^{-1}), $q_{comb,s,o}^{\cdot}$ is the combined convective and radiative heat flux to the outside surface (kJh^{-1}), $h_{conv,s,o}$ is the convective heat transfer coefficient at the outside surface, $T_{a,s}$ is the outside surface temperature (°C), $T_{s,o}$ is the ambient temperature (°C), $\varepsilon_{s,o}$ is the longwave emissivity of the outside surface from the WINDOW library, σ is the Stefan–Boltzmann constant, T_{fsky} is the fictive temperature difference between the ground and sky (°C), $f_{s,sky}$ is the view factor of the sky, T_{sgrd} is the fictive ground temperature (°C), $1 - f_{s,sky}$ is the view factor of each external surface, and T_{sky} is the fictive sky temperature (°C).

Compared with the standard mode, the artificial star node is not considered in the detailed mode because the convective heat flux is calculated separately. The detailed mode used the Gebhart factor ($G_{ir,1-2}$) for longwave radiation exchange, considering multi-reflection from surfaces. The Gebhart factor is the fraction of emission from surface A_1 that reaches surface A_2 and is absorbed. $G_{ir,1-2}$ includes all paths (direct and multiple reflections) to reach A_2. The Gebhart matrix for longwave radiation is given by Equation (22):

$$G_{ir} = (I - F\rho_{ir})^{-1} F\varepsilon_{ir}, \tag{22}$$

where ρ_{ir} and ε_{ir} are diagonal matrices describing reflectivity and emissivity, I is an identity matrix, and F is the view factor, which is the fraction of diffusely radiated energy leaving surface 1 that is incident on surface 2; 'ir' represents the longwave range of the radiation spectrum (infrared).

The auxiliary matrix ($G_{ir}{}^*$), given by Equation (23), was introduced to determine the longwave radiation ($\dot{Q}_{ir}$) in the enclosure:

$$G_{ir}{}^* = \left(I - G_{ir}{}^T\right) A\varepsilon\sigma, \tag{23}$$

where $G_{ir}{}^T$ is the transpose of G_{ir}, A is the diagonal matrix describing the surface areas, σ is the Stefan–Boltzmann constant, and ε is the longwave emissivity.

Then,

$$\dot{Q}_{ir} = G_{ir}{}^* T^4, \tag{24}$$

where T is the temperature vector of the enclosure.

Considering that thermal screens and roofs are opened during the day to harness solar radiation, the effects of the beam and diffuse radiation modes are assumed to be negligible in greenhouses compared with residential buildings. However, during the night, the thermal screens and roofs are fully closed when energy is supplied to the conditioned zone of the greenhouse, and the properties of the thermal screens and covering materials determine the conditions of other zones. Therefore, different longwave radiation exchange modes impact the thermal comfort of the greenhouse.

The simple and standard longwave radiation modes cannot consider radiation exchange over more than one air node. The greenhouse was modelled as a single air node per zone as shown in Figure 11, to investigate the effects of the three longwave radiation modes. The shape of a building must be convex and closed to generate its view factor matrix for the detailed longwave radiation mode. Farm A is a concave-shaped greenhouse, and farm B is a convex-shaped greenhouse (Figure 12). Consequently, for farm A, the simple and standard longwave radiation models were utilised, whereas for farm B, the simple, standard, and detailed modes were adopted.

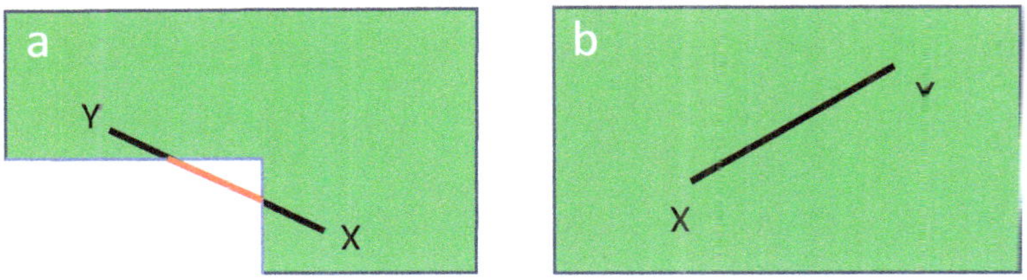

Figure 12. Plan view: (**a**) concave-shaped farm A; (**b**) convex-shaped farm B.

2.4. BES Model Validation

The proposed BES model was validated using the Nash–Sutcliffe efficiency coefficient (NSE) represented by Equation (25). This coefficient quantifies the fit of the experimentally measured data with the simulated data in a 1:1 plot. Its value ranges from $-\infty$ to 1, with values closer to 1 indicating substantial predictive potential of the model:

$$\mathrm{NSE} = 1 - [\frac{\sum_{i=0}^{n}({X_i}^{exp} - X^{\,sim})^2}{\sum_{i=0}^{n}({X_i}^{exp} - {X_i}^{mean})^2}], \quad (25)$$

where ${X_i}^{exp}$ is the experimentally measured data, ${X_i}^{sim}$ is the simulated data, and ${X_i}^{mean}$ is the mean of the experimentally measured data.

2.5. Sensitivity Analysis

A sensitivity analysis was performed to determine the impact of the longwave radiation modes on the total monthly heating demand. In mathematical modelling, sensitivity analysis quantifies the effects of several independent variables on the dependent variable under a specific condition, and the sensitivity coefficient (SC) is a dimensionless factor widely used to characterise the error assessment. The simplest form of the sensitivity coefficients for comparative energy studies, as given by Lam and Hui [41], is presented in Equation (26):

$$\mathrm{SC} = \frac{\Delta \mathrm{OP}}{\Delta \mathrm{IP}}, \quad (26)$$

where OP is the output, and IP is the input.

In addition to the radiation modes, several other factors, including infiltration, moisture change, and capacitance, affect the heating requirement of a greenhouse in the BES model. Rather than changing these parameters, the output of the heating requirements from the radiation modes is considered as the OP, whereas the measured heating consumption from the greenhouse was considered as the IP.

3. Results and Discussion

3.1. Radiometric Properties of the Novel Greenhouse Materials

The experiment based on RBM was conducted for several nights, and the results were compared with those by spectrophotometry. Figures 13 and 14 show the longwave radiative properties of the fluorine film and Obscura measured from 18:00 to 06:00 based on RBM. The reflectance, transmittance, and emittance of the fluorine film were 0.43, 0.94, and 0.02, respectively. The reflectance, transmittance, and emittance of Obscura were 0.96, 0.001, and 0.045, respectively. Because the Obscura has approximately zero transmittance in the longwave spectrum, spectrophotometry was used only for the fluorine film within the waveband of 2.5 to 25 µm (Figure 15a). The longwave transmittance result by spectrophotometry was the same as that by RBM. The consistency of the results using the two methods validates the accuracy in the determination of the radiative properties of the materials used in the BES modelling (Table 4).

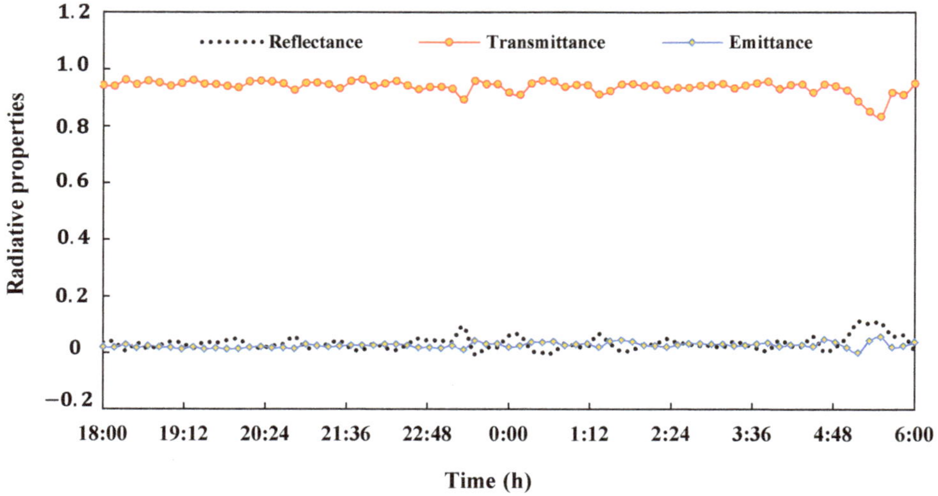

Figure 13. Longwave radiative properties of fluorine film.

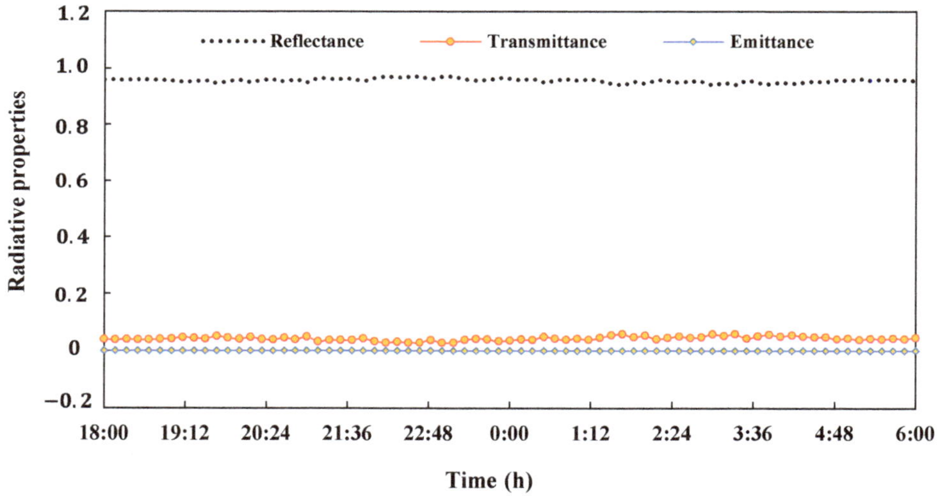

Figure 14. Longwave radiative properties of Obscura.

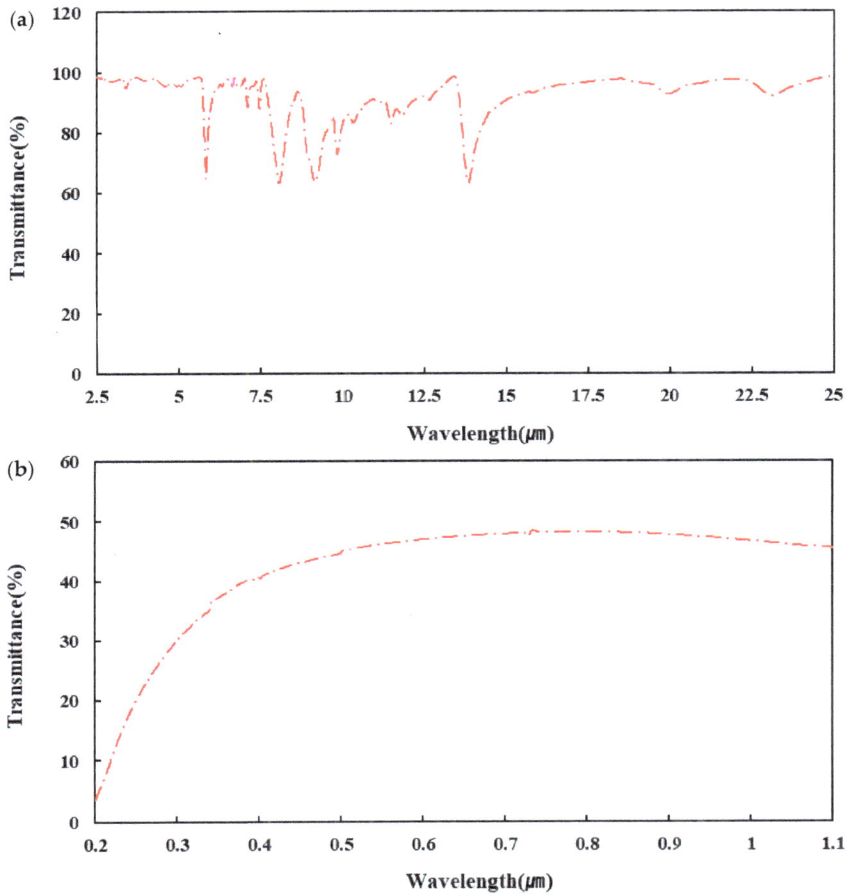

Figure 15. (**a**) Variation in fluorine film with wavelength (**a**) longwave (**b**) shortwave.

Table 4. Properties of the novel greenhouse materials.

Cover Characteristics	Fluorine Film	Obscura
Thickness (mm)	0.08	0.34
Solar transmittance (front)	0.92	0.01
Solar transmittance (back)	0.92	0.01
Solar reflectance (front)	0.06	0.64
Solar reflectance (back)	0.06	0.64
Visible radiation transmittance (front)	0.92	0.01
Visible radiation transmittance (back)	0.92	0.01
Visible radiation reflection (front)	0.06	0.64
Visible radiation reflection (back)	0.06	0.64
Thermal radiation transmittance	0.94	0.001
Thermal radiation emission (front)	0.02	0.045
Thermal radiation emission (back)	0.03	0.045
Thermal conductivity ($Wm^{-1}K^{-1}$)	0.15	0.35
Infiltration ($m^3h^{-1}m^2$)	-	-

The diffuse shortwave transmittance of the fluorine film, measured using an ultraviolet-visible spectrophotometer (Jasco V-730), is shown in Figure 15b. The measured diffuse transmittance is 45.86%. However, owing to the limitations of spectrophotometry, the total transmittance was measured using a haze meter. The total transmittance was 92.49% with the beam and diffuse transmissions contributing to 46.74% and 45.75%, respectively. The solar transmittance of the fluorine film indicates that it transmits more than 90% of the shortwave radiation, which is vital for plant growth. This value is higher than the transmittance of the commonly used greenhouse covering materials, polyethylene (71%) [34] and HG (89%) [33]. The solar transmittance obtained is similar to the longwave transmittance; this does not agree with Choab et al. [42], who recommended high and low transmittances in the shortwave and longwave spectra, respectively, for covering materials. However, the installed thermal screens have very low transmittance in the longwave spectrum (Tempa, 5%; Luxous, 33%) to limit heat loss before transmission through the film.

3.2. Comparison of Results of the TRNSYS Model with the Experimental Measurements

The daily average outside temperature and average solar radiation on the horizontal surface are shown in Figure 16. The dataset represents meteorological conditions in Yeoju, South Korea, during the winter season between December 2021 and March 2022. The lowest and the highest hourly temperatures (−17.8 °C and 20.2 °C, respectively) were recorded on 26 December 07:00 and 12 March 15:00, respectively. For temperature, the daily average low (of −12.2 °C) and high (of 13.3 °C) were recorded on 26 December and 13 March, respectively. The maximum hourly solar radiation (1103.6 Wm^{-2}) was on 28 March, 12:00. Both minimum and maximum daily average solar radiations (15.1 Wm^{-2} and 344 Wm^{-2}, respectively) were recorded in March. The coldest month was January, with a monthly average temperature of −3.6 °C.

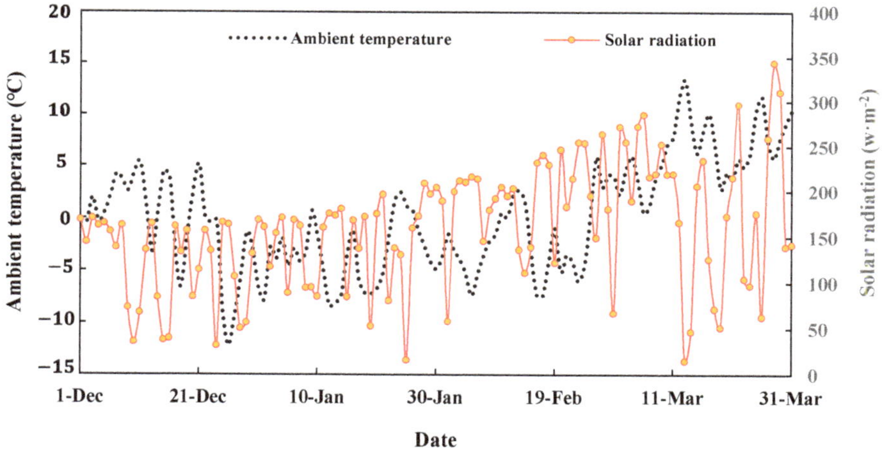

Figure 16. Daily average outside temperature and solar radiation in Yeoju-si, South Korea.

The simulated internal temperatures in the different zones for both farms were compared with the experimentally measured temperatures of the greenhouse. Because the daily greenhouse conditioning is approximately the same during the winter season, only a few days of the analysis are discussed. Figure 17 shows the simulated zone 1 temperatures compared with the experimentally measured temperatures from 26 to 31 January. On most days, the greenhouse's experimental daytime temperature was approximately 30 °C for both the farms. For farm A, the simulated daytime temperatures differ from the observed temperatures by a maximum of 3.0 °C (above the observed temperature) and 3.0 °C (below the observed temperature) for the simple and standard modes, respectively. For farm B, the simulated maximum daytime temperatures differ from the observed temperatures by a maximum of 3.0 °C (above the observed temperature), 2.3 °C (below the observed temperature), and 0.94 °C (below the observed temperature) for the simple, standard, and detailed modes, respectively. The simple and standard modes over-predicted and under-predicted the daytime temperatures, respectively, for both greenhouses. The controlled simulated night temperature for the three modes was 15 °C. So, it was not possible to determine the impact of the modes on the night temperature in zone 1. The calculated NSE values for farm A were 0.85 and 0.78 for the simple and standard modes, respectively. The calculated NSE values for farm B were 0.73, 0.78, and 0.83 for the simple, standard, and detailed modes, respectively.

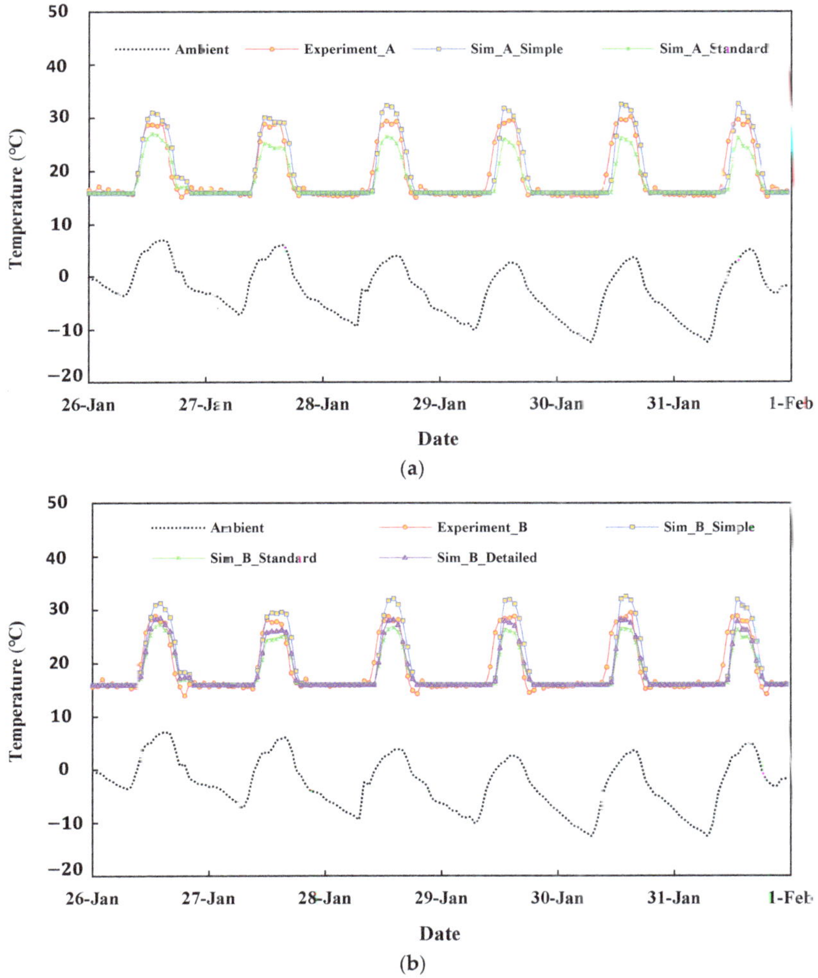

Figure 17. Comparison of greenhouse zone 1 temperature from experiment and simulation: (**a**) Farm A; (**b**) Farm B.

Figure 18 shows the simulated zone 2 temperatures compared with the experimentally measured temperatures. The thermal screens placed between zones 1 and 2 caused an approximately 5 °C temperature difference between the two zones during the night when zone 1 was being heated. For farm A, the simple and standard modes over-predicted the daytime temperatures by a maximum of 8 °C and 3 °C, respectively. Similarly, for farm B, the simulated daytime temperatures differ from the observed temperatures by a maximum of 3.4 °C, 1.72 °C, and 0.8 °C above the observed temperature for the simple, standard, and detailed modes, respectively. However, the simple and standard modes under-predicted the night temperature by a maximum of 5.34 °C and 2.54 °C, respectively, in Farm A. By contrast, for farm B, the maximum difference of the simple, standard, and detailed modes was 3.5 °C, 3.7 °C, and 3.4 °C below the observed night temperatures, respectively. The calculated NSE values for farm A were 0.79 and 0.89 for the simple and standard modes, respectively. The calculated NSE values for farm B were 0.89, 0.88, and 0.9 for the simple, standard, and detailed modes, respectively.

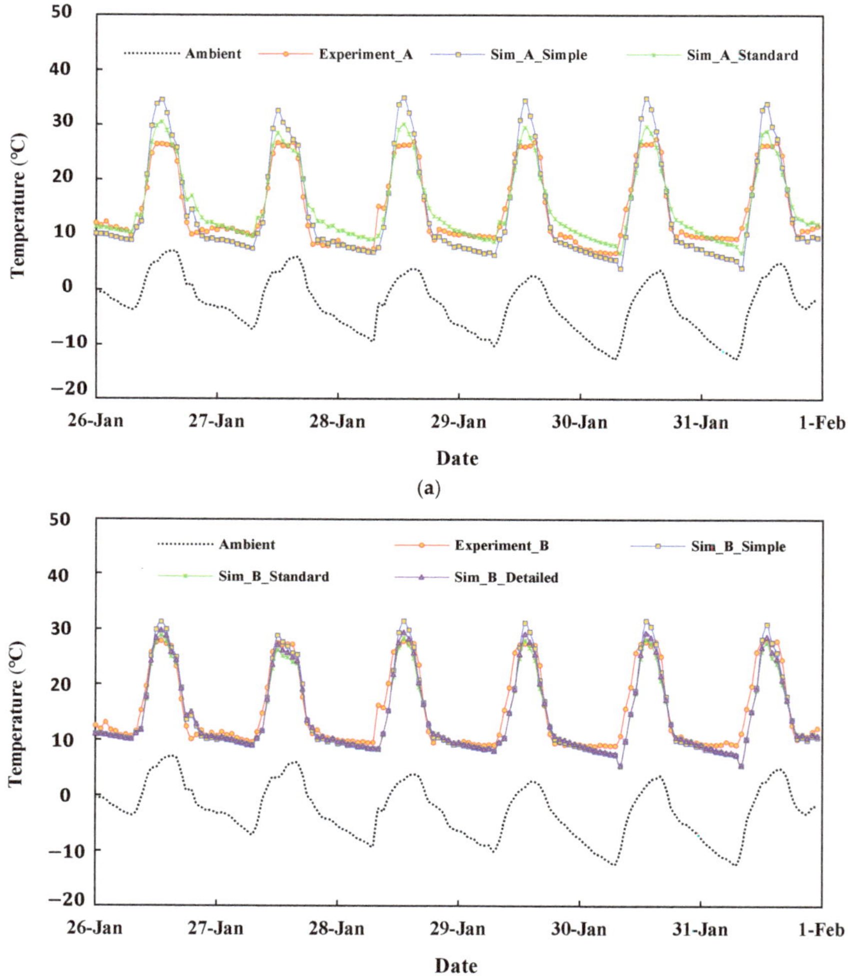

Figure 18. Comparison of greenhouse zone 2 temperature from experiment and simulation: (**a**) Farm A; (**b**) Farm B.

Figure 19 shows the simulated zone 3 temperatures compared with the experimentally measured temperatures. The thermal screens placed between zones 2 and 3 caused an 8 °C temperature difference between the two zones during the night. This temperature difference is higher than between Zone 1 and 2 owing to the low transmissive properties of Tempa compared with Luxous. The average temperature difference between zone 3 and the ambient was 10 °C. The calculated NSE values for farm A were 0.91 and 0.93 for the simple and standard modes, respectively. The calculated NSE values for farm B were 0.92, 0.91, and 0.92 for the simple, standard, and detailed modes, respectively.

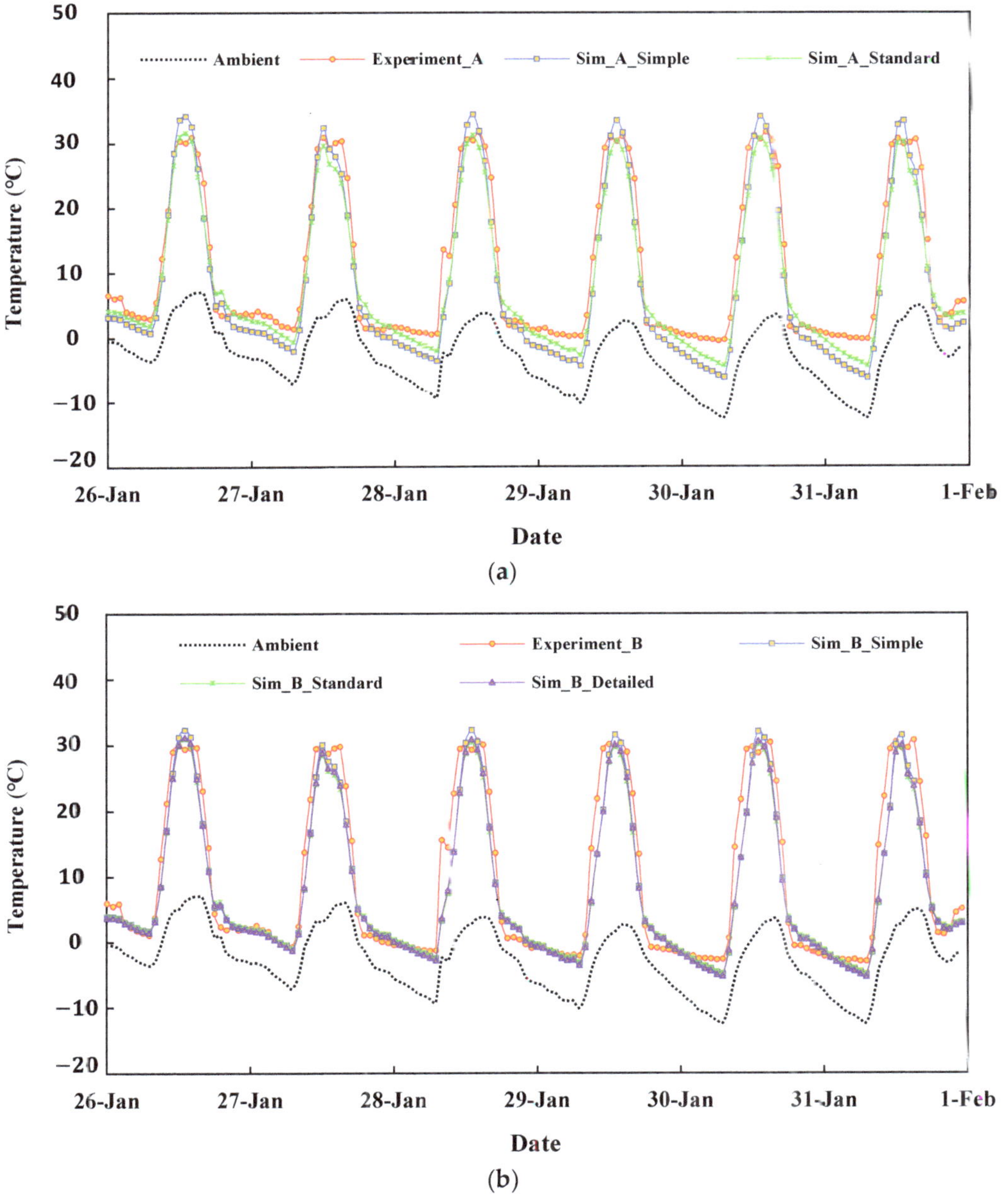

Figure 19. Comparison of greenhouse zone 3 temperature from experiment and simulation: (**a**) Farm A; (**b**) Farm B.

As the NSE values for all radiation modes were higher than 0.5, the BES model was valid, and the standard and detailed models were recommended for predicting the internal greenhouse temperature for concave and convex-shaped greenhouses. However, the user-defined emissivities of the surface for longwave radiation can only be handled using a detailed model. Therefore, a detailed model should be adopted to accurately model the internal greenhouse temperature.

Because the overall performance of the standard and detailed radiation modes for farms A and B was superior, both were used to analyse the heating demand of the two greenhouses. However, the maximum heating demand varied for the different radiation modes (Table 5). Figure 20 compares the simulated heating demand with the experimental energy consumption during the period of the lowest ambient temperature from 21 to 27 December 2021. There were significant hourly changes between the simulated and measured results owing to the transient state of the experimental greenhouse and the intermittent application of the energy control systems. The maximum heating loads for farms A and B were 113.5 kcal/hm^2 and 120.7 kcal/hm^2, respectively, on 26 December when the ambient temperature (−17.8 °C) was minimum and the greenhouse set point temperature was 15 °C. For the same greenhouse with similar hybrid heat pump systems using geothermal sources and solar heat, Jeon et al. [43] designed a greenhouse (1015 m^2) with a maximum heating load of 148.8 kcal/hm^2 for the lowest ambient temperature of −19 °C and greenhouse set point temperature of 23 °C. The simulated maximum heating demands for both the farms with the thermal screens were high compared with Rasheed et al. [23], who obtained 109.9 kcal/hm^2, 98 kcal/hm^2, and 81.3 kcal/hm^2 using single-, double-, and triple layered-thermal screens, respectively, for a greenhouse floor area of 7572.6 m^2. However, the simulated heating demand was lower than that of a greenhouse with a floor area of 391.2 m^2 and with a maximum heating load of 250 kcal/hm^2 [44]. This trend indicates that greenhouse heating demand per square meter decreases with increasing floor area. Heat is lost through all greenhouse surfaces, including walls, roofs, and floors, and the amount of heat lost per square meter increases as the wall-to-total surface area ratio increases.

Table 5. Heating loads using TRNSYS for different radiation modes.

Farm	Greenhouse Set Point Temperature (°C)	Maximum Outside Temperature (°C)	Greenhouse Heating Area (m^2)	Radiation Mode	Maximum Heating Load (kcal/hm^2)
A	15	−17.8	2160	Simple Standard	101.3 113.5
B	15	−17.8	1782	Simple Standard Detailed	116.4 123.4 120.7

The total daily heating demand was compared with the total daily energy consumed for BES model performance. The maximum heating demand was on 19 January (farm A: 3547.8 Mcal/day, and farm B: 3118.9 Mcal/day) when the daily average temperature was −7.3 °C. In the study by Ahamed et al. [22] using the CSGHEAT models with MATLAB, the average daily heating requirement in a CSG integrated with thermal screens was 900 MJ/day (215.2 Mcal/day). The CSG heating area was 210 m^2. This is approximately ten times lower than that of the experimental glass greenhouse. The calculated NSE values for farms A and B were 0.89 and 0.9, respectively, indicating the model's ability to predict the greenhouse energy load.

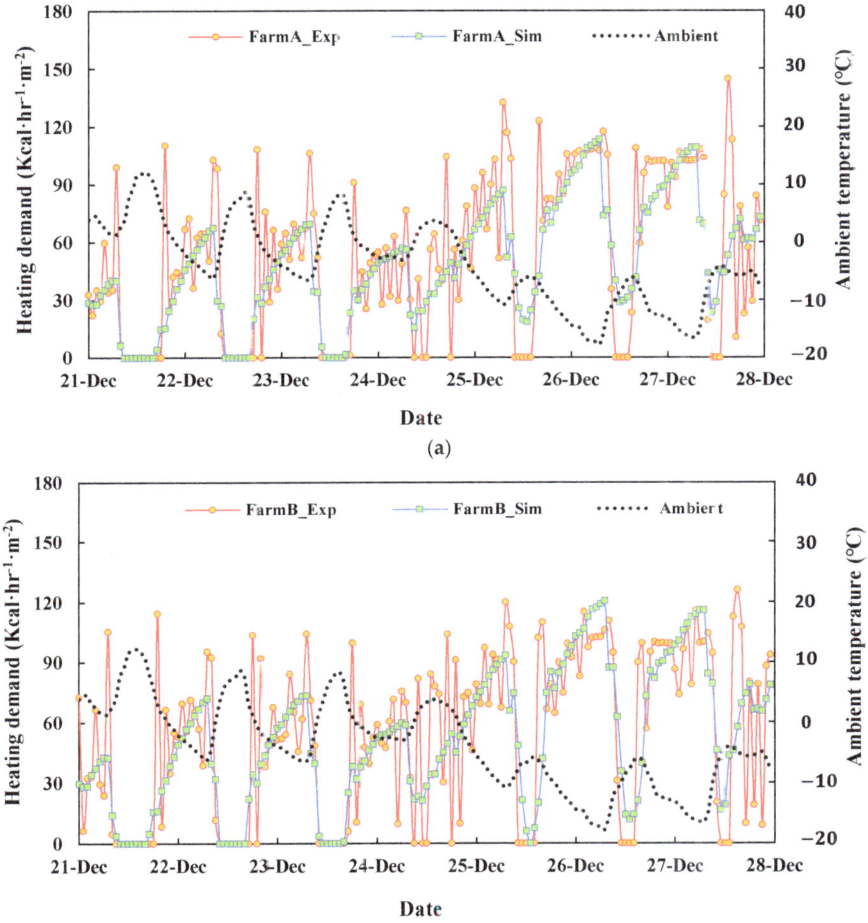

Figure 20. Comparison of greenhouse hourly heating demand from experiment and simulation: (a) Farm A; (b) Farm B.

Figure 21 shows the monthly heating demands predicted by the radiation modes and the experimental heating energy consumption. The radiation modes affected the simulated heating requirement of the greenhouse. The monthly heating demand predicted by the simple and standard radiation modes for farm A matched the experimental measurements with deviations within 27.7% and 7.6%, respectively. The monthly heating loads predicted by the simple, standard, and detailed radiation modes for farm B were similar to the experimental measurements with deviations within 10.5%, 6.7%, and 2.9%, respectively. For both greenhouses, the monthly heating requirements predicted by the simple mode were less than the experimental measurements. However, the standard mode under- and over-predicted the monthly heating requirements in farms A and B, respectively.

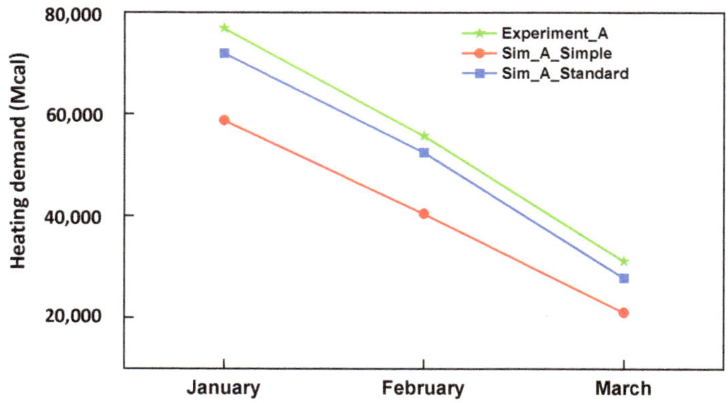

(**a**)

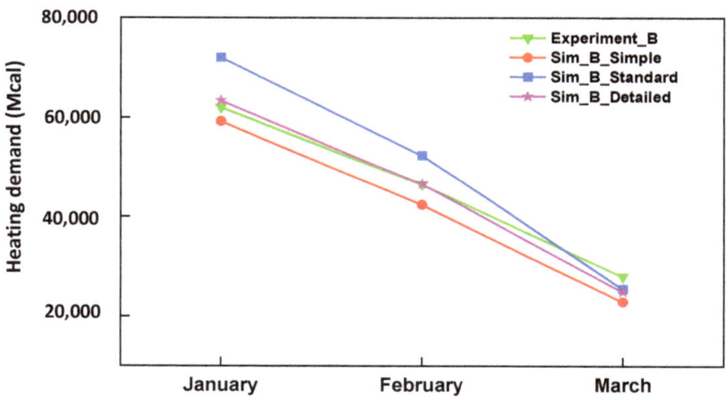

(**b**)

Figure 21. Greenhouse monthly heating demands: (**a**) Farm A; (**b**) Farm B.

Figure 22 shows the SC for both the farms. The simple and standard modes for farm A had an average SC of 0.7 and 0.9, respectively, whereas the simple, standard, and detailed modes for farm B had an average SC of 0.9, 1.1, and 1.0, respectively. In addition, for farm B, as the monthly ambient temperature increases, the heating demand predicted by the standard and detailed modes converged, indicating that the differences in the predictions between the radiation modes were more pronounced during cold months for a convex-shaped greenhouse.

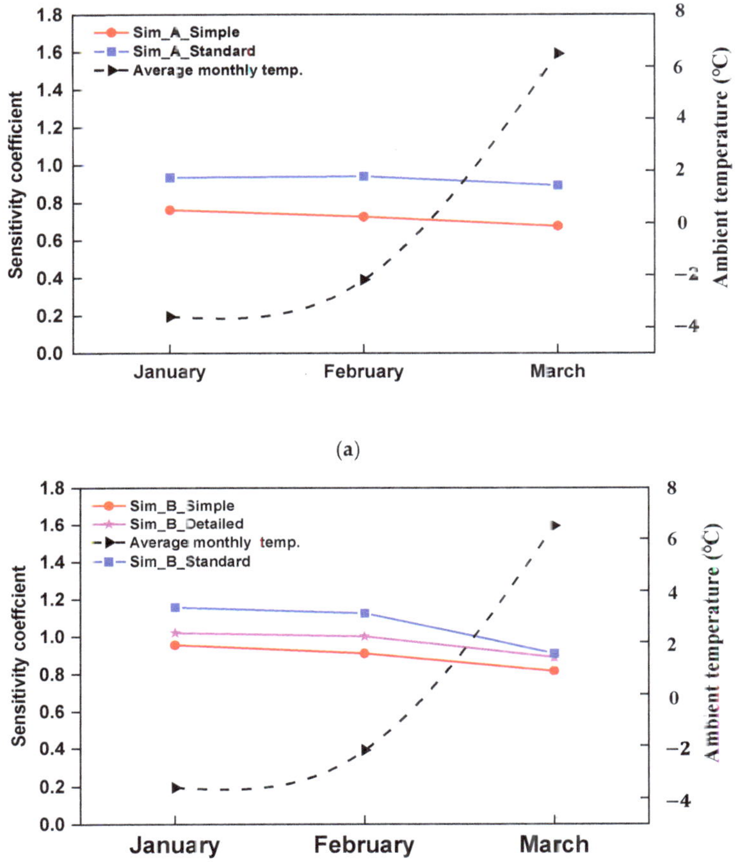

(a)

(b)

Figure 22. Sensitivity coefficients of TRNSYS radiation modes for the greenhouse heating requirement: (**a**) Farm A; (**b**) Farm B.

3.3. Limitation of the Study and Future Work

The proposed BES model has some limitations. Firstly, three weather data (windspeed, wind direction, and ambient pressure) used in the simulation were downloaded at a neighbouring station, 18 km from the experimental greenhouse. The results obtained could lead to a slight variation if the model is to be adapted for year-round simulation, as the greenhouse solely depends on wind forces for natural ventilation in the summer. However, the division of the greenhouse based on the geometry shows the variation in the radiation mode on the internal temperature and heating demand of the greenhouse. More so, in the absence of on-site data, several researchers have used neighbouring stations weather data. Choi et al. [45] had studied the windspeed trend of 31 KMA stations from 1985–2019 and concluded that the average wind speed in Korea does not change that much, while Kim et al. [46] developed a TRNSYS model for Jincheon using Cheongju weather data, 23.9 km away from Jincheon. Secondly, the planted crops were not considered in the simulation. Although they are the most important factor in greenhouse simulation [47], the absence of a stand-alone component in TRNSYS limited their consideration in our study.

The present TRNSYS model could only access the greenhouse thermal performance based on radiation modes without including the internal heat gains due to evapotranspiration from the crop. Modified crop models to calculate the latent mass flow and latent power due to the presence of crop will be implemented in future.

4. Conclusions

The greenhouse industry is faced with challenges to increase production and reduces the use of resources. Mathematical modelling has been seen as alternative to identify potential solutions. In this study, TRNSYS was used to investigate the thermal performance of a glass greenhouse integrated with movable thermal screens. For the novel greenhouse covering materials and thermal screens considered in the study, the thermo-physical, aerodynamics, and radiative properties were experimentally determined for accurate BES modelling. The internal greenhouse temperature and heating demand were simulated using different longwave radiation modes, and the simulated results were validated using transient experimental data. The performances of the standard and detailed modes were superior in concave- and convex-shaped greenhouses as the radiation modes were sensitive to ambient temperature. The findings of this study will aid greenhouse growers, researchers, and engineers in selecting the appropriate longwave radiation mode for modelling the thermal performance of a greenhouse based on the geometry of the greenhouse.

Author Contributions: Simulation, M.A.A. and A.R. (Adnan Rasheed); validation, M.A.A.; formal analysis, Q.O.O., T.D.A., A.R. (Anis Rabiu) and W.-H.N.; investigation, M.A.A. and H.-W.L.; resources, H.-W.L., W.-H.N. and Y.-C.Y.; data curation, Q.O.O., T.D.A., A.R. (Anis Rabiu) and Y.-C.Y.; writing; M.A.A.; supervision, H.-W.L. and W.-H.N.; project administration, H.-W.L.; funding acquisition, H.-W.L. All authors have read and agreed to the published version of the manuscript.

Funding: This work was supported by the Korea Institute of Planning and Evaluation for Technology in Food, Agriculture, and Forestry (IPET) through the Agricultural Energy Self-Sufficient Industrial Model Development Program, funded by the Ministry of Agriculture, Food, and Rural Affairs (MAFRA) (120096-3). This research was supported by the Basic Science Research Program through the National Research Foundation of Korea (NRF), funded by the Ministry of Education (NRF-2019R1I1A3A01051739).

Data Availability Statement: Data available on request due to restrictions.

Conflicts of Interest: The authors declare no conflict of interest.

Nomenclature

Symbols

$A_{s,i}$	Inside surface area, m^2
C_m	Thermal capacitance of the zone masses, kJK^{-1}
C_p	Specific heat capacity of water, $kcalkg^{-1}K^{-1}$
E_s	Emissive power of the material, Wm^{-2}
$f_{s,sky}$	View factor of the sky
G_{ir}	Gebhart factor
$G_{ir}{}^*$	Auxiliary matrix
$G_{ir}{}^T$	Transpose of G_{ir}
h	Greenhouse height, m
$h_{conv,s,o}$	Convective heat transfer coefficient at the outside surface, $Wm^{-2}K^{-1}$
h_o	Reference height, m
h_v	Heat of vaporisation of water, $kJkg^{-1}$
$\dot{m}$	Mass flow rate of water, kgh^{-1}
$\dot{m}_{ig}$	Mass flow rate due to couplings of two zones, kgm^{-3}
$\dot{m}_{inf}$	Mass flow rate of infiltration air, kgm^{-3}
$\dot{m}_{vent}$	Mass flow rate of ventilation air, kgm^{-3}

M_{eff}	Effective moisture capacitance, kg
$\dot{q}_{comb,s,i}$	Combined convective and radiative heat flux in the inner surface, kJh^{-1}
$\dot{q}_{comb,s,o}$	Combined convective and radiative heat flux to the outside surface, kJh^{-1}
$\dot{q}_{c,s,o}$	Convective heat flux to the outside surface, kJh^{-1}
$\dot{q}_{r,s,o}$	Radiative heat flux to the outside surface, kJh^{-1}
Q	Energy consumed, $kcalh^{-1}$
Q_a	Inward sky radiation toward the material, Wm^{-2}
Q_b	Upward longwave radiation from the material to the sky, Wm^{-2}
Q_c	Upward longwave radiation from the material to the black fabric during the night, Wm^{-2}
$\dot{Q}_{cond}$	Heat conduction through the building envelope, kJh^{-1}
$\dot{Q}_{conv,in}$	Convective heat flux between the zone and the inner surface, kJh^{-1}
$\dot{Q}_{conv,out}$	Convective heat flux between the external surface and ambient, kJh^{-1}
$\dot{Q}_{cplg,i}$	Gains due to inter-connected air nodes kJh^{-1}
$\dot{Q}_{g,c,i}$	Internal convective gains, $kJ\ h^{-1}$
Q_d	Inward longwave radiation from the black fabric toward the material, Wm^{-2}
$\dot{Q}_{ig,rad}$	Radiative heat flux, $kJ\ h^{-1}$
$\dot{Q}_{inf,i}$	Infiltration heat flux, $kJ\ h^{-1}$
$\dot{Q}_{ISHCC,i}$	Solar radiation absorbed on all internal shading devices of the zone, kJh^{-1}
$\dot{Q}_{lat,i}$	Latent energy flux of the zone, kJh^{-1}
$\dot{Q}_{lw,in}$	Longwave radiation exchange between two inner surfaces, kJh^{-1}
$\dot{Q}_{lw,out}$	Longwave radiation emitted by the outside surfaces to the Surroundings, kJh^{-1}
$\dot{Q}_{sens,i}$	Sensible heat flux of the zone, kJh^{-1}
$\dot{Q}_{solar,i}$	Solar radiation entering an air node through external windows, kJh^{-1}
$\dot{Q}_{solx}$	Absorbed solar gain on the outside opaque surfaces, kJh^{-1}
$\dot{Q}_{surf,i}$	Convective gain from surfaces, kJh^{-1}
$\dot{Q}_{ven,i}$	Ventilation heat flux, $kJ\ h^{-1}$
$R_{star,i}$	Resistance of each surface, Ohms
$R_{equiv,i}$	Equivalent resistance of all the surfaces, Ohms
S_a	Downward shortwave radiation from the sky, Wm^{-2}
S_b	Outward shortwave radiation from the material toward the sky, Wm^{-2}
S_c	Outward shortwave radiation from the material toward the black material, Wm^{-2}
S_d	Radiation from the black fabric toward the material, Wm^{-2}
Δt	Temperature difference, K
ΔT	Change in simulation time step
T	Temperature vector of the enclosure
$T_{a,s}$	Outside surface temperature, K
T_b	Surface temperature, K
T_{fsky}	Fictive temperature difference between the ground and sky, K
T_i	Equivalent air node temperatures, K
T_m	Temperature of the zone masses, K
T_{sgrd}	Fictive ground temperature, K
T_{sky}	Fictive sky temperature, K
$T_{s,i}$	Surface air node temperatures, K
$T_{s,o}$	Ambient temperature, K
T_{star}	Artificial temperature of the air node, K
V_h	Calculated wind speed, ms^{-1}
V_o	Reference wind speed, ms^{-1}
w_a	Ambient humidity ratio, $kg_{water}kg_{air}^{-1}$
w_j	Adjacent air node humidity ratio, $kg_{water}kg_{air}^{-1}$
w_i	Air node humidity ratio, $kg_{water}kg_{air}^{-1}$
w_{vent}	Humidity ratio of ventilation air $kg_{water}kg_{air}^{-1}$
W_g	Internal humidity gain, $kg_{water}h^{-1}$
X_i^{exp}	Experimentally measured data
X_i^{sim}	Simulated data and
X_i^{mean}	Mean of the experimentally measured data

Greek symbols	
$\propto$	Power law exponent
ρ_L	Longwave reflectance of the material
τ_L	Longwave transmittance of the material
σ	Stefan–Boltzmann constant
ε_b	Emissivity of the black fabric
ρ_s	Reflectance of the material
τ_S	Transmittance of the material
$\varepsilon_{s,o}$	Longwave emissivity of the outside surface from the WINDOW library
ρ_{ir}	Diagonal matrices describing reflectivity
ε_{ir}	Diagonal matrices describing emissivity
Abbreviations	
A	Diagonal matrix describing the surface areas
I	Identity matrix
F	View factor
ir	Longwave range of the radiation spectrum (infrared)
BES	Building Energy Simulations
TRNSYS	Transient System Simulation
NSE	Nash–Sutcliffe Efficiency Coefficient
CSG	Chinese-style Solar Greenhouse
HG	Horticultural Glass
KMA	Korean Meteorological Administration
QTM	Quick Thermal Meter
RBM	Radiation Balance Method
SC	Sensitivity Coefficient
OP	Output
IP	Input

References

1. Baudoin, W.; Nono-Womdim, R.; Lutaladio, N.; Hodder, A.; Castilla, N.; Leonardi, C.; De Pascale, S.; Qaryouti, M. Cultural Practices and Environment. In *Hobby Hydroponics*; CRC Press: Boca Raton, FL, USA, 2013; pp. 42–55, ISBN 9789251076491.
2. Akpenpuun, T.D.; Na, W.H.; Ogunlowo, Q.O.; Rabiu, A.; Adesanya, M.A.; Addae, K.S.; Kim, H.T.; Lee, H.-W. Effect of Glazing Configuration as an Energy-Saving Strategy in Naturally Ventilated Greenhouses for Strawberry (*Seolhyang Sp.*) Cultivation. *J. Agric. Eng.* **2021**, *52*, 1177. [CrossRef]
3. Ogunlowo, Q.O.; Akpenpuun, T.D.; Na, W.H.; Rabiu, A.; Adesanya, M.A.; Addae, K.S.; Kim, H.T.; Lee, H.W. Analysis of Heat and Mass Distribution in a Single-and Multi-Span Greenhouse Microclimate. *Agriculture* **2021**, *11*, 891. [CrossRef]
4. Banakar, A.; Montazeri, M.; Ghobadian, B.; Pasdarshahri, H.; Kamrani, F. Energy Analysis and Assessing Heating and Cooling Demands of Closed Greenhouse in Iran. *Therm. Sci. Eng. Prog.* **2021**, *25*, 101042. [CrossRef]
5. Mazzeo, D.; Baglivo, C.; Panico, S.; Congedo, P.M. Solar Greenhouses: Climates, Glass Selection, and Plant Well-Being. *Sol. Energy* **2021**, *230*, 222–241. [CrossRef]
6. Akpenpuun, T.D.; Na, W.H.; Ogunlowo, Q.O.; Rabiu, A.; Adesanya, M.A.; Addae, K.S.; Kim, H.T.; Lee, H.W. Effect of Greenhouse Cladding Materials and Thermal Screen Configuration on Heating Energy and Strawberry (Fragaria Ananassa Var. "Seolhyang") Yield in Winter. *Agronomy* **2021**, *11*, 2498. [CrossRef]
7. Baglivo, C.; Mazzeo, D.; Panico, S.; Bonuso, S.; Matera, N.; Congedo, P.M.; Oliveti, G. Complete Greenhouse Dynamic Simulation Tool to Assess the Crop Thermal Well-Being and Energy Needs. *Appl. Therm. Eng.* **2020**, *179*, 115698. [CrossRef]
8. Zhang, Y.; Gauthier, L.; De Halleux, D.; Dansereau, B.; Gosselin, A. Effect of Covering Materials on Energy Consumption and Greenhouse Microclimate. *Agric. For. Meteorol.* **1996**, *82*, 227–244. [CrossRef]
9. Rasheed, A.; Na, W.H.; Lee, J.W.; Kim, H.T.; Lee, H.W. Optimization of Greenhouse Thermal Screens for Maximized Energy Conservation. *Energies* **2019**, *12*, 3592. [CrossRef]
10. Shukla, A.; Tiwari, G.N.; Sodha, M.S. Thermal Modeling for Greenhouse Heating by Using Thermal Curtain and an Earth-Air Heat Exchanger. *Build. Environ.* **2006**, *41*, 843–850. [CrossRef]
11. Ahamed, M.S.; Guo, H.; Tanino, K. Energy Saving Techniques for Reducing the Heating Cost of Conventional Greenhouses. *Biosyst. Eng.* **2019**, *178*, 9–33. [CrossRef]
12. Rasheed, A.; Lee, J.W.; Lee, H.W. Development of a Model to Calculate the Overall Heat Transfer Coefficient of Greenhouse Covers. *Spanish J. Agric. Res.* **2017**, *15*, e0208. [CrossRef]
13. Guo, Y.; Zhao, H.; Zhang, S.; Wang, Y.; Chow, D. Modeling and Optimization of Environment in Agricultural Greenhouses for Improving Cleaner and Sustainable Crop Production. *J. Clean. Prod.* **2021**, *285*, 124843. [CrossRef]

14. Lee, S.Y.; Lee, I.B.; Lee, S.N.; Yeo, U.H.; Kim, J.G.; Kim, R.W.; Decano-Valentin, C. Dynamic Energy Exchange Modelling for a Plastic-Covered Multi-Span Greenhouse Utilizing a Thermal Effluent from Power Plant. *Agronomy* **2021**, *11*, 1461. [CrossRef]
15. Zhang, G.; Ding, X.; Li, T.; Pu, W.; Lou, W.; Hou, J. Dynamic Energy Balance Model of a Glass Greenhouse: An Experimental Validation and Solar Energy Analysis. *Energy* **2020**, *198*, 117281. [CrossRef]
16. Baneshi, M.; Gonome, H.; Maruyama, S. Wide-Range Spectral Measurement of Radiative Properties of Commercial Greenhouse Covering Plastics and Their Impacts into the Energy Management in a Greenhouse. *Energy* **2020**, *210*, 118535. [CrossRef]
17. Ahamed, M.S.; Guo, H.; Tanino, K. A Quasi-Steady State Model for Predicting the Heating Requirements of Conventional Greenhouses in Cold Regions. *Inf. Process. Agric.* **2018**, *5*, 33–46. [CrossRef]
18. Katzin, D.; van Henten, E.J.; van Mourik, S. Process-Based Greenhouse Climate Models: Genealogy, Current Status, and Future Directions. *Agric. Syst.* **2022**, *198*, 103388. [CrossRef]
19. Fabrizio, E. Energy Reduction Measures in Agricultural Greenhouses Heating: Envelope, Systems and Solar Energy Collection. *Energy Build.* **2012**, *53*, 57–63. [CrossRef]
20. Ahamed, M.S.; Guo, H.; Tanino, K. Development of a Thermal Model for Simulation of Supplemental Heating Requirements in Chinese-Style Solar Greenhouses. *Comput. Electron. Agric.* **2018**, *150*, 235–244. [CrossRef]
21. Mashonjowa, E.; Ronsse, F.; Milford, J.R.; Pieters, J.G. Modelling the Thermal Performance of a Naturally Ventilated Greenhouse in Zimbabwe Using a Dynamic Greenhouse Climate Model. *Sol. Energy* **2013**, *91*, 381–393. [CrossRef]
22. Ahamed, M.S.; Guo, H.; Tanino, K. Modeling Heating Demands in a Chinese-Style Solar Greenhouse Using the Transient Building Energy Simulation Model TRNSYS. *J. Build. Eng.* **2020**, *29*, 101114. [CrossRef]
23. Rasheed, A.; Kwak, C.S.; Na, W.H.; Lee, J.W.; Kim, H.T.; Lee, H.W. Development of a Building Energy Simulation Model for Control of Multi-Span Greenhouse Microclimate. *Agronomy* **2020**, *10*, 1236. [CrossRef]
24. Sharma, P.K.; Tiwari, G.N.; Sorayan, V.P.S. Temperature Distribution in Different Zones of the Micro-Climate of a Greenhouse: A Dynamic Model. *Energy Convers. Manag.* **1999**, *40*, 335–348. [CrossRef]
25. Asa'd, O.; Ugursal, V.I.; Ben-Abdallah, N. Investigation of the Energetic Performance of an Attached Solar Greenhouse through Monitoring and Simulation. *Energy Sustain. Dev.* **2019**, *53*, 15–29. [CrossRef]
26. Rabiu, A.; Na, W.; Denen, T.; Rasheed, A.; Aderemi, M. ScienceDirect Determination of Overall Heat Transfer Coefficient for Greenhouse Energy-Saving Screen Using Trnsys and Hotbox. *Biosyst. Eng.* **2022**, *217*, 83–101. [CrossRef]
27. Castellucci, S.; Carlini, M. Modelling and Simulation for Energy Production Parametric Dependence in Greenhouses. *Math. Probl. Eng.* **2010**, *2010*, 590943. [CrossRef]
28. Choab, N.; Allouhi, A.; El Maakoul, A.; Kousksou, T.; Saadeddine, S.; Jamil, A. Effect of Greenhouse Design Parameters on the Heating and Cooling Requirement of Greenhouses in Moroccan Climatic Conditions. *IEEE Access* **2020**, *9*, 2986–3003. [CrossRef]
29. Rasheed, A.; Kim, H.T.; Lee, H.W. Modeling-Based Energy Performance Assessment and Validation of Air-To-Water Heat Pump System Integrated with Multi-Span Greenhouse on Cooling Mode. *Agronomy* **2022**, *12*, 1374. [CrossRef]
30. Yeoju Climate, Weather by Month, Average Temperature (South Korea)—Weather Spark. Available online: https://weatherspark.com/y/142307/Average-Weather-in-Yeoju-South-Korea-Year-Round (accessed on 4 March 2022).
31. Weather Data Opening Portal. Available online: https://data.kma.go.kr/data/grnd/selectAsosRltmList.do?pgmNo=36 (accessed on 18 June 2022).
32. Jung, S.-H.; Lee, J.-W.; Lee, S.-Y.; Lee, H.-W. Analysis of Wind Velocity Profile for Calculation of Wind Pressure on Greenhouse. *Prot. Hortic. Plant Fact.* **2015**, *24*, 135–146. [CrossRef]
33. Valera, M.D.; Molina, A.F.; Alvarez, M. *Protocolo de Auditoría Energética En Invernaderos Auditoría Energética de Un Invernadero Para Cultivo de Flor Cortada*; Instituto para la diversificacion y ahorro de la Energia: Madrid, Spain, 2008; ISBN 9788496680265.
34. Rafiq, A.; Na, W.H.; Rasheed, A.; Lee, J.W.; Kim, H.T.; Lee, H.W. Measurement of Longwave Radiative Properties of Energy-Saving Greenhouse Screens. *J. Agric. Eng.* **2021**, *52*. [CrossRef]
35. Nijskens, J.; Deltour, J.; Coutisse, S.; Nisen, A. Radiometric and thermal properties of the new plastic films for greenhouse covering. *Acta Hortic.* **1989**, *77*, 7–38. [CrossRef]
36. Rasheed, A.; Lee, J.W.; Lee, H.W. Development and Optimization of a Building Energy Simulation Model to Study the Effect of Greenhouse Design Parameters. *Energies* **2018**, *11*, 2001. [CrossRef]
37. Solar Energy Laboratory. *TRNSYS 18 Manual Documentation*; University of Wisconsin: Madison, WI, USA, 2018; Volume 5, Available online: http://www.trnsys.com (accessed on 4 March 2022).
38. Transsolar. TRNSYS 18 Manual Documentation (TRNFLOW Manual). 2009. Available online: http://www.transsolar.com (accessed on 4 March 2022).
39. Lim, A. A Comparative Study between TRNSYS and RC Thermal Models to Simulate a District Thermal Demand. Master's Thesis, Eindhoven University of Technology, Eindhoven, The Netherlands, 2020. Available online: https://research.tue.nl/en/studentTheses/a-comparative-study-between-trnsys-and-rc-thermal-models-to-simul (accessed on 4 March 2022)
40. Seem, J.E. Modeling of Heat in Buildings. Ph.D. Thesis, Solar Energy Laboratory, University of Wisconsin Madison, Madison, WI, USA, 2011.
41. Lam, J.C.; Hui, S.C.M. Sensitivity Analysis of Energy Performance of Office Buildings. *Build. Environ.* **1996**, *31*, 27–39. [CrossRef]
42. Choab, N.; Allouhi, A.; El Maakoul, A.; Kousksou, T.; Saadeddine, S.; Jamil, A. Review on Greenhouse Microclimate and Application: Design Parameters, Thermal Modeling and Simulation, Climate Controlling Technologies. *Sol. Energy* **2019**, *191*, 109–137. [CrossRef]

43. Jeon, J.G.; Lee, D.G.; Paek, Y.; Kim, H.G. Study on Heating Performance of Hybrid Heat Pump System Using Geothermal Source and Solar Heat for Protected Horticulture. *J. Korean Sol. Energy Soc.* **2015**, *35*, 49–56. [CrossRef]
44. Rasheed, A.; Na, W.H.; Lee, J.W.; Kim, H.T.; Lee, H.W. Development and Validation of Air-to-Water Heat Pump Model for Greenhouse Heating. *Energies* **2021**, *14*, 4714. [CrossRef]
45. Adjustment, A.H. Long-Term Trend of Surface Wind Speed in Korea: Anemometer Height Adjustment. *Atmosphere* **2021**, *31*, 101–112.
46. Kim, M.H.; Kim, D.; Heo, J.; Lee, D.W. Techno-Economic Analysis of Hybrid Renewable Energy System with Solar District Heating for Net Zero Energy Community. *Energy* **2019**, *187*, 115916. [CrossRef]
47. Cooper, P.I.; Fuller, R.J. A Transient Model of the Interaction between Crop, Environment and Greenhouse Structure for Predicting Crop Yield and Energy Consumption. *J. Agric. Eng. Res.* **1983**, *28*, 401–417. [CrossRef]

sustainability

MDPI

Article

Influence of the Height in a Colombian Multi-Tunnel Greenhouse on Natural Ventilation and Thermal Behavior: Modeling Approach

Edwin Villagrán [1], Jorge Flores-Velazquez [2,*], Mohammad Akrami [3,*] and Carlos Bojacá [1]

1 Department of Biological and Environmental Sciences, Faculty of Natural Sciences and Engineering, Jorge Tadeo Lozano University, Bogotá 111321, Colombia; edwina.villagranm@utadeo.edu.co (E.V.); carlos.bojaca@utadeo.edu.co (C.B.)
2 Coordination of Hydrosciences, Postgraduate Collage, Carr Mex Tex km 36.5, Montecillo Edo de Mexico 62550, Mexico
3 Department of Engineering, University of Exeter, Exeter EX4 4QF, UK
* Correspondence: jorgelv@colpos.mx (J.F.-V.); m.akrami@exeter.ac.uk (M.A.)

Abstract: The dimensions of a passive greenhouse are one of the decisions made by producers or builders based on characteristics of the available land and the economic cost of building the structure per unit of covered area. In few cases, the design criteria are reviewed and the dimensions are established based on the type of crop and local climate conditions. One of the dimensions that is generally exposed to greater manipulation is the height above the gutter and the general height of the structure, since a greenhouse with a lower height has a lower economic cost. This has led some countries in the tropical region to build greenhouses that, due to their architectural characteristics, have inadequate microclimatic conditions for agricultural production. The objective of this study was to analyze the effect on air flows and thermal distribution generated by the increase of the height over gutter of a Colombian multi-tunnel greenhouse using a successfully two-dimensional computational fluid dynamics (CFD) model. The simulated numerical results showed that increasing the height of the greenhouse allows obtaining temperature reductions from 0.1 to 11.7 °C depending on the ventilation configuration used and the external wind speed. Likewise, it was identified that the combined side and roof ventilation configuration (RS) allows obtaining higher renovation indexes (RI) in values between 144 and 449% with respect to the side ventilation (S) and roof ventilation (R) configurations. Finally, the numerical results were successfully fitted within the surface regression models responses.

Keywords: microclimate; response surface; renovation index; CFD simulation; airflow

Citation: Villagrán, E.; Flores-Velazquez, J.; Akrami, M.; Bojacá, C. Influence of the Height in a Colombian Multi-Tunnel Greenhouse on Natural Ventilation and Thermal Behavior: Modeling Approach. *Sustainability* **2021**, *13*, 13631. https://doi.org/10.3390/su132413631

Academic Editors: Muhammad Sultan, Yuguang Zhou, Walter Den and Uzair Sajjad

Received: 25 October 2021
Accepted: 6 December 2021
Published: 9 December 2021

1. Introduction

In Colombia, plasticulture and protected agriculture have promoted a method of agricultural production that generates higher crop yields per unit of available land area, compared to open field production [1]. The use of passive greenhouses, where the microclimate is managed by means of natural ventilation, has become more widespread. This type of greenhouse is also quite common in other regions of the world [2]. This cultivation method allows, among others, the intensification of horticultural or ornamental production [3], better management of water resources and fertilizer applications [4,5], partial or total control of the micro-climatic variables that affect crop growth and development [6]. In recent years, this production technology has also become a crop protection tool used by producers in the face of increasingly frequent and severe weather events due to climate change [3,7,8].

The use of greenhouses or protected agriculture structures with plastic roofing worldwide has been estimated at more than 3.5 million hectares, which have been established mainly in countries such as China, Korea, Spain, Mexico, France and Italy [9]. In the specific case of Colombia, statistics show that there are approximately 11,000 hectares destined for the ornamental and horticultural sectors. The main production areas are located mostly in regions with altitudes above 1500 m above sea level, where cold and humid mountain climate conditions predominate [10–12]. In Colombia, as in the other countries mentioned above, different types of greenhouses have been built and a high percentage of them are light structures with a low technological level [13,14].

These typologies of geometric designs of plastic roofing greenhouses already conceived, originated from some initial concerns of the producers such as land area, easily available structural and roofing materials, versatile structure designs for its construction, but mainly the final economic cost of the greenhouse or structure used. [15,16]. However, an adequate design process should include an analysis of the local climatic conditions so that, based on these conditions, the builders or decision makers can propose the appropriate architecture and geometry of the greenhouse that, together with the selected roofing material, will generate adequate microclimate conditions for the growth and development of the crops [17,18].

In terms of climatic information, there are already local or regional information tools or systems where it is possible to obtain data series of the main meteorological variables of a specific site. In the last three decades, significant progress has been made in improving the optical and thermal properties of polyethylene roofing material, seeking greater benefits for the development of the main horticultural and ornamental crops [19,20]. Therefore, in each country the producer has the possibility of selecting a type of plastic roofing according to his needs. This facilitates the selection of an optimal roofing material or the most suitable for the local climate conditions, so that inside the structures the behavior of temperature, relative humidity and vapor pressure deficit can be managed within the optimal range for agricultural production [21].

On the other hand, there is the architectural aspect and the geometrical dimensions of the greenhouse to be built. Although there are some design parameters established for naturally ventilated greenhouses, such as the width and length [22]. The information available on other parameters such as height is variable and many of the related studies recommend increasing the height of structures used in tropical and subtropical climate conditions [10,23]. However, the height level of the structure is not defined to a specific limit but varies between types of greenhouses and types of climatic conditions [17,24]. It would be ideal, therefore, to have at the availability of the greenhouse structure designers, a software or a design methodology that allows them to evaluate this characteristic and to be able to define the adequate dimensions. This design tool should be able to be implemented in different types of greenhouses and under different operation scenarios.

One of the methodologies that can offer quick solutions is computational fluid dynamics (CFD) simulation. This simulation tool has been implemented to analyze aspects related to the characteristics of multiple types of greenhouses [25–27]. In addition, when a CFD model is successfully validated experimentally, it is possible to analyze unbuilt scenarios, which promotes the efficient use of resources and avoids the construction of unsuitable greenhouse structures or undesirable microclimate conditions [28,29].

Regarding the use of CFD studies applied to the optimization of passive greenhouse structures, it should be mentioned that, for example, in a two-dimensional study for a Chinese solar greenhouse, it was determined that, as longer greenhouse sections were generated, higher temperatures were produced inside the structure [30]. Similarly, Villagrán et al. [10] determined that by increasing the roof ventilation surface with respect to the covered floor area (S_{VC}/S_C) from 2.5 to 20.1% in a traditional Colombian greenhouse, it was possible to obtain ventilation rates higher than 0.04 $m^3\ s^{-1}\ m^{-2}$ for the prevailing conditions of the local climate.

Regarding the greenhouse width, it is known that in Almeria type structures [31], which include the arc type [32] and gothic tunnel type [33], as the more spans that are joined laterally, the greater the width of the greenhouse, the ventilation rate is reduced exponentially, which in turn generates higher temperature values inside the greenhouse [31–33]. For the height dimension, Boulard and Fatnassi [34] using a CFD-2D study on an 8-span arch-type greenhouse found that when the height above the greenhouse gutter increases from 3 to 5 m, the reduction of the thermal gradient between the interior and the exterior is reduced from 3 5 to 2.0 °C. This allowed the authors to conclude that the greenhouse height positively affects the natural ventilation phenomenon and allows to generate better thermal conditions inside the greenhouse [34].

Recently Fatnassi et al. [35] conducted a study in a tunnel type greenhouse with butterfly roof vents, where the thermal behavior of the greenhouse was evaluated numerically under four different gutter heights, 4, 5, 6 and 7 m, respectively. It was found that when the height is increased from 4 to 5 m, the temperature inside the greenhouse decreases by approximately 2 °C, while when it is increased from 6 to 7 m the temperature value only decreases by 0.2 °C. According to these results, the authors concluded that the effect of increasing greenhouse height on air temperature is clearly asymptotic, therefore, the increase in height should be analyzed as a climatic benefit versus economic cost ratio.

In this study, an experimentally validated CFD-2D model has been implemented under the climatic conditions of a region of the savanna of Bogota, Colombia. The CFD model was used to evaluate the effects on the air flow velocity and the interior temperature of a gothic-type greenhouse as the height level increases from 2.5 m to 5.0 m, under three ventilation configurations, side ventilation (R), roof vents (S) and combined ventilation (RS). Finally, the numerical data obtained in each of the developed simulation scenarios were grouped by ventilation configuration and fitted to response surface regression models. This was done to generate a more integrated and simpler method of analysis of the response variables analyzed and their relationship with the microclimate generated

2. Materials and Methods

2.1. Description of Prototype Proposed for Analysis

The analysis proposed in this research was carried out on a gothic multi-tunnel greenhouse with a 200 μm thick polyethylene cover located in the Savannah of Bogota, Colombia. The greenhouse was composed of a total of six spans, each span had a width length of 9.33 m, for a total greenhouse width length of 55.98 m (Figure 1). The greenhouse had a gutter height of 4.0 m and a maximum roof height of 8.3 m. It was equipped with ventilation areas on the side walls, with an effective opening of 3.7 m, which equals a lateral ventilation surface with respect to the covered floor area of 13.21%. Likewise, the ventilation surface was complemented with a roof ventilation area of 1.5 m of total opening in each of the spans, therefore, the roof ventilation surface with respect to the covered surface is 16.1%.

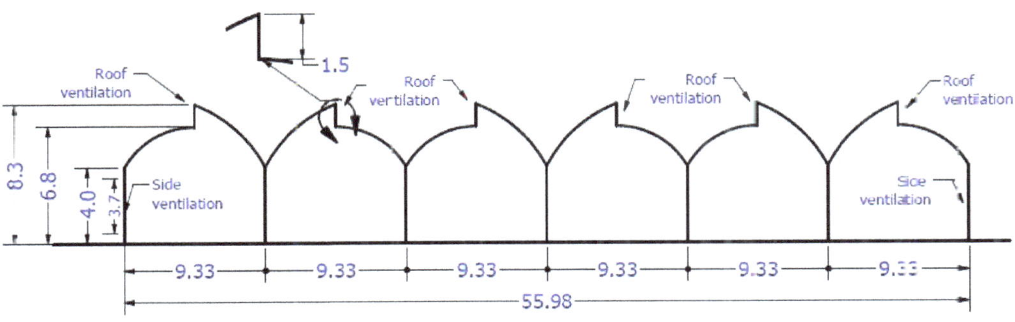

Figure 1. Geometric diagram of the cross section of the greenhouse studied.

2.2. Computational Fluid Dynamics Modeling

The numerical calculation of the physical processes involving fluid movement and related variables such as pressure, force, density, temperature and their associated changes in heat and mass transfer phenomena can be solved using computational fluid dynamics [36]. Currently, CFD is one of the most widely used analysis, design or redesign methodologies in different sectors of the economy, therefore, it is possible to find CFD studies approached from mechanical, environmental, aeronautical and automotive engineering [37]. On the other hand, its use in the agricultural sector has not been insignificant and in the last two decades it has been implemented as a tool for process analysis involving natural ventilation processes or energy optimization of greenhouse structures [22,38].

CFD analysis is a resolution technique that solves a set of nonlinear partial equations from algebraic discretization by means of numerical simulation using the finite volume method. A CFD simulation process is composed of three phases of development, preprocessing, processing and post-processing [39]. In the preprocessing phase, the computational domain, and the geometry of the structure on which the natural ventilation process will be analyzed are designed. In this phase the numerical meshing process applied to the whole computational domain including the greenhouse geometry is also generated, these processes must be performed based on criteria of good CFD simulation practices [22,40].

In the processing phase, the numerical solution of the evaluated problem is performed, when the air and energy flow in the model is calculated by solving the governing equations based on the physical laws of conservation of energy, mass and momentum. In this phase, the sub models considered as source terms of the governing equations are selected, such as turbulence, buoyancy or free convection, solar radiation, porous media and phenomena associated with mass transfer [41]. Finally, in the post-processing phase, it is possible to develop a qualitative and quantitative analysis of the numerical results obtained for each scenario evaluated [42].

It should be noted that the CFD-2D model used in this research has been previously successfully validated experimentally, verifying its high predictive capacity for air flow patterns and thermal distribution inside the greenhouse analyzed. The validation results can be verified in the works developed by Villagrán and Bojacá [16,43] and by Villagrán et al. [1]. Therefore, for this research work, details related to the validation process will not be discussed, although relevant aspects of each of the CFD simulation phases will be mentioned.

2.2.1. Pre-Processing

The commercial software ANSYS ICEM CFD (v. 18.0, ANSYS Inc., Canonsburg, PA, USA, EE. UU.) was used, by means of which a large computational domain was created in a two-dimensional configuration. It included the geometry of the cross section of the analyzed greenhouse, as well as the process of meshing the computational domain was performed by means of this software. The two-dimensional CFD studies of natural ventilation are useful and have a capacity to predict the ventilation rate and the thermal distribution, inside a structure when the dominant external wind currents blow perpendicular to the ventilation areas [27,44].

It was determined that the dimensions of the computational domain, establishing as the reference for the maximum height of the structure (H), should have dimensions from the region of the airflow inlet to the windward sidewall of the structure of 15H. From the leeward sidewall to the airflow outlet boundary of 20H and a minimum height from ground level of 10H (Figure 2). These dimensions are established following the guidelines given in numerical studies of natural ventilation of greenhouses and are the dimensions that allow to adequately describe the physical phenomena associated with the analyzed problem within the atmospheric boundary layer [1,45].

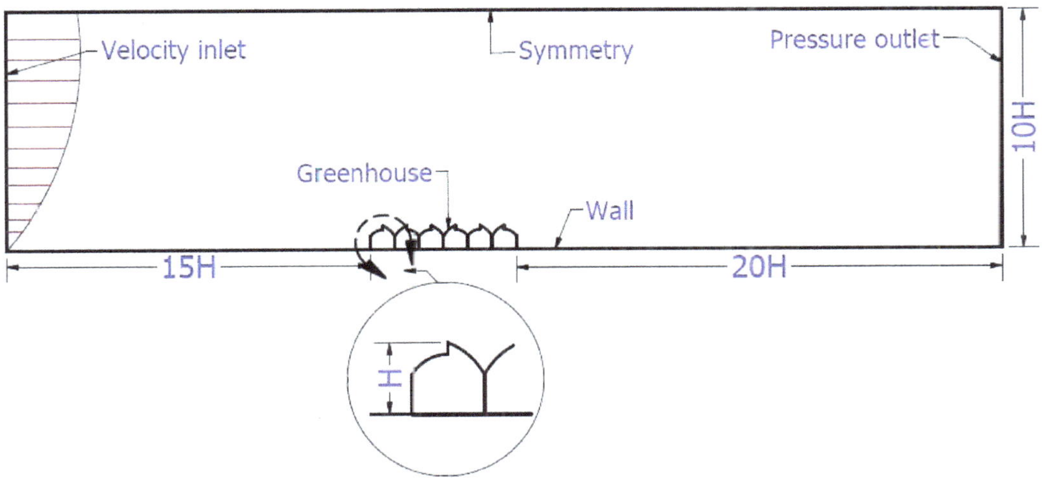

Figure 2. Overall dimensions of the computational domain.

In the meshing process, the computational domain volume was discretized into an unstructured grid of square or rectangular elements (Figure 3). The size of the numerical grid was defined by a mesh independence test as described in the work done by Villagran et al. [1]. at the end of the analysis process, the selected numerical grid was composed of a total of 1,104,369 elements. The quality parameters evaluated were cell size and cell-to-cell size variation; it was found that 95.2% of the cells of the mesh were within the high-quality interval (0.95–1). The orthogonality criterion was also evaluated, where the minimum value obtained was 0.92, a value considered as high quality [10].

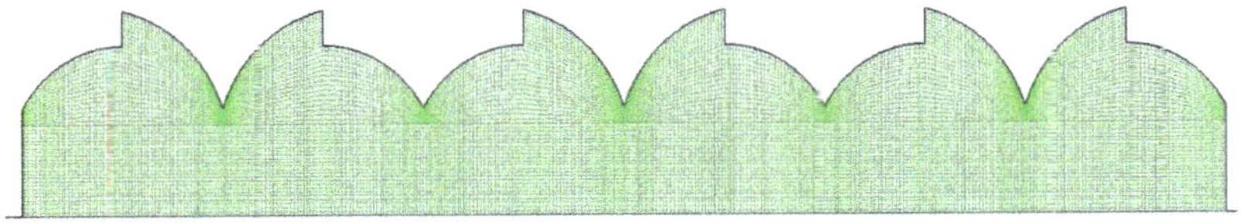

Figure 3. Detail of the cross-sectional grid of the greenhouse evaluated.

Once the geometry has been constructed, it is possible to define the boundary conditions to be established in the limits of the computational domain and in the geometry of the greenhouse (Figure 1). In this specific case the right side was defined as the air flow inlet boundary, for which a condition determined by a logarithmic wind speed profile was established, this logarithmic profile depends on the climatic characteristics and the type of soil in the study region, factors that have been previously determined in the work developed by Villagrán et al. [1].

The left boundary on the other hand was determined as a pressure and airflow outflow boundary, while the upper boundary of the computational domain was set as a wall boundary condition with an imposed solar radiation flux. The wall-type boundary condition was also established for the lower boundary of the computational domain, the floor, the walls and the roof of the greenhouse, while the ventilation areas were set to the indoor boundary condition depending on the ventilation configuration analyzed. The physical and optical properties of the materials within the computational domain were also defined with the values summarized in Table 1.

Table 1. Physical and optical properties used in the CFD simulation.

Property	Soil	Air	Polyethylene
Density (ρ, kg m^{-3})	1350	1.225	923
Thermal conductivity (k, W m^{-1} K^{-1})	1.3	0.0242	0.4
Specific heat (Cp, J K^{-1} kg^{-1})	800	1006.43	2300
Absorptivity	0.90	0.19	0.06
Scattering coefficient	−15	0	0
Refractive index	1.92	1	1.53
Emissivity	0.95	0.9	0.7

Subsequently, six greenhouse models (M1–M6) were created with the only variations being the height under the gutter (H1) and the lateral ventilation dimension (SV), as described in Table 2. The meshing process for models M1 to M6 was performed on the same geometric model already discretized from model M4, simply decreasing the greenhouse height for models M1 to M3 and increasing it for models M5 and M6.

Table 2. Geometric details of the analyzed greenhouse models.

Model	H1 (m)	SV (m)	Scheme
Model 1 (M1)	2.5	1.2	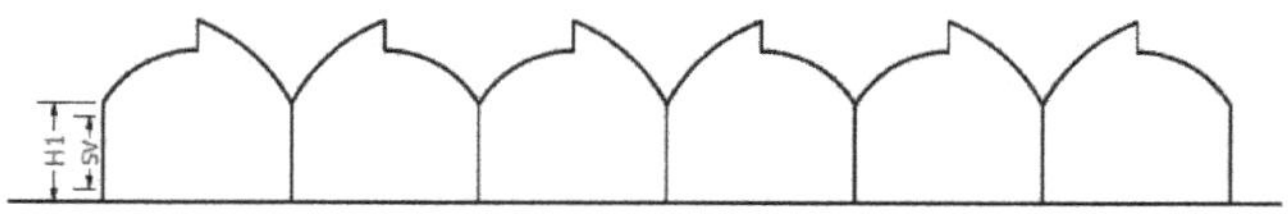
Model 2 (M1)	3.0	1.7	
Model 3 (M3)	3.5	2.2	
Model 4 (M4)	4.0	2.7	
Model 5 (M5)	4.5	3.2	
Model 6 (M6)	5.0	3.7	

Finally, the simulation scenarios to be analyzed were defined, these were built from the ventilation configurations used locally, which are: ventilation by the lateral sides (S), by the roof vents (R) and combined ventilation (RS). After this, the greenhouse model to be evaluated was defined from M1 to M6 and finally the climatic conditions to be used as starting conditions for each simulation were determined. These conditions were known from previous studies developed in the same study region and where it was established that the maximum daytime temperature averaged 22.3 °C, the maximum solar radiation level was 853 Wm-2 and wind conditions can vary between values of 0.2 and 1.6 ms^{-1}. Therefore, it was determined to perform the evaluations under four wind speeds S1 (0.2 ms^{-1}), S2 (0.5 ms^{-1}), S3 (1 ms^{-1}) and S4 (1.6 ms^{-1}). According to the above, 72 simulations were performed and coded as shown in Table 3.

Table 3. Coding of the 72 simulated scenarios.

Scenarios					
S-M1S1	S-M4S1	R-M1S1	R-M4S1	RS-M1S1	RS-M4S1
S-M1S2	S-M4S2	R-M1S2	R-M4S2	RS-M1S2	RS-M4S2
S-M1S3	S-M4S3	R-M1S3	R-M4S3	RS-M1S3	RS-M4S3
S-M1S4	S-M4S4	R-M1S4	R-M4S4	RS-M1S4	RS-M4S4
S-M2S1	S-M5S1	R-M2S1	R-M5S1	RS-M2S1	RS-M5S1
S-M2S2	S-M5S2	R-M2S2	R-M5S2	RS-M2S2	RS-M5S2
S-M2S3	S-M5S3	R-M2S3	R-M5S3	RS-M2S3	RS-M5S3
S-M2S4	S-M5S4	R-M2S4	R-M5S4	RS-M2S4	RS-M5S4
S-M3S1	S-M6S1	R-M3S1	R-M6S1	RS-M3S1	RS-M6S1
S-M3S2	S-M6S2	R-M3S2	R-M6S2	RS-M3S2	RS-M6S2
S-M3S3	S-M6S3	R-M3S3	R-M6S3	RS-M3S3	RS-M6S3
S-M3S4	S-M6S4	R-M3S4	R-M6S4	RS-M3S4	RS-M6S4

2.2.2. Processing

For the numerical solution of the Reynolds-averaged Navier-Stokes (RANS) equations, the ANSYS FLUENT processing software was used. The air flow in the greenhouse was considered to be turbulent and with a density change associated with temperature changes that can be modeled by the Boussinesq approximation. The method of resolution selected for the CFD model was semi-implicit for the pressure and velocity bound equations using the SIMPLE algorithm. Finally, residual convergence criteria were established in 10^{-6} for the energy equation and in 10^{-3} for the other variables such as continuity, turbulence and momentum [41].

The influence of solar radiation intensity on the spatial behavior of temperature was considered using the discrete order (DO) radiation model, which allows modeling the radiation and calculating the convective exchanges that occur in the computational domain, [46].

It was also decided not to include any type of crop, because the purpose of the study is not to analyze the heat and mass transfer flows that occur between the indoor environment and any species of plants. On the contrary, the main objective is to make an analysis of the ventilation phenomenon under the most critical condition that can occur in a real scenario and that is in empty conditions and under the most extreme climatic conditions [37,47]. This is an approach that is still valid and continues to be implemented in a significant number of studies carried out in various countries [48–51]. In addition, as different types of horticultural, ornamental, aromatic and medicinal crops are grown in the region, the analysis developed in this study can be applied to any type of crop.

2.2.3. Post-Processing

For this phase, where the main objective is to perform the qualitative and quantitative analysis of the evaluated scenarios, ANSYS CFD-Post software was used. Therefore, for each of the 72 simulations developed, two-dimensional plots of temperature and airflow patterns were generated and numerical values of air velocity and temperature in the greenhouse cross section at a height of 1.4 m above ground level were extracted. These data sets were analyzed for each simulation scenario and as a whole using response surface regression models.

2.2.4. Response Surface Modeling

Response surface regression models were fitted to each of the three ventilation scenarios under consideration (roof (R), side (S), and roof—side (RS)). For each scenario, models were fitted to evaluate the response of varying levels of exterior wind speed and greenhouse height on the internal temperature and air velocity. The objectives of applying response surface models on top of the CFD simulated scenarios was threefold: (1) to establish an approximate relationship between the response variables and the exterior wind speed and greenhouse height that can be used for predicting temperature or internal wind speed values for a given combination of the predictors; (2) to determine the statistical significance of the explanatory variables through hypothesis testing; and (3) to determine the optimum combination of the explanatory variables that result in the maximum response over the region under consideration.

Response surface models are an extension of linear models and works similarly; however, extra arguments should be added to consider the response-surface portion of the model [52]. In the present study, second order models including first-order response, two-way interactions and quadratic terms for the explanatory variables were included in the model's formulation. Once calibrated, the models were tested for their goodness of fit through measures such as the coefficient of determination, lack of fit and pure error [53,54].

The output of the models was plotted as perspective plots to depict the combined effect of the experimental factors (external wind speed and greenhouse height) over the dependent variables (internal temperature and air velocity). The models were fitted using the rsm package (v. 2.10.2; [52]) included in the statistical software R (v. 4.0.4; [55]).

3. Results and Discussion

3.1. Airflow Patterns

3.1.1. Side Ventilation (S)

The airflow patterns under the ventilation configuration through the sidewall (S) areas of the greenhouse can be seen in Figure 4. In general terms, it is observed that the airflow pattern enters the greenhouse through the ventilation area on the windward side and moves horizontally in an airflow pattern that has a height very similar to the height under the gutter of each of the greenhouses evaluated (M1-M6), this horizontal flow leaves the greenhouse again through the ventilation area arranged on the leeward wall.

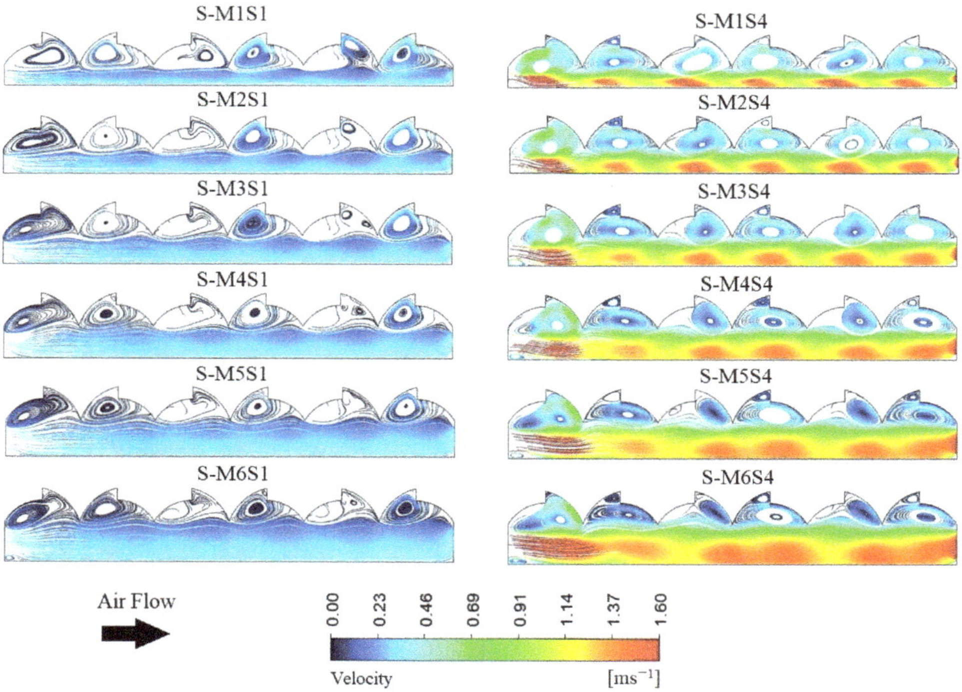

Figure 4. Air distribution patterns simulated with ventilation configuration S and for the six greenhouse models (M1 to M6) under wind speeds S1 (0.2 ms^{-1}) and S4 (1.6 ms^{-1}).

Another common feature is the air circulation loops that are generated in the upper part of each of the greenhouse spans, this occurs because due to the buoyancy phenomenon and the associated pressure components, the warm air inside the greenhouse moves vertically towards the roof area and its space in the lower part of each span is occupied by the fresh air that enters from the outside environment [3,56]. It should also be noted that qualitatively it is observed that as the height of the greenhouse increases, the air flow that moves over the nearby area where the crops are grown has a higher velocity; this same behavior is observed in the air flow that leaves the greenhouse through the leeward ventilation.

To quantitatively analyze the airflows, velocity data were extracted from a cross profile at a height of 1.4 m above ground level. In general, the airflows show a behavior for velocity that oscillates between a minimum of 0.25 ± 0.05 ms^{-1} for S-M1S1 and a maximum of 1.42 ± 0.16 ms^{-1} for S-M6S4. It was also observed that the air flows inside the greenhouse show a direct relationship with the wind speed outside, therefore, the higher the wind speed, the higher the air velocity inside the greenhouse. This occurs in greenhouses where this type of ventilation is analyzed and where porous insect-proof screens are not contemplated in the ventilation areas [29,57].

On the other hand, it is also observed that as the height of the greenhouse increases, the average air flow velocity inside the greenhouse increases for the wind speed scenarios analyzed (S1 to S4). In summary, the increase in velocity between the extreme simulation scenarios was 42.3, 20.9, 19.4 and 16.2% for M1S1, M1S2, M1S3, M1S4 compared to M6S1, M6S2, M6S3, M6S4, respectively. These velocity increases in the higher greenhouse models occur because the effect of air friction is reduced as the ratio (L/H1) between the greenhouse length (L) and the height above the gutter (H1) decreases, which coincides with what was previously reported in the study developed by Chu et al. [44].

Another factor that showed a direct relationship with height was the uniformity of the air flow velocity behavior along the cross axis of the greenhouse (Figure 5). In general terms, it is observed that as the height of the structure increases, the velocity behavior curves show less oscillation between the points of high and low air velocity. This can also be verified with the standard deviation (sd) values. In S-M1S4 the sd value was ± 0.27 ms^{-1} while for S-M6S4 it was $\pm$ 0.16 ms^{-1}. This will undoubtedly promote a more homogeneous microclimate behavior inside the greenhouse, helping to limit the negative impacts on growth and development generated by non-uniform microclimates [58].

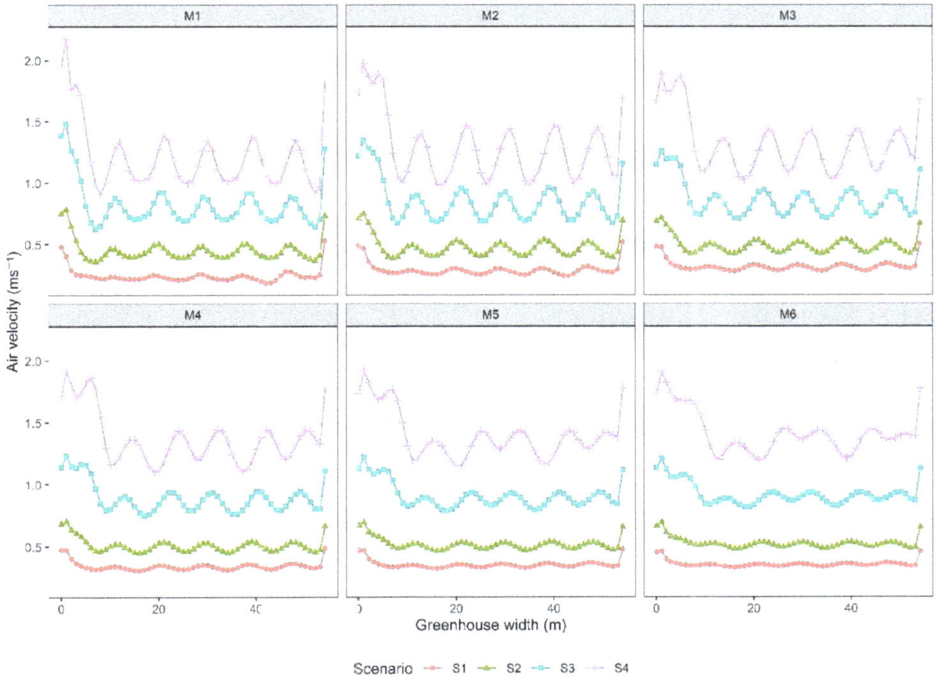

Figure 5. Spatial distribution of air velocity inside the greenhouse for the simulated scenarios under the S configuration.

3.1.2. Roof Ventilation (R)

Figure 6 shows the simulated airflow patterns for this ventilation configuration. In this case, since the side windows are closed, the airflow patterns are forced in and out through the windows in the roof region [16]. For the low velocity scenarios S1 where the thermal component of natural ventilation dominates, a behavior where three flow patterns are generated is observed. In span 1, an air flow was identified that enters through the ventilation area and becomes a convective movement loop between the floor and the roof of this span.

Subsequently, between span 2 and the middle of span 3 there is an airflow pattern that moves in the direction of the outside wind at ground level and a flow that moves in the opposite direction to the outside wind at the top of the roof of these spans. Finally, through the ventilation area of span 3 an airflow enters and mixes with another airflow entering from the ventilation area of span 4, flows that move towards the leeward wall and exits through the ventilation areas of spans 4 and 5. It should also be noted that the confluence of flows over the central area of span 3 generates a small low velocity loop near the floor region, being this an inadequate movement pattern that may cause the generation of a heat spot over this region [29,33,35].

For the case of the S4 scenario wind speed, the natural ventilation of the greenhouse will depend on the thermal and wind components together [59,60]. In this case, we were able to identify an air flow pattern that enters through span 1 and forms a recirculation pattern between the roof of the span, the floor and the wall of the windward side, where the highest air flow velocity occurs at ground level, which coincides with that reported by Kwon et al. [61]. Likewise, part of the air flow entering through the window of span 1 is mixed with another air flow entering through the window of span 2, which generates an acceleration of the air flow above ground level; the same behavior is repeated with the air flows entering through spans 3 and 4; finally, the air flows leave the greenhouse through spans 4 and 5.

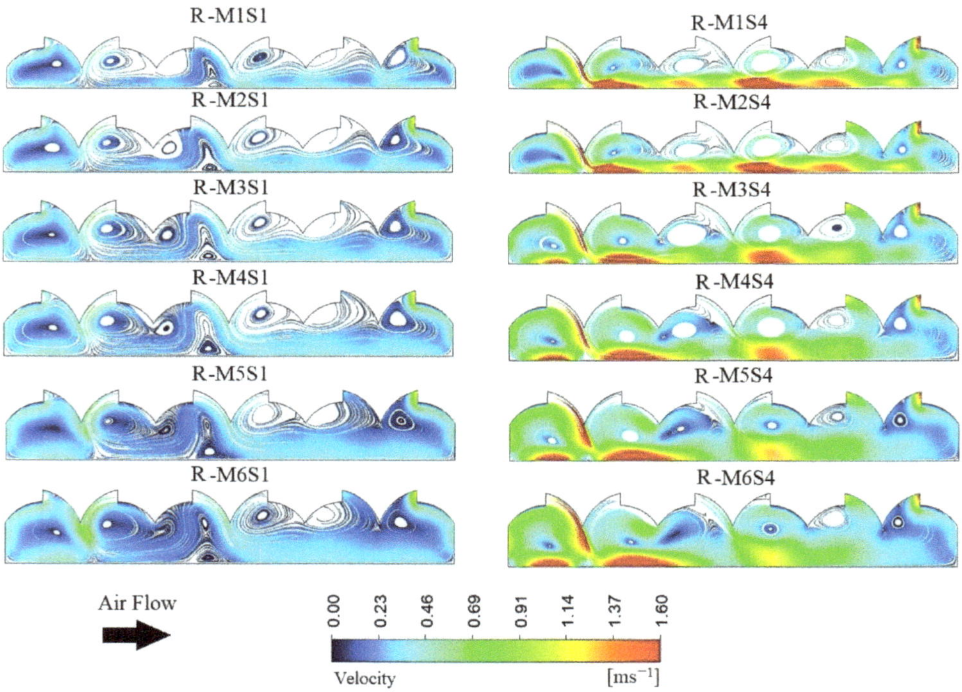

Figure 6. Simulated air distribution patterns with ventilation configuration R and for the six greenhouse models (M1 to M6) under wind speeds S1 (0.2 ms^{-1}) and S4 (1.6 ms^{-1}).

In numerical terms, the air flow velocity inside the greenhouse over the area where the crops are grown ranged from a minimum of 0.26 $\pm$ 0.07 ms^{-1} for R-M1S1 to a maximum of 0.92 $\pm$ 0.36 ms^{-1} in R-M1S4. In this specific case no increase in air flow velocity is observed as the greenhouse height increases in the case of the low velocity scenario S1 it is observed that between M1 and M6 there is only an increase of 11.5% but this same value is obtained between M1 compared to M2, M3 and M4.

For the case of S2 the air velocity increased between M1 and M2 by 2.5% but compared to the other greenhouse models it decreased by up to 12.5% with M6 specifically. In the S3 scenarios the air flow velocity decreased up to 20.5% compared M1 with M5 and M6 and finally in S4 the air velocity also decreased up to 15% compared M1 with M6.

Regarding the behavior of the air flow in the cross axis of the greenhouse, it was found that for models M1, M2 and M3 there is a very similar distributed behavior in space, while for M4, M5 and M6 there is a small change in this spatial distribution, which may be caused by the change in greenhouse heights (Figure 7). Likewise, the inside air velocities increase as the wind speed increases, which has already been demonstrated in several research works [22,33,49,62].

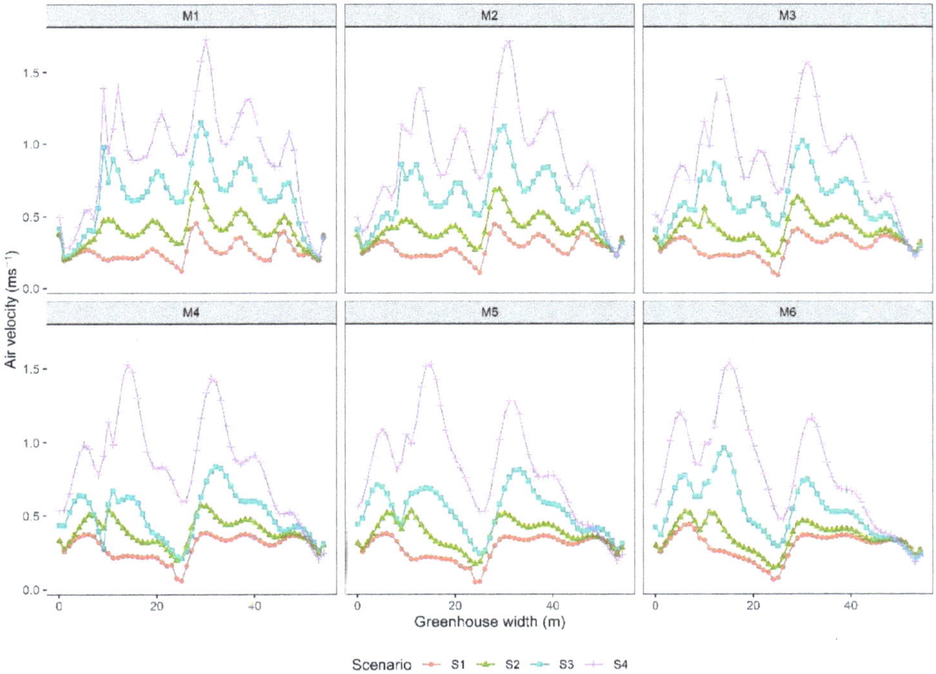

Figure 7. Spatial distribution of air velocity inside the greenhouse for the simulated scenarios under the R configuration.

On the other hand, there is a quite heterogeneous behavior among each of the six spans, finding that the smallest standard deviation among these scenarios was ± 0.07 ms^{-1} in the low velocity scenarios S1, while for the high velocity scenarios S4 this standard deviation was up to $\pm$ 0.36 ms^{-1}. The lowest air speed points were the regions near the leeward and windward wall and the highest air speed points were the areas below spans 2 and 4.

3.1.3. Side and Roof Ventilation (RS)

The spatial distribution of the airflow patterns for the combined ventilation configuration can be seen in Figure 8. Qualitatively, more continuous and intense airflows can be observed over the entire cross-sectional area analyzed, both in the crop and canopy regions, which should positively impact the greenhouse renovation rates and directly the thermal performance of the greenhouse [62].

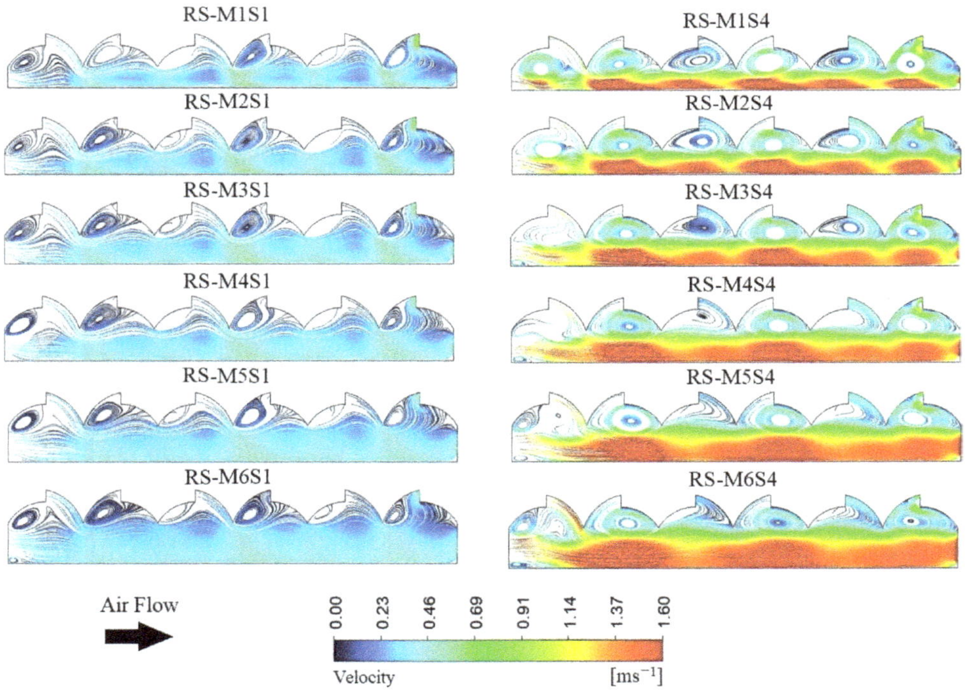

Figure 8. Simulated air distribution patterns with the RS ventilation configuration and for the six greenhouse models (M1 to M6) under wind speeds S1 (0.2 ms^{-1}) and S4 (1.6 ms^{-1}).

Numerically, the air flows presented a mean velocity that ranged from a minimum value of 0.31 ± 0.09 ms^{-1} in RS-M1S1 to a maximum of 1.53 ± 0.17 in RS-M6S4. For the case of S1, it is observed that the airflow velocity increases up to 29.1% in M5 and M6 compared to M1. Likewise, in S4 the increase in velocity was 15% in the M6 greenhouse with respect to the value obtained in M1, therefore, the greenhouse height has a significant impact on the airflow velocity. This is mainly due to the fact that, according to the analysis of natural ventilation carried out in previous studies, the greater the distance between the central axis of the side ventilation with respect to the central axis of the roof ventilation, the more dynamic the wind and thermal component is and therefore the better the air flow patterns are obtained [56,63].

On the other hand, the spatial behavior of the air flow velocity inside the greenhouse cross section can be found for some of the simulated scenarios in Figure 9. In general terms, it can be observed that as the height of the greenhouse increases, the less oscillation between the points of higher and lower velocity exists in the spatial distribution curves. At the same time, it should be noted again that these changes in air flow velocity are mainly due to the dynamics of incoming and outgoing air flows and to the interaction of warm and cold air masses that mix and generate buoyancy flows [10,64]. Although these values are less critical in the M3 to M5 greenhouse scenarios under the RC combined ventilation configuration.

Regarding air flow velocities, in general, under the three configurations evaluated, the values obtained are within the ranges reported in passive greenhouse ventilation studies [5,65]. Although for the low velocity condition S1, 100% of the airflows present velocities lower than the minimum recommended value of 0.5 ms^{-1} for plant growth inside greenhouses [42,65].

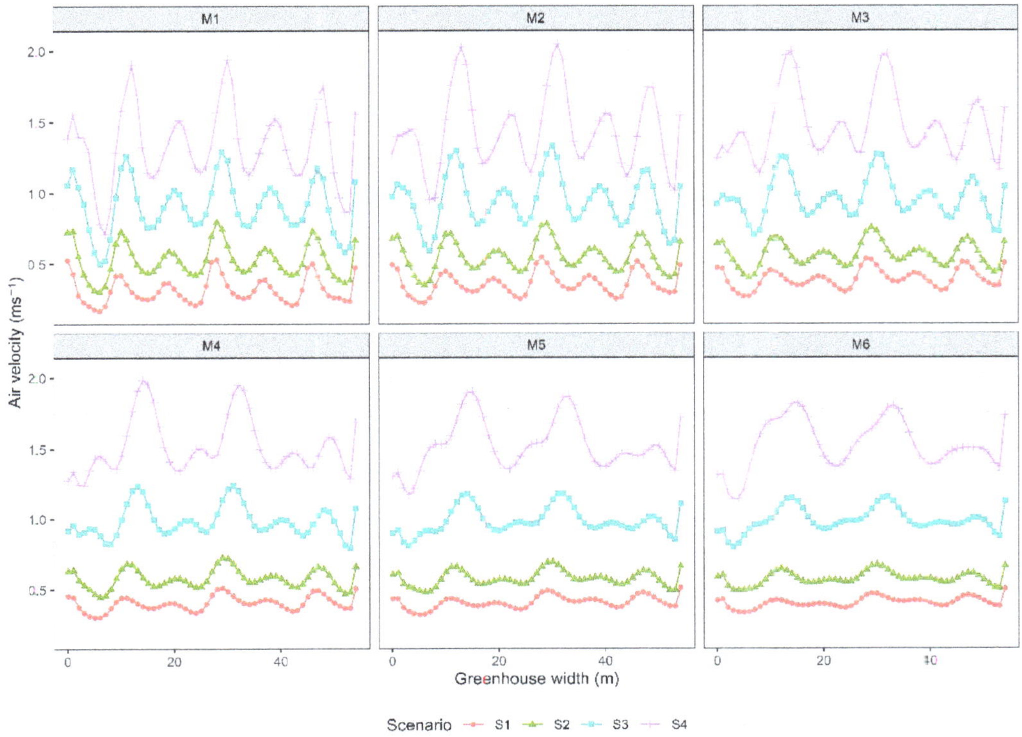

Figure 9. Spatial distribution of air velocity inside the greenhouse for the simulated scenarios under the RS configuration.

3.2. Renewal Index (RI)

The renewal index (RI) was calculated using the method of integration of the volumetric flow of air leaving the ventilation areas of the greenhouse, the results obtained can be seen in Figure 10. The RI values ranged from a minimum of 5.89 Vol h^{-1} obtained in S-M1S1 to a maximum of 72.3 Vol h^{-1} obtained in S-M6S4. It is important to highlight that these are the contrasting scenarios, therefore, the minimum RI was obtained in the greenhouse with the lowest height and ventilation area, evaluated at the lowest wind speed (S1), while the maximum RI was obtained in the greenhouse with the highest height and ventilation area evaluated at the highest wind speed (S4).

For the lowest wind speed (S1), RI values between 5.89 Vol h^{-1} and 19.1 Vol h^{-1} were obtained in the side ventilation configuration S for M1 and M6, respectively. Likewise, for the ventilation configuration through the vents of the roof region R, RI values of 19.4 and 19.7 Vol h^{-1} were obtained in M1 and M6, which are values compared to those obtained in S of 329% and 3.68%, respectively. Finally, for the RS combined ventilation configuration, an RI value was obtained for M1 of 26.5 Vol h^{-1} which is a value 449% higher compared to the S ventilation configuration and 36.6% higher compared to the R ventilation configuration. For M6, a value of 36.8 Vol h^{-1} was obtained, which is 92.7% and 86.8% higher than the S and R ventilation configurations, respectively.

Regarding the S4 velocity scenario, RI values were obtained for the S ventilation configuration of 19.5 and 50.2 Vol h^{-1} for M1 and M6, respectively. For the R ventilation configuration, values of 43.8 and 46.7 Vol h^{-1} were obtained for M1 and M6, respectively, which represents an increase in M1 of 224.6% with respect to the S configuration; on the other hand, for M6 there was a reduction of 14.6% in RI with respect to S. In the case of the RS configuration, a value of 59.7 Vol h^{-1} was obtained in M1, which is equivalent to a higher value of RI by 306.1% with respect to the S configuration and 136.6% with respect to R. While in M6 an RI value of 72.3 Vol h^{-1} was obtained, this value being 144 and 154.8% higher than those obtained in S and R, respectively.

To highlight the scenarios where RI values above or equal to the recommended minimum (RI $\geq$ 40 Vol h^{-1}) are achieved for naturally ventilated agricultural structures [11,66]. Under ventilation configuration S, this was only obtained under greenhouse models M4, M5 and M6 and under external wind speed conditions of 1.5 ms^{-1}, with this same speed condition for ventilation configuration R adequate RI values are obtained in all greenhouse models (M1–M6).

While for the RS combined ventilation configuration, the RI values are adequate for all greenhouses (M1–M6) under wind speed scenarios higher than 1 ms^{-1}, the same is true for the M4, M5 and M6 models under wind speeds higher than 0.5 ms^{-1}. It is therefore relevant for greenhouse growers or greenhouse builders or decision makers to seriously analyze the prevailing wind speed conditions in the study region [62,67].

These results reaffirm some conclusions already determined in previous works related to the natural ventilation of greenhouses, where it was identified that the renewal indexes are dependent on the ventilation configuration used [68–71]. It was also identified that there is a linear relationship between the renewal index and wind speed [38,56,72,73], the ventilation configuration that allows to obtain the highest renewal index values is the combined configuration of roof and side vents [10,16,74,75].

On the other hand, in the side ventilation configuration, the increase of renewal indexes as a function of the increase of the side ventilation area is relevant and significant only in narrow greenhouses (width < 60 m) or with few attached spans (<6 spans) [76–79]. Finally, the IR in low external wind speed conditions are more stable and higher in greenhouse structures with relevant ventilation areas in the roof region [14,29,80].

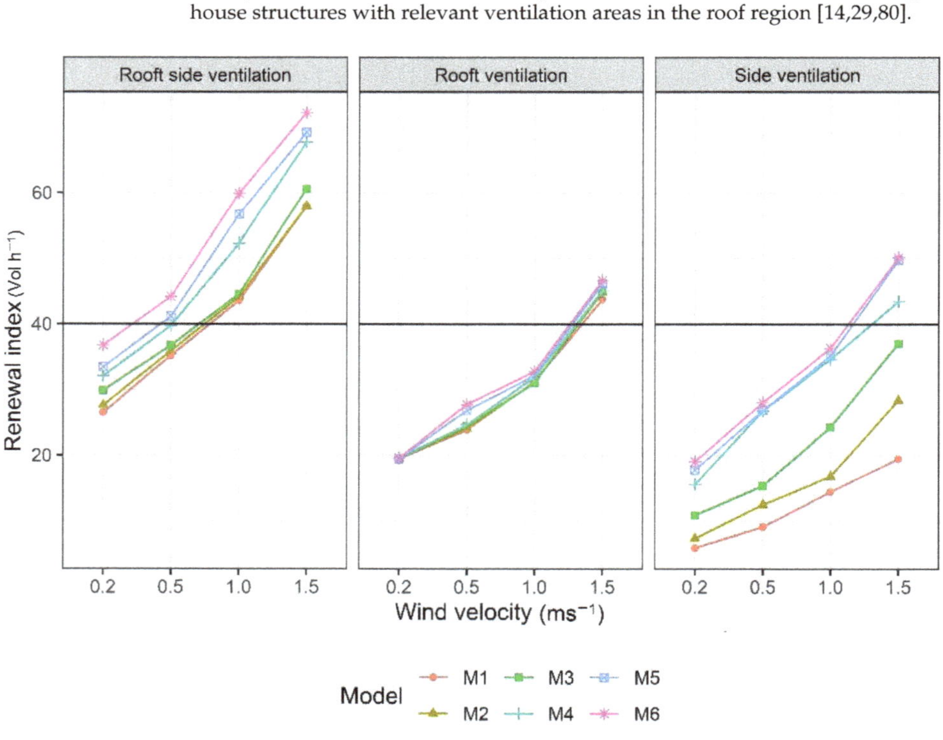

Figure 10. Calculated values of the simulated renewal indexes for each of the scenarios considered.

3.3. Spatial Distribution of Temperature

3.3.1. Side Ventilation (S)

The distribution of the thermal behavior for each of the evaluated scenarios can be seen in Figure 11. In general terms, irrespective of the outside wind speed, it was found that the cool zones correspond to the ventilation regions where the air enters the greenhouse, while the regions of higher temperature are located in the region where the air flow exits from the interior of the structure. This behavior occurs mainly because the air that enters the greenhouse through the windward window of span 1, as it crosses the cross section, mixes with the warm air and increases its energy level by heat transfer [10,45].

On the other hand, the height of the greenhouse directly influences the magnitude and spatial distribution of the temperature; it was observed that as the height of the greenhouse increases, the magnitude of the temperature and the percentage of the area of the structure with high temperature values decrease. This is an effect generated by the higher renovation index and the higher level of thermal inertia obtained in higher greenhouses [1,35]. Likewise, it is possible to observe the effects of the external wind speed, where for the highest speed scenario S4 it was found that the temperature value inside the greenhouse was lower and also presented a greater homogeneous behavior, coinciding with what was reported Flores-Velázquez [81].

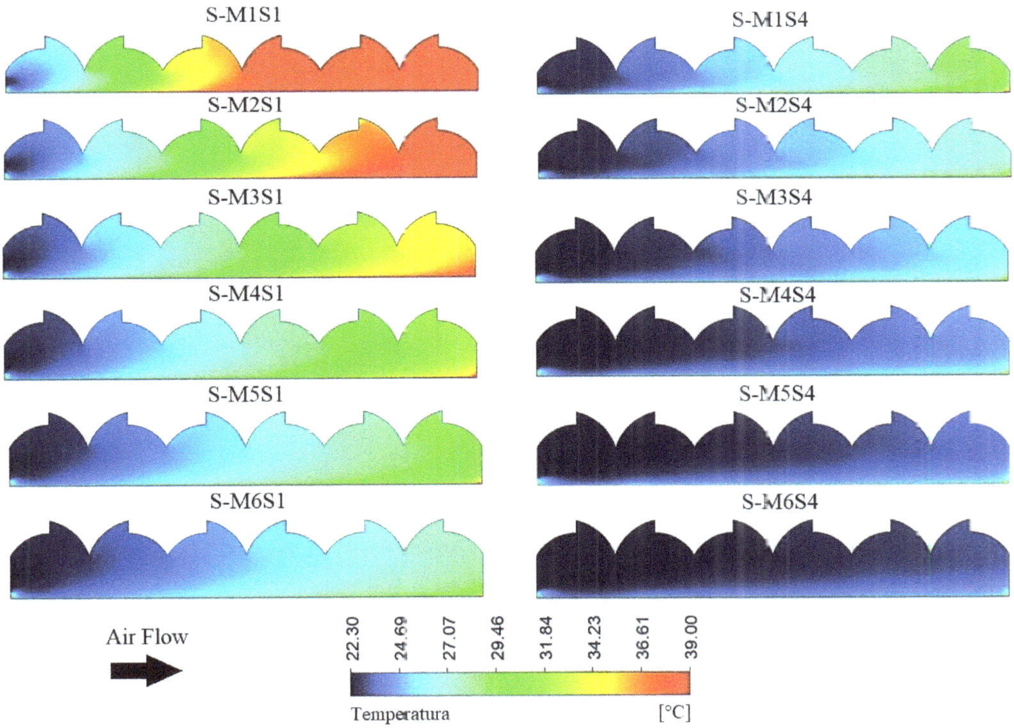

Figure 11. Simulated thermal distribution patterns with ventilation configuration S and for the six greenhouse models (M1 to M6) under wind speeds S1 (0.2 ms^{-1}) and S4 (1.6 ms^{-1}).

In numerical terms, the average air temperature inside the greenhouse ranged between maximum values of 37.5 ± 8.9 °C for S-M1S1 and a minimum of 23.3 ± 0.6 °C for S-M6S4. For the S1 scenario it can be observed that the temperature decreases 11.7 °C comparing greenhouses M1 and M6 respectively, likewise for the S4 scenario this temperature reduction between M1 and M6 is only 2.5 °C, therefore, the increase in greenhouse height will be more relevant in regions where calms or low wind speeds predominate. Although under these conditions it should also be analyzed up to what level it is convenient from the technical and economic point of view to increase the greenhouse height; since for the M4 and M5 models under this condition no less important reductions of 9.7 and 10.8 °C, respectively, would be obtained.

Another relevant criterion to be analyzed is the uniform distribution of temperature in the cross axis of the greenhouse. This is undoubtedly a factor that has received more attention in recent years, since it has a direct influence on the physiological behavior of the plants and on the final yield of the crops [58,82]. Therefore, for this study, the values obtained in the cross section of each of the greenhouses were plotted as a function of wind speed (Figure 12).

In general, it is again observed the difference that exists between span 1 on the windward side (cool area) and span 9 on the leeward side (warm area) with relevant thermal differentials higher than 15 °C. Likewise, in some critical scenarios, the temperature in some areas of the cross section of the greenhouse exceeds 40 °C, a value that is quite inadequate for the growth and development of any vegetal species [83].

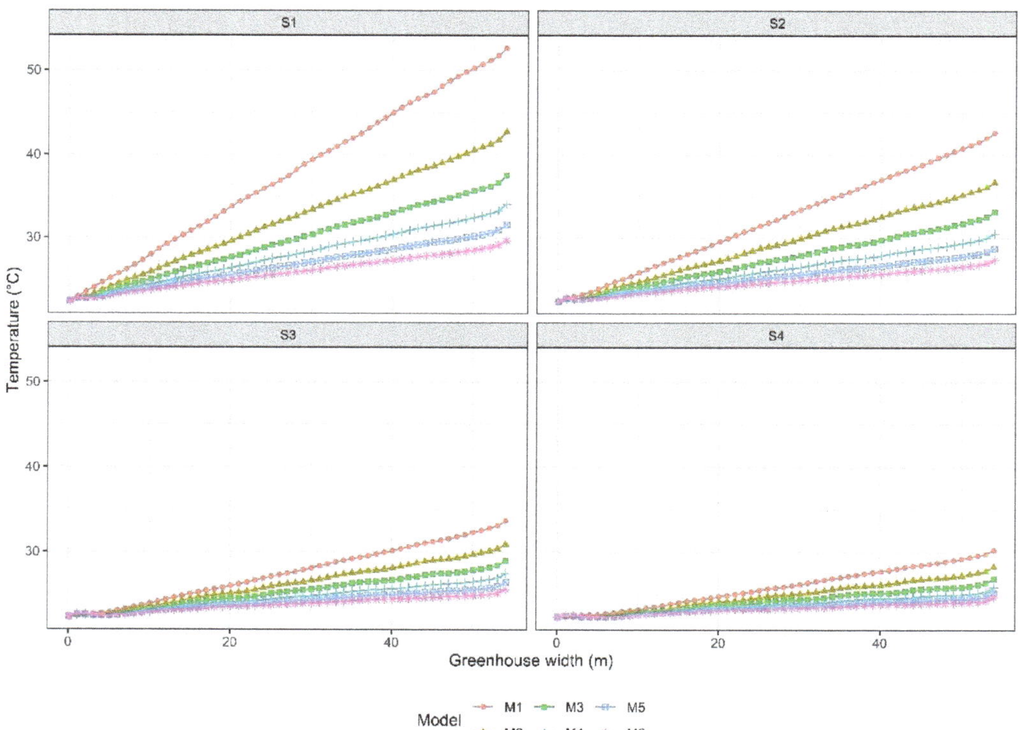

Figure 12. Spatial distribution of air temperature inside the greenhouse for the simulated scenarios under the S configuration.

3.3.2. Roof Ventilation (R)

In qualitative terms, it is observed that under ventilation configuration R, lower temperatures are obtained with respect to configuration S (Figure 13). In general, it is observed that the highest temperature areas are located on the windward and leeward sides of the span, respectively. This behavior is related to the ventilation configuration and to the characteristics of the airflows of these regions near the greenhouse walls previously analyzed. a similar behavior was reported in a work on natural ventilation of four types of multi-tunnel greenhouses by Park et al. [84].

For this configuration it is also observed that there is a positive effect of wind speed and greenhouse height on the magnitude and distribution of the temperature inside the structure. In the most critical case S-M1S1 it is observed that there are relatively important areas of high temperature in spans 1, 3 and 5, while in the S-M6S1 scenario these areas are significantly reduced. In S-M1S4 the high temperature area disappears in span 3 and the hot areas in spans 1 and 6 are reduced to an area very close to the side wall of each side and, finally, in S-M6S4 these high temperature areas disappear completely in the spans already described.

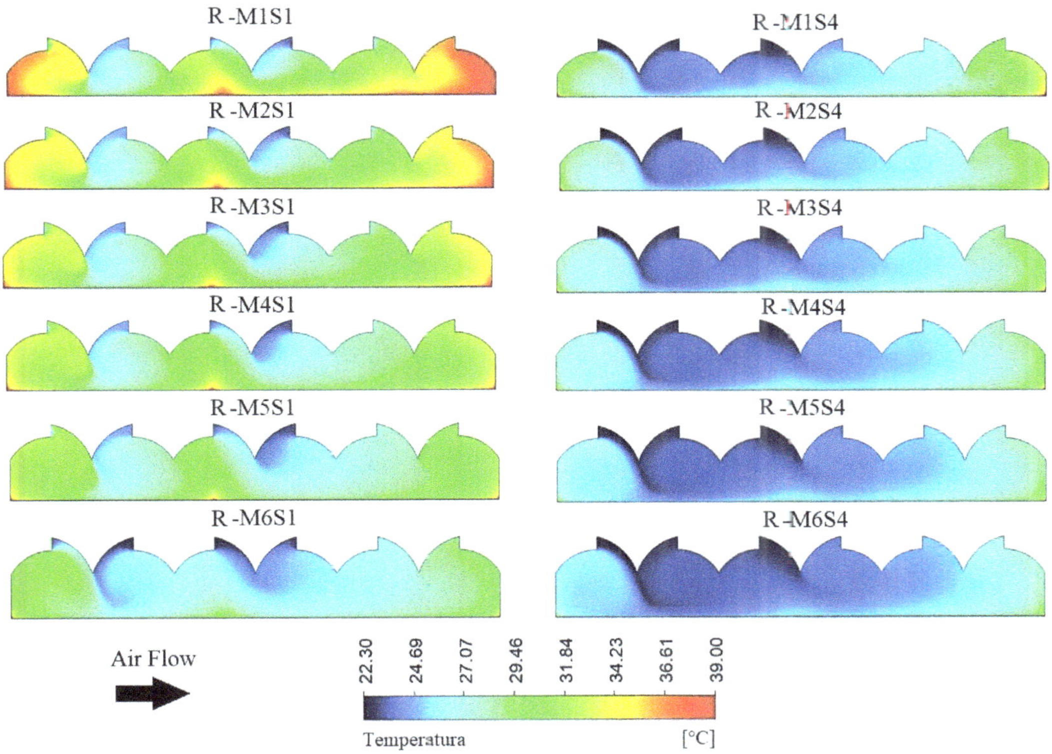

Figure 13. Simulated thermal distribution patterns with ventilation configuration R and for the six greenhouse models (M1 to M6) under wind speeds S1 (0.2 ms^{-1}) and S4 (1.6 ms^{-1}).

In numerical terms the average air temperature ranged between maximum value of 32.7 ± 3.1 °C for R-M1S1, value that is lower by 4.8 °C with respect to S-M1S1, the minimum value obtained was 25.0 ± 1.3 °C in R-M6S4 value that is higher by 1.7 °C with respect to S-M6S4. For the S1 scenario the temperature reduction obtained was 1.3, 2.2, 2.8, 3.6 and 4.3 °C in M2 to M6 compared to M1 respectively, while in S4 these reductions with respect to M1 were 0.5, 0.7, 1.1, 1.1, 1.4, 1.7 °C in M2 to M6, successively, results that continue to confirm the importance of roof ventilation in low wind speed conditions (S1).

For this case, the thermal behavior on the greenhouse cross axis in each scenario evaluated shows a totally different spatial temperature distribution than that observed in the previous ventilation configuration (Figure 14). Due to the air inlet and outlet flows through the roof vents and the airflow distribution patterns discussed in Figure 6, it is possible to observe the temperature variations occurring between span 1, 3 and 6 successively. It can also be seen how this temperature distribution tends to be more uniform and smaller in magnitude as the height of the greenhouse and the outside wind speed increase.

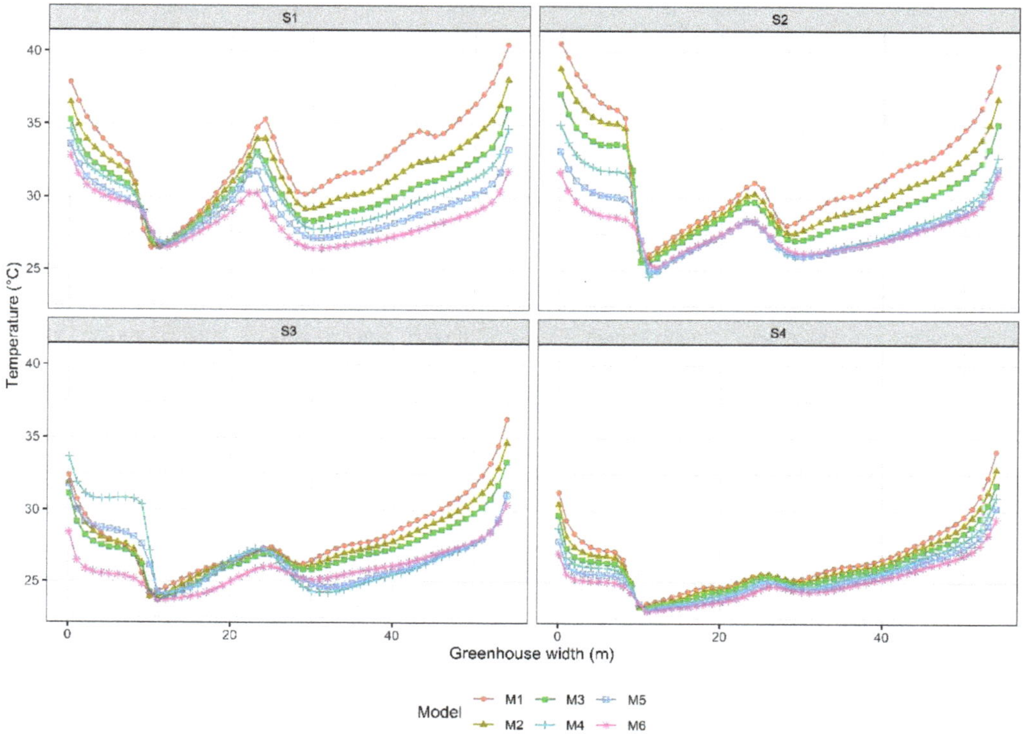

Figure 14. Spatial distribution of air temperature inside the greenhouse for the simulated scenarios under the R configuration.

3.3.3. Side Ventilation and Roof (RS)

The spatial distribution of the temperature for this configuration allows us to observe qualitatively that the temperature value in each of the simulated scenarios has a lower magnitude compared to the S and R configurations (Figure 15). For these scenarios, it is observed that the regions of lower temperature coincide with the ventilation areas where the air flow enters, both in the lateral sides and in the roof ventilation areas, which coincides with what was analyzed in the work by Villagran et al. [10].

For scenario S1, it can be observed that the high temperature regions are in the spans located on the leeward side. This high temperature region gradually disappears as the height of the greenhouse increases. In the case of M1, the highest temperature region occurs in spans 4, 5 and 6, with a heat spot in the middle of the area of span 6. However, in M6, this heat patch disappears from the span 6 and only a small region of higher temperature is observed near the ventilation area on the leeward side of the greenhouse. Similar behavior is observed in scenario S4, although in this case the temperature values appear to be qualitatively lower than those obtained in S1.

Numerically, it was found that the temperature value inside the structure presented a maximum value of 29.4 ± 3.8 °C in RS-M1S1, being this value lower by 8.1 and 3.3 °C compared to S and R, respectively. Therefore, the RS ventilation configuration also allows obtaining a higher cooling efficiency under the same climatic conditions, which is a very relevant factor for the management of the microclimate in greenhouses located in the savanna of Bogota [1].

On the other hand, the minimum mean inside air temperature value was 23.2 ± 0.6 °C in RS-M6S4, a value that is lower by 1.7 and 0.1 °C with respect to those obtained in R and S, respectively. Therefore, again it can be observed that the greatest benefits in terms of thermal distribution are obtained for low wind speed conditions. It is also important to note that the temperature values obtained in most of the scenarios are within the range of 25 and 30 °C, ranges where species of commercial and food interest such as Tomatoes tend to develop adequately. As well as some ornamental species of commercial interest for the international market such as the Rose or Carnation [1,85].

Regarding the spatial distribution, it was again found that as the greenhouse height increases and the external wind speed increases, more stable and uniform temperature values are obtained in the cross section of the greenhouse (Figure 16). The thermal gradients for the same moment inside the greenhouse were reduced from 13.2 °C in M1S1 to 1.94 °C in M6S4, the latter value being a recommended limit to guarantee the homogeneity of a naturally ventilated greenhouse [39,86].

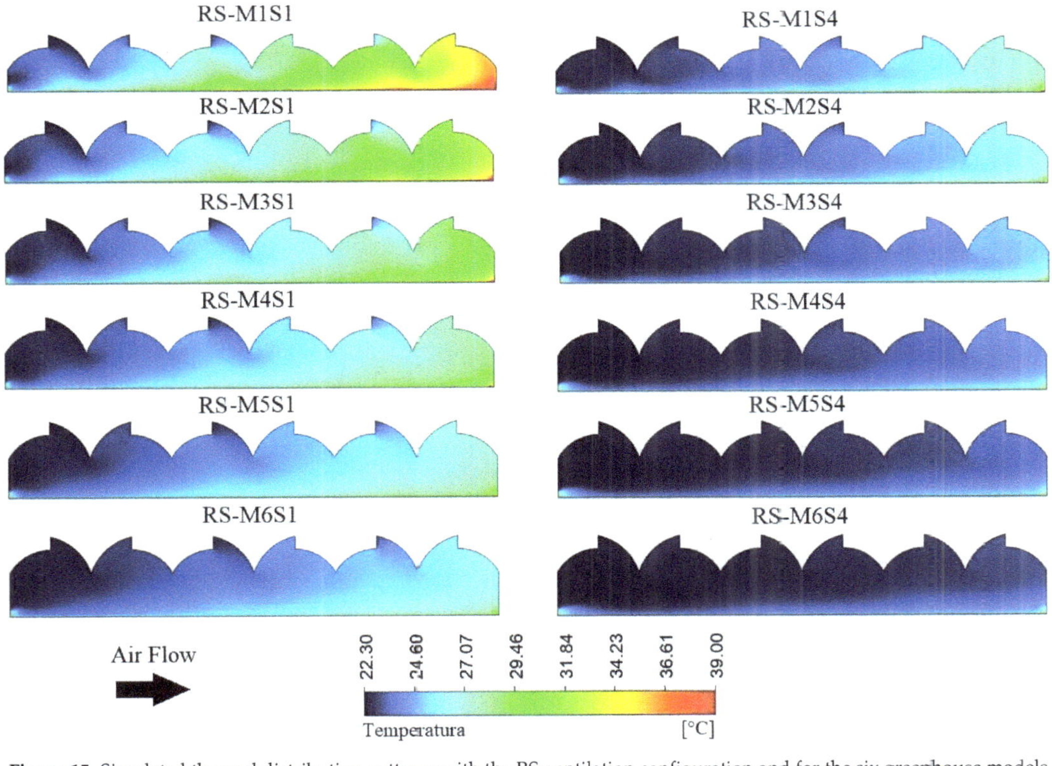

Figure 15. Simulated thermal distribution patterns with the RS ventilation configuration and for the six greenhouse models (M1 to M6) under wind speeds S1 (0.2 ms^{-1}) and S4 (1.6 ms^{-1}).

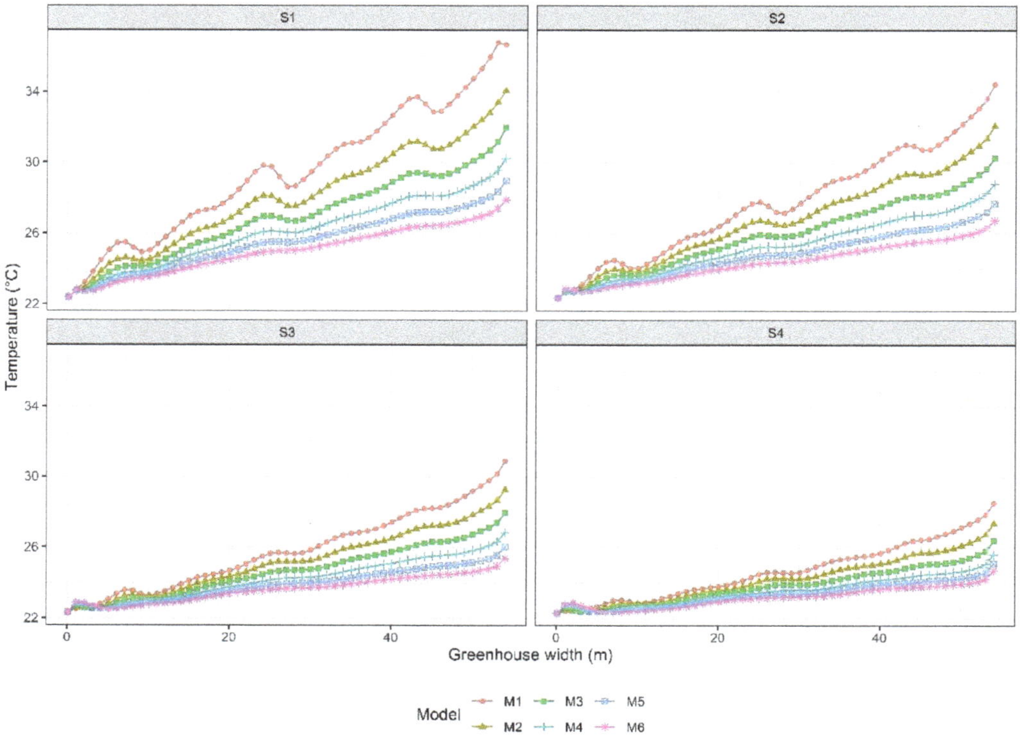

Figure 16. Spatial distribution of air temperature inside the greenhouse for the scenarios simulated under the RS configuration.

3.4. Response Surface Modeling

Finally, looking for a better and simpler interpretation of the results obtained by CFD simulation, these numerical results of the evaluated scenarios were adjusted to response surface models. This methodology allows observing the behavior of the response variable in scenarios not simulated numerically, if these scenarios are within the limiting ranges of the simulated boundary conditions

The results for the fitted response surface models are summarized in Table 4. The second order model alternative gave the best results for all models. The lack of fit test showed that the models accurately fit the data exception made for the sidewalls scenario and with temperature as the response variable. The adjusted R-squared coefficient of determination indicated varying results whether the outcome variable was the internal air velocity or the temperature. Highest R-squared values were obtained for the models with wind speed as the dependent variable as compared to their temperature pairs under the same scenario.

Despite the varying results for the adjusted R-squared coefficient, all terms included in the models were considered significant. Since most of the results indicated a good fit of the model to the calibration data, we evaluated the effect of varying levels of greenhouse height and external wind speed for each ventilation scenario through response surface plots. The response surface plots indicated a similar behavior for internal temperature and air velocity irrespective of the ventilation scenario (Figure 17). The individual effects of the predictors showed that increasing the greenhouse height effectively resulted in lower temperatures, while the internal air velocity was slightly affected by the greenhouse height. The major effect of the external wind speed on both, the temperature, and the internal air velocity, is clearly depicted in Figure 17. Increasing the greenhouse height under increasing levels of external wind speed resulted in lower temperature values within the greenhouse.

Table 4. Response surface models and goodness of fit measures applied to the three ventilation scenarios and considering internal wind speed and temperature as dependent variables.

Scenario—Ventilation Type	Model	Adjusted R-Squared	Lack of Fit Test
Roof (R)	$IAV \sim H1 + EWS - H1^2 + EWS^2 + H1 : EWS$	0.6405	F = 0.6145; $p = 0.891$
	$T \sim H1 + EWS + H^2 + EWS^2 + H1 : EWS$	0.5663	F = 1.0693; $p = 0.378$
Side (S)	$IAV \sim H1 + EWS - H1^2 + EWS^2 + H1 : EWS$	0.8908	F = 0.5115; $p = 0.954$
	$T \sim H1 + EWS + H1^2 + EWS^2 + H1 : EWS$	0.5893	F = 2.2616; $p = 0.002$
Roof + Side (RS)	$IAV \sim H1 + EWS - H1^2 + EWS^2 + H1 : EWS$	0.8986	F = 0.6915; $p = 0.822$
	$T \sim H1 + EWS + H1^2 + EWS^2 + H1 : EWS$	0.5114	F = 0.2596; $p = 0.999$

IAV: internal air velocity; *T*: temperature; *H*1: greenhouse height; *EWS*: external wind speed.

On the other hand, lower greenhouse heights combined with decreased external wind speeds resulted in lower internal air velocity. However, while all scenarios depicted the same trends, differences arise when looking at the scale of variation of the internal temperatures and air velocity. The sidewalls scenario showed that increasing the greenhouse height under increased external wind speeds have a more dramatic effect in terms of lowering the internal temperature of the greenhouse (Figure 17c), compared to roof or combined ventilation (Figure 17a,e).

The option to apply only the roof ventilation decreased the internal wind speed under almost null external wind speed no matter what the greenhouse height is. However, when the external wind speed increases along with the greenhouse height, internal air velocity also increases but the highest internal air velocity is reached with the lowest greenhouse height and the maximum external wind speed (Figure 17b). This results is opposed to the other scenarios where the top internal air velocity was reached when the maximum greenhouse height and external wind speed were evaluated (Figure 17d,f).

The optimal combination of greenhouse height and external wind speed that minimized the temperature and achieved an adequate internal wind speed seemed to be located around the maximum height and external wind speed for all scenarios. These results implied that higher values for both explanatory variables should be considered to find the lowest internal temperature and the highest internal air velocity. However, going beyond and increasing the region of analysis will yield unrealistic results due to greenhouse construction restrictions and maximum local wind speeds.

The results presented here are in line with those reported by Villagran et al. [1], and Bustamante et al. [37], indicating that increased greenhouse heights and the combination of roof and sidewalls ventilation yields improved internal climate conditions, particularly for greenhouse established in high altitude tropical regions.

It is also important to recommend that a future study regarding this greenhouse typology should focus on the structural design of the model evaluated and an economic analysis of the height alternatives proposed. This to be able to dimension the robustness of the structure and an associated cost per m^2 of covered area for each of the 6 models proposed, although in Colombia there are currently no regulations in force for the construction of greenhouses, a work of this magnitude can provide economic and structural tools for decision making. This information can be used for the generation of public policy management documents that can promote and regulate the use of greenhouses conceived with design criteria.

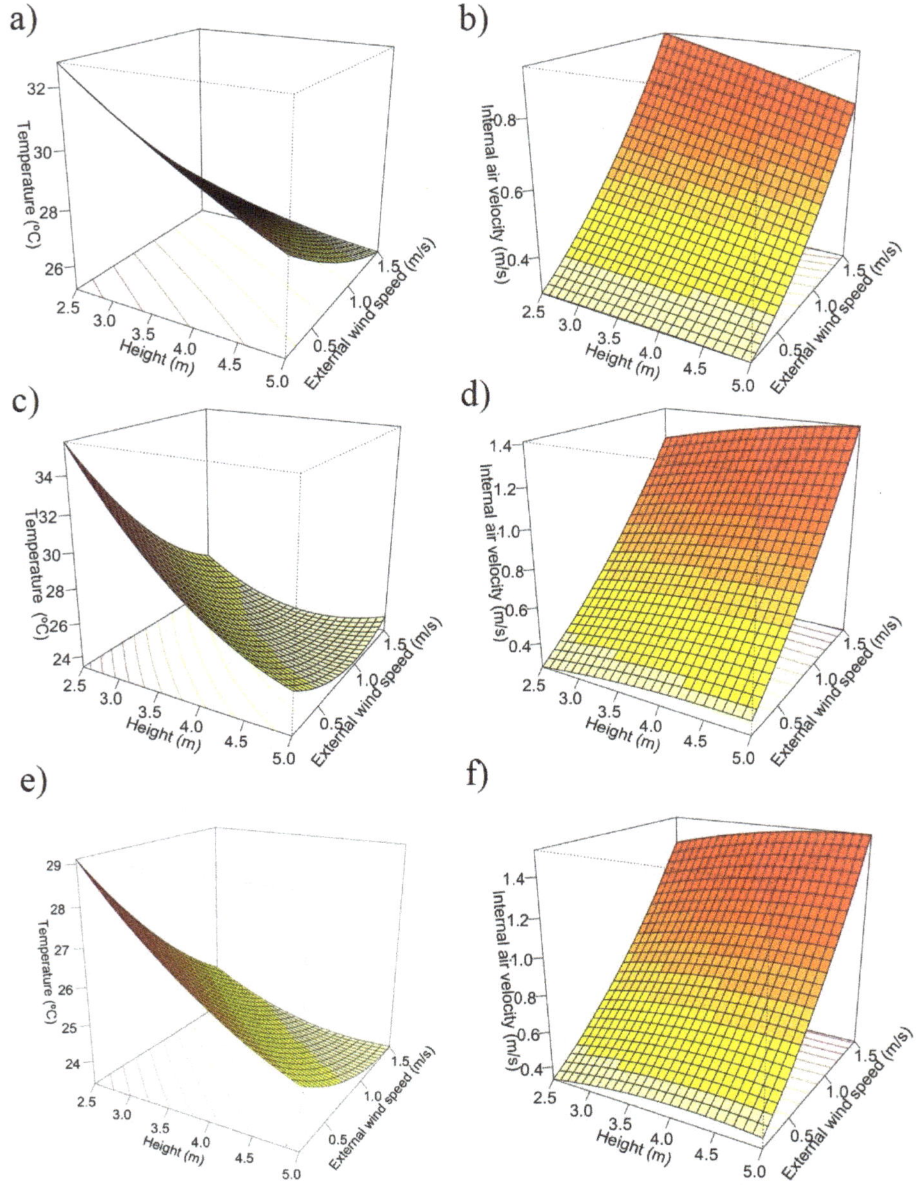

Figure 17. Effect of greenhouse height and external wind speed on internal air velocity and temperature for the roof (**a**,**b**), sidewalls (**c**,**d**), and roof + sidewalls (**e**,**f**) ventilation scenarios.

4. Conclusions

The numerical simulation tool is still an agile and accurate alternative to determine the renewal indexes and thermal distribution in passive greenhouses. It is also analyze unconstructed scenarios, facilitating decision making for farmers and providing an alternative analysis that can contribute to improve the technical and economic sustainability of protected agriculture.

The use of previously validated CFD models allows generating simulated data series of temperature and air velocity inside a greenhouse and associating them to response variables such as the height of the structure evaluated or others. This facilitated the association of the obtained data to other analysis methodologies such as regression with response surface models obtaining acceptable adjustments in the analyzed variables, therefore, it was possible to generate response surface graphs which facilitates the interpretation of the numerical results as a whole for each ventilation configuration analyzed (S, R, RS).

The natural ventilation of a greenhouse is highly dependent on the behavior of the wind speed of the study region. For the three ventilation configurations analyzed (S, R and RS), it was found that the renewal indexes were maximized as a function of the increase in the external wind speed. Although it should be noted that regardless of the ventilation configuration for the lowest wind speed S1 (0 2 ms^{-1}), none of the greenhouse models exceeded the recommended minimum value of 40 hourly renovations, although these IR values are less critical in the RC configuration. The optimization of the IR in the RS ventilation configuration allowed obtaining a higher cooling efficiency inside the greenhouse and a uniformity in the spatial distribution of the temperature. Maximum temperature values of 29.4 ± 3.8 °C were obtained in RS-M1S1, being this value lower by 8.1 and 3.3 °C with respect to S-M1S1 and R-M1S1 respectively.

The crop production in greenhouse in the future should be with the efficient use of resources, so that numerical simulation techniques will have the goal of adapting the climatic environment in the design and operation of protected agriculture. In addition to the greenhouse dimensions, the position and size of windows are among the other influential factors to be considered in future studies

Author Contributions: Conceptualization, E.V., M.A., J.F.-V.; Methodology, E.V, C.B., J.F.-V., M.A.; Software, E.V.; Validation, E.V., C.B.; Investigation, E.V., C.B.; Writing—original draft preparation, E.V., C.B., J.F.-V., M.A.; Writing—review and editing, E.V., C.B., J.F.-V., M.A. All authors have read and agreed to the published version of the manuscript.

Funding: The present study was funded by the Servicio Nacional de Aprendizaje (SENA), la Asociación Colombiana de Exportadores de Flores (Asocolflores) y al Centro de Innovación de la Floricultura Colombiana (Ceniflores) through the project "Generación de una herramienta de diseño u optimización de ventilación natural de los invernaderos dedicados a la producción de flores de corte en cuatro subregiones de la Sabana de Bogotá, mediante el uso de herramientas de simulación basadas en la técnica Dinámica de Fluidos Computacional (CFD)".

Data Availability Statement: Not applicable.

Acknowledgments: To the Universidad Jorge Tadeo Lozano for the support for the development of the doctoral research project called "*Development of a methodological proposal for the design, evaluation and microclimatic optimization of passive greenhouses through the use of numerical simulation tools*".

Conflicts of Interest: The authors declare no conflict of interest.

References

1. Villagrán, E.A.; Romero, E.J.B.; Bojacá, C.R. Transient CFD analysis of the natural ventilation of three types of greenhouses used for agricultural production in a tropical mountain climate. *Biosyst. Eng.* **2019**, *188*, 288–304. [CrossRef]
2. Sun, H.; Wang, L.; Lin, R.; Zhang, Z.; Zhang, B. Mapping Plastic Greenhouses with Two-Temporal Sentinel-2 Images and 1D-CNN Deep Learning. *Remote Sens.* **2021**, *13*, 2820. [CrossRef]
3. Villagran, E.; Leon, R.; Rodriguez, A.; Jaramillo, J. 3D Numerical Analysis of the Natural Ventilation Behavior in a Colombian Greenhouse Established in Warm Climate Conditions. *Sustainability* **2020**, *12*, 8101. [CrossRef]
4. Nikolaou, G.; Neocleous, D.; Katsoulas, N.; Kittas, C. Irrigation of Greenhouse Crops. *Horticulturae* **2019**, *5*, 7. [CrossRef]
5. Pakari, A.; Ghani, S. Airflow assessment in a naturally ventilated greenhouse equipped with wind towers: Numerical simulation and wind tunnel experiments. *Energy Build.* **2019**, *199*, 1–11. [CrossRef]
6. Holcman, E.; Sentelhas, P.C.; Mello, S.D.C. Cherry tomato yield in greenhouses with different plastic covers. *Ciência Rural* **2017**, *47*. [CrossRef]
7. Revathi, S.; Sivakumaran, N.; Radhakrishnan, T. Design of solar-powered forced ventilation system and energy-efficient thermal comfort operation of greenhouse. *Mater. Today Proc.* **2021**, *46*, 9893–9900. [CrossRef]

8. Salah, A.H.; Hassan, G.E.; Fath, H.; Elhelw, M.; Elsherbiny, S. Analytical investigation of different operational scenarios of a novel greenhouse combined with solar stills. *Appl. Therm. Eng.* **2017**, *122*, 297–310. [CrossRef]
9. Jiménez-Lao, R.; Aguilar, F.; Nemmaoui, A.; Aguilar, M. Remote Sensing of Agricultural Greenhouses and Plastic-Mulched Farmland: An Analysis of Worldwide Research. *Remote Sens.* **2020**, *12*, 2649. [CrossRef]
10. Munar, E.; Aldana, C. CFD Simulation of the Increase of the Roof Ventilation Area in a Traditional Colombian Greenhouse: Effect on Air Flow Patterns and Thermal Behavior. *Int. J. Heat Technol.* **2019**, *37*, 881–892. [CrossRef]
11. Villagrán, E.; Flores-Velazquez, J.; Bojacá, C.; Akrami, M. Evaluation of the Microclimate in a Traditional Colombian Green-house Used for Cut Flower Production. *Agronomy* **2021**, *11*, 1330. [CrossRef]
12. Diaz, D.; Bojacá, C.; Schrevens, E. Modeling the suitability of the traditional plastic greenhouse for tomato production across Colombian regions. *Acta Hortic.* **2018**, 857–864. [CrossRef]
13. Fernández-García, M.; Vidal-López, P.; Rodríguez-Robles, D.; Villar-García, J.; Agujetas, R. Numerical Simulation of Multi-Span Greenhouse Structures. *Agriculture* **2020**, *10*, 499. [CrossRef]
14. Villagran, E.; Bojacá, C.; Akrami, M. Contribution to the Sustainability of Agricultural Production in Greenhouses Built on Slope Soils: A Numerical Study of the Microclimatic Behavior of a Typical Colombian Structure. *Sustainability* **2021**, *13*, 4748. [CrossRef]
15. Flores-Velazquez, J. Analisis de la Ventilación en los Principales Modelos de Invernaderos en Mejico Mediante Dinámica de Fluidos Computacional (CFD). Ph.D. Thesis, University of Almería, Almería, Spain, 2010.
16. Munar, E.A.V.; Aldana, C.R.B. Study of natural ventilation in a Gothic multi-tunnel greenhouse designed to produce rose (Rosa spp.) in the high-Andean tropic. *Ornam. Hortic.* **2019**, *25*, 133–143. [CrossRef]
17. Choab, N.; Allouhi, A.; El Maakoul, A.; Kousksou, T.; Saadeddine, S.; Jamil, A. Review on greenhouse microclimate and application: Design parameters, thermal modeling and simulation, climate controlling technologies. *Sol. Energy* **2019**, *191*, 109–137. [CrossRef]
18. Bot, G. Physical modeling of greenhouse climate. *IFAC Proc. Vol.* **1991**, *24*, 7–12. [CrossRef]
19. Seven, S.A.; Taştan, Ö.F.; Tas, C.E.; Ünal, H.; Ince, I.A.; Menceloglu, Y.Z. Insecticide-releasing LLDPE films as greenhouse cover materials. *Mater. Today Commun.* **2019**, *19*, 170–176. [CrossRef]
20. Babaghayou, M.; Mourad, A.-H.I.; Lorenzo, V.; Chabira, S.; Sebaa, M. Anisotropy evolution of low density polyethylene greenhouse covering films during their service life. *Polym. Test.* **2018**, *66*, 146–154. [CrossRef]
21. Alsadon, A.; Al-Helal, I.; Ibrahim, A.; Abdel-Ghany, A.; Al-Zaharani, S.; Ashour, T. The effects of plastic greenhouse covering on cucumber (*Cucumis sativus* L.) growth. *Ecol. Eng.* **2016**, *87*, 305–312. [CrossRef]
22. Bournet, P.-E.; Boulard, T. Effect of ventilator configuration on the distributed climate of greenhouses: A review of experimental and CFD studies. *Comput. Electron. Agric.* **2010**, *74*, 195–217. [CrossRef]
23. Cemek, B.; Atiş, A.; Küçüktopçu, E. Evaluation of temperature distribution in different greenhouse models using computational fluid dynamics (CFD). *Anadolu J. Agric. Sci.* **2017**, *32*, 54. [CrossRef]
24. Ghoulem, M.; El Moueddeb, K.; Nehdi, E.; Boukhanouf, R.; Calautit, J.K. Greenhouse design and cooling technologies for sustainable food cultivation in hot climates: Review of current practice and future status. *Biosyst. Eng.* **2019**, *183*, 121–150. [CrossRef]
25. Teitel, M.; Montero, J.; Baeza, E. Greenhouse Design: Concepts and Trends. *Acta Hortic.* **2012**, *952*, 605–620. [CrossRef]
26. Akrami, M.; Javadi, A.A.; Hassanein, M.J.; Farmani, R.; Dibaj, M.; Tabor, G.R.; Negm, A. Study of the Effects of Vent Configuration on Mono-Span Greenhouse Ventilation Using Computational Fluid Dynamics. *Sustainability* **2020**, *12*, 986. [CrossRef]
27. Benni, S.; Santolini, E.; Barbaresi, A.; Torreggiani, D.; Tassinari, P. Calibration and comparison of different CFD approaches for airflow analysis in a glass greenhouse. *J. Agric. Eng.* **2017**, *48*, 49–52. [CrossRef]
28. Kim, R.-W.; Kim, J.-G.; Lee, I.-B.; Yeo, U.-H.; Lee, S.-Y.; Decano-Valentin, C. Development of three-dimensional visualisation technology of the aerodynamic environment in a greenhouse using CFD and VR technology, part 1: Development of VR a database using CFD. *Biosyst. Eng.* **2021**, *207*, 33–58. [CrossRef]
29. Villagrán, E.A.; Bojacá, C.R. Effects of surrounding objects on the thermal performance of passively ventilated greenhouses. *J. Agric. Eng.* **2019**, *50*, 20–27. [CrossRef]
30. Tong, G.; Christopher, D.M.; Zhang, G. New insights on span selection for Chinese solar greenhouses using CFD analyses. *Comput. Electron. Agric.* **2018**, *149*, 3–15. [CrossRef]
31. Molina-Aiz, F.; Valera, D. Configuration by evaluating ventilation efficiency based on computational fluid dynamics. *Acta Hortic.* **2011**, *893*, 669–677. [CrossRef]
32. He, K.; Chen, D.; Sun, L.; Huang, Z.; Liu, Z. Effects of Vent Configuration and Span Number on Greenhouse Microclimate under Summer Conditions in Eastern China. *Int. J. Vent.* **2015**, *13*, 381–396. [CrossRef]
33. Kacira, M.; Sase, S.; Okushima, L. Effects of Side Vents and Span Numbers on Wind-Induced Natural Ventilation of a Gothic Multi-Span Greenhouse. *Jpn. Agric. Res. Q. JARQ* **2004**, *38*, 227–233. [CrossRef]
34. Boulard, T.; Fatnassi, H. Greenhouse Aeration and Climate Optimization Based on CFD Studiess. *Plasticulture* **2005**, *124*, 38.
35. Fatnassi, H.; Boulard, T.; Benamara, H.; Roy, J.; Suay, R.; Poncet, C. Increasing the height and multiplying the number of spans of greenhouse: How far can we go? *Acta Hortic.* **2017**, *1170*, 137–144. [CrossRef]
36. Xu, F.-Y.; Lu, H.-F.; Chen, Z.; Guan, Z.-C.; Chen, Y.-W.; Shen, G.-W.; Jiang, Z. Selection of a computational fluid dynamics (CFD) model and its application to greenhouse pad-fan cooling (PFC) systems. *J. Clean. Prod.* **2021**, *302*, 127013. [CrossRef]

37. Villagran, E.; Henao-Rojas, J.; Franco G. Thermo-Environmental Performance of Four Different Shapes of Solar Greenhouse Dryer with Free Convection Operating Principle and No Load on Product. *Fluids* **2021**, *6*, 183. [CrossRef]
38. Akrami, M.; Salah, A.H.; Javadi, A.A.; Fath, H.E.; Hassanein, M.J.; Farmani, R.; Dibaj, M.; Negm, A. Towards a Sustainable Greenhouse: Review of Trends and Emerging Practices in Analysing Greenhouse Ventilation Requirements to Sustain Maximum Agricultural Yield. *Sustainability* **2020**, *12*, 2794. [CrossRef]
39. Villagran, E. Implementation of ventilation towers in a greenhouse established in low altitude tropical climate conditions: Numerical approach to the behavior of the natural ventilation. *Rev. Ceres* **2021**, *68*, 10–22. [CrossRef]
40. Tominaga, Y.; Mochida, A.; Yoshie, R.; Kataoka, H.; Nozu, T.; Yoshikawa, M.; Shirasawa, T. AIJ guidelines for practical applications of CFD to pedestrian wind environment around buildings. *J. Wind. Eng. Ind. Aerodyn.* **2008**, *96*, 1749–1761. [CrossRef]
41. Villagran, E.; Ramirez, R.; Rodriguez, A.; Pacheco, R.; Jaramillo, J. Simulation of the Thermal and Aerodynamic Behavior of an Established Screenhouse under Warm Tropical Climate Conditions: A Numerical Approach. *Int. J. Sustain. Dev. Plan.* **2020**, *15*, 487–499. [CrossRef]
42. Villagran, E.A.; Noreña, J.E.J. Microclimatic behavior of a screen house proposed for horticultural production in low-altitude tropical climate conditions. *Comun. Sci.* **2020**, *11*, e3350. [CrossRef]
43. Munar, E.A.V.; Aldana, C.R.B. Numerical evaluation of passive strategies for nocturnal climate optimization in a greenhouse designed for rose production (Rosa spp.). *Ornam. Hortic.* **2019**, *25*, 351–364. [CrossRef]
44. Chu, C.-R.; Lan, T.-W.; Tasi, R.-K.; Wu, T.-R.; Yang, C.-K. Wind-driven natural ventilation of greenhouses with vegetation. *Biosyst. Eng.* **2017**, *164*, 221–234. [CrossRef]
45. Senhaji, A.; Mouqallid, M.; Majdoubi, H. CFD Assisted Study of Multi-Chapels Greenhouse Vents Openings Effect on Inside Airflow Circulation and Microclimate Patterns. *Open J. Fluid Dyn.* **2019**, *9*, 119–139. [CrossRef]
46. Baxevanou, C.; Fidaros, D.; Bartzanas, T.; Kittas, C. Yearly numerical evaluation of greenhouse cover materials. *Comput. Electron. Agric.* **2018**, *149*, 54–70. [CrossRef]
47. Rasheed, A.; Lee, J.W.; Kim, H.T.; Lee, H.W. Efficiency of Different Roof Vent Designs on Natural Ventilation of Single-Span Plastic Greenhouse. *Prot. Hortic. Plant Fact.* **2019**, *28*, 225–233. [CrossRef]
48. Aguilar-Rodríguez, C.E.; Flores-Velázquez, J.; Rojano, F.; Flores-Magdaleno, H.; Panta, E.R. Simulation of Water Vapor and Near Infrared Radiation to Predict Vapor Pressure Deficit in a Greenhouse Using CFD. *Processes* **2021**, *9*, 1587. [CrossRef]
49. Rocha, G.A.O.; Pichimata, M.A.; Villagran, E. Research on the Microclimate of Protected Agriculture Structures Using Numerical Simulation Tools: A Technical and Bibliometric Analysis as a Contribution to the Sustainability of Under-Cover Cropping in Tropical and Subtropical Countries. *Sustainability* **2021**, *13*, 10433. [CrossRef]
50. Montaño-Rodríguez, S.; Villagrán-Munar, E.A.; Osorio-Fiaga, D.F.; Bojacá-Aldana, C.R.; Velásquez-Vargas, W.L. Numeric Simulation of Macrotunnel Greenhouse Thermic Behavior Used for Edible Mushrooms under Tropical Weather Conditions. *Tecnol. Marcha J.* **2020**, *32*, 78–85.
51. Sun, X.; Yang, H.; Liu, Q.-F.; Liu, Y.-H. Event-triggered control for greenhouse temperature under natural ventilation based on computational fluid dynamics. *Syst. Sci. Control. Eng.* **2021**, *9*, 93–101. [CrossRef]
52. Lenth, R.V. Response-Surface Methods inR, Usingrsm. *J. Stat. Softw.* **2009**, *32*, 1–17. [CrossRef]
53. Kalita, K.; Shivakoti, I.; Ghadai, R.K. Optimizing process parameters for laser beam micro-marking using genetic algorithm and particle swarm optimization. *Mater. Manuf. Process.* **2017**, *32*, 1101–1108. [CrossRef]
54. Kalita, K.; Dey, P.; Joshi, M.; Haldar, S. A Response Surface Modelling Approach for Multi-Objective Optimization of Com-posite Plates. *Steel Compos. Struct.* **2019**, *32*, 455–466. [CrossRef]
55. R Core Team. *R: A Language and Environment for Statistical Computing*; R Foundation for Statistical Computing: Vienna, Austria, 2020.
56. Baeza, E.J.; Pérez-Parra, J.J.; Montero, J.I.; Bailey, B.J.; López, J.C.; Gázquez, J.C. Analysis of the role of sidewall vents on buoyancy-driven natural ventilation in parral-type greenhouses with and without insect screens using computational fluid dynamics. *Biosyst. Eng.* **2009**, *104*, 86–96. [CrossRef]
57. Villagrán, E.A.; Gil, R.; Acuña, J.F.; Bojacá, C.R. Optimization of Ventilation and Its Effect on the Microclimate of Acolombian Multispan Greenhouse. *Agron. Colomb.* **2012**, *30*, 282–288.
58. Ma, D.; Carpenter, N.; Maki, H.; Rehman, T.U.; Tuinstra, M.R.; Jin, J. Greenhouse environment modeling and simulation for microclimate control. *Comput. Electron. Agric.* **2019**, *162*, 134–142. [CrossRef]
59. Molina-Aiz, F.D.; Valera, D.L.; Álvarez, A.J. Measurement and simulation of climate inside Almería-type greenhouses using computational fluid dynamics. *Agric. For. Meteorol.* **2004**, *125*, 33–51. [CrossRef]
60. López, A.; Valera, D.L.; Molina-Aiz, F. Sonic Anemometry to Measure Natural Ventilation in Greenhouses. *Sensors* **2011**, *11*, 9820–9838. [CrossRef] [PubMed]
61. Kwon, J.-K.; Lee, S.-H.; Seong, J.-H.; Moon, J.-P.; Lee, S.-J.; Choi, B.-M.; Kim, K.-J. Analysis of Natural Ventilation Characteristics of Venlo-type Greenhouse with Continuous Roof Vents. *J. Biosyst. Eng.* **2011**, *36*, 444–452. [CrossRef]
62. Reynafarje, X.; Villagrán, E.; Bojacá, C.; Gil, R.; Schrevens, E. Simulation and validation of the airflow inside a naturally ventilated greenhouse designed for tropical conditions. *Acta Hortic.* **2020**, 55–62. [CrossRef]
63. Villagrán-Munar, E.A.; Bojacá-Aldana, C.R.; Acuña-Caita, J.F. *Diseño y Evaluación Climatica de Un Invernadero Para Condiciones de Clima Intertropical de Montaña*; Universidad Nacional de Colombia: Bogota, Colombia, 2015; p. 138.

64. Mesmoudi, K.; Meguellati, K.; Bournet, P.-E. Thermal analysis of greenhouses installed under semi arid climate. *Int. J. Heat Technol.* **2017**, *35*, 474–486. [CrossRef]
65. McCartney, L.; Orsat, V.; Lefsrud, M.G. An experimental study of the cooling performance and airflow patterns in a model Natural Ventilation Augmented Cooling (NVAC) greenhouse. *Biosyst. Eng.* **2018**, *174*, 173–189. [CrossRef]
66. ASAE. Heating, Ventilating and Cooling Greenhouses. *ANSI/ASAE* **2008**, *3*, 9. [CrossRef]
67. Villagran, E. Two-Dimensional Numerical Study of the Microclimate Generated in Three Screenhouses for the Climatic Conditions of the Colombian Caribbean. *Int. J. Heat Technol.* **2021**, *39*, 460–468. [CrossRef]
68. Espejel Trujano, D.; López Cruz, I.L. Determinación de Las Tasas de Ventilación Natural En Un Invernadero Mediante Modelos Teóricos y Gases Trazadores. *Rev. Mex. Cienc. Agrícolas* **2013**, *4*, 185–198. [CrossRef]
69. Baeza, E.J.; Pérez-Parra, J.; Lopez, J.C.; Kacira, M.; Gázquez, J.C.; Montero, J.I. Validation of CFD Simulations for Three Dimensional Temperature Distributions of a Naturally Ventilated Multispan Greenhouse Obtained by Wind Tunnel Measurements. *Acta Hortic.* **2011**, *893*, 571–580. [CrossRef]
70. Mimar, E.A.V.; Aldana, C.R.B. Microclimate i Simulation in a Greenhouse Used for Roses Production under Conditions of Intertropical Climate. *Chil. J. Agric. Anim. Sci.* **2019**, *35*, 137–150. [CrossRef]
71. Kacira, M.; Sase, S.; Okushima, L. Optimization of vent configuration by evaluating greenhouse and plant canopy ventilation rates under wind-induced ventilation. *Trans. ASAE* **2004**, *47*, 2059–2067. [CrossRef]
72. Bartzanas, T.; Boulard, T.; Kittas, C. Numerical simulation of the airflow and temperature distribution in a tunnel greenhouse equipped with insect-proof screen in the openings. *Comput. Electron. Agric.* **2002**, *34*, 207–221. [CrossRef]
73. Molina-Aiz, F.; Fatnassi, H.; Boulard, T.; Roy, J.; Valera, D. Comparison of finite element and finite volume methods for simulation of natural ventilation in greenhouses. *Comput. Electron. Agric.* **2010**, *72*, 69–86. [CrossRef]
74. He, K.; Chen, D.; Sun, L.; Huang, Z.; Liu, Z. Analysis of the Climate inside Multi-span Plastic Greenhouses under Different Shade Strategies and Wind Regimes. *Korean J. Hortic. Sci. Technol.* **2014**, *32*, 473–483. [CrossRef]
75. Sase, S. Air movement and climate uniformity in ventilated greenhouses. *Acta Hortic.* **2006**, *719*, 313–324. [CrossRef]
76. Baeza, E.; Montero, J.I.; Pérez-Parra, J.; Bailey, B.J.; López, J.C.; Vázquez, J.C. *Avances En El Estudio de La Ventilación Natural*; Cajamar Caja Rural: Almeria, España, 2012.
77. Kittas, C.; Boulard, T.; Papadakis, G. Natural ventilation of a greenhouse with ridge and side openings: Sensitivity to temperature and wind effects. *Trans. ASAE* **1997**, *40*, 415–425. [CrossRef]
78. Gil, R.; Bojacá, C.; Casilimas, H.; Schrevens, E.; Suay, R. Assessment of sidewall and roof vents opening configurations to improve airflow inside greenhouses. *Acta Hortic.* **2012**, *952*, 141–146. [CrossRef]
79. Villagrán, E.A.; Gil, R.; Acuña, J.F.; Bojacá, C.R. *Agronomía Colombiana*; Universidad Nacional de Colombia, Facultad de Agronomía: Bogota, Colombia, 2012; Volume 30.
80. Katsoulas, N.; Bartzanas, T.; Boulard, T.; Mermier, M.; Kittas, C. Effect of Vent Openings and Insect Screens on Greenhouse Ventilation. *Biosyst. Eng.* **2006**, *93*, 427–436. [CrossRef]
81. Flores-Velázquez, J.; López-Cruz, I.L.; Mejía-Sáenz, E.; Montero, J.I. Evaluación Del Desempeño Climático de Un Inver-nadero Baticenital Del Centro de México Mediante Dinámica de Fluidos Computacional (CFD). *Agrociencia* **2014**, *48*, 131–146.
82. Villagran, E.; Bojacá, C. Experimental evaluation of the thermal and hygrometric behavior of a Colombian greenhouse used for the production of roses (Rosa spp.). *Ornam. Hortic.* **2020**, *26*, 205–219. [CrossRef]
83. Saberian, A.; Sajadiye, S.M. The effect of dynamic solar heat load on the greenhouse microclimate using CFD simulation. *Renew. Energy* **2019**, *138*, 722–737. [CrossRef]
84. Park, M.J.; Choi, D.K.; Son, J.K.; Yoon, S.-W.; Kim, H.T.; Lee, S.-K.; Kang, D.H. Field Survey and Analysis of Natural Ventilation Characteristics of Multi-span Greenhouse with Different Roof Vent. *Prot. Hortic. Plant Fact.* **2020**, *29*, 36–42. [CrossRef]
85. Villagran, E.; Bojacá, C. Analysis of the microclimatic behavior of a greenhouse used to produce carnation (*Dianthus caryophyllus* L.). *Ornam. Hortic.* **2020**, *26*, 190–204. [CrossRef]
86. Zorzeto, T.Q.; Leal, P.A.M.; Nunes, E.F. Homogeneity of Temperature and Relative Humidity of Air in Greenhouse. *Environment* **2014**, *1*, 12.
87. Bustamante, N.; Acuña, C.J.F.; Valera, D.L. Effect of the Height of the Greenhouse on the Plant-Climate Relationship as a Development Parameter in Mint (Mentha Spicata) Crops in Colombia. *Ing. Investig.* **2016**, *36*, 6–13. [CrossRef]

sustainability

MDPI

Article

Long-Term Fertilizer Reduction in Greenhouse Tomato-Cucumber Rotation System to Assess N Utilization, Leaching, and Cost Efficiency

Yucong Geng [1,†], Muhammad Amjad Bashir [1,2,†], Ying Zhao [3], Jianhang Luo [3], Xiaotong Liu [3], Feng Li [4], Hongyuan Wang [1], Qurat-Ul-Ain Raza [5], Abdur Rehim [2,5], Xuejun Zhang [3,*] and Hongbin Liu [1]

1 Key Laboratory of Nonpoint Source Pollution Control, Ministry of Agriculture, Institute of Agricultural Resources and Regional Planning, Chinese Academy of Agricultural Sciences (CAAS), Beijing 100081, China; gengyucong@caas.cn (Y.G.); amjad.bashir941@gmail.com (M.A.B.); wanghongyuan@caas.cn (H.W.); liuhongbin@caas.cn (H.L.)
2 College of Agriculture, Bahadur Sub-Campus Layyah, Bahauddin Zakariya University, Multan 60800, Pakistan; abdur.rehim@bzu.edu.pk
3 Institute of Agricultural Resources and Environment, NingXia Academy of Agricultural and Forestry Sciences, 590 Huanghe Donglu Road, Yinchuan 750002, China; tony029@163.com (Y.Z.); daguolitz@sina.com (J.L.); tz_azzfighting@163.com (X.L.)
4 Institute of Agricultural Economy and Information Technology, NingXia Academy of Agricultural and Forestry Sciences, 590 Huanghe Donglu Road, Yinchuan 750002, China; nxlfeng@163.com
5 Department of Soil Science, FAS&T, Bahauddin Zakariya University, Multan 60800, Pakistan; quratulain1111@yahoo.com
* Correspondence: zhxjun2002@163.com
† The authors contributed equally and should be considered co-first authors.

Citation: Geng, Y.; Bashir, M.A.; Zhao, Y.; Luo, J.; Liu, X.; Li, F.; Wang, H.; Raza, Q.-U.-A.; Rehim, A.; Zhang, X.; et al. Long-Term Fertilizer Reduction in Greenhouse Tomato-Cucumber Rotation System to Assess N Utilization, Leaching, and Cost Efficiency. *Sustainability* **2022**, *14*, 4647. https://doi.org/10.3390/su14084647

Academic Editor: Domenico Ronga

Received: 20 March 2022
Accepted: 8 April 2022
Published: 13 April 2022

Publisher's Note: MDPI stays neutral with regard to jurisdictional claims in published maps and institutional affiliations.

Abstract: Technology-oriented approaches to reduce chemical fertilization in agroecosystems without influencing the vegetable yield is a proficient method for sustainable agriculture and environmental safety. However, short-term studies are not capable to indicate the effects of various inputs in a long run; therefore, a six-year experiment was conducted in Ningxia, China. The experimental units were: no fertilizer control (CK: 0 kg N ha^{-1}), chicken manure (M: organic N 362 kg ha^{-1}), reduced chemical fertilizer (RCF: chemical N 992 kg ha^{-1} + organic N 362 kg ha^{-1}), and conventional fertilizer (CF: chemical N 1515 kg ha^{-1} + organic N 362 kg ha^{-1}). The study aimed to assess the effects of reduced fertilization on yield, nitrogen (N) use efficiency (NUE), N leaching, and the economic benefits. The results achieved herein indicate that RCF has significantly improved the NUE, reduced N leaching (23.7%), and improved the N economic benefit (NEB, 41.8%) as compared to the CF. Based on yield, net benefit and NUE conditions, the optimum N application range was 634–821 kg N ha^{-1} for tomato and 556–778 kg N ha^{-1} for cucumber. The study concluded that reduced chemical application of N is an important factor to control environmental pollution and improve fertilizer use efficiency. Further experiments are suggested to examine the optimum N rate provision from chemical fertilizer and its ratio with organic fertilization.

Keywords: vegetable yield; nitrogen use efficiency; nutrient leaching; leaching-to-input ratio; nitrogen fertilizer economic benefit

1. Introduction

Greenhouse vegetable production has the advantages of less planting area requirement and relatively higher yield [1]. Increased vegetable consumption and farmers' increased per-capita earning expectations have promoted greenhouse vegetable production globally [2,3]. In 2009, the area under greenhouse vegetable cultivation in China was 3.35 million hectares, and the country had the highest yield of greenhouse vegetables globally [4]. By 2013, Chinese production accounted for 50% of world vegetable production, and the area under cultivation in China for greenhouse vegetables had increased to more than 3.7 million hectares [5]. In 2016, the area under greenhouse vegetable cultivation in China was 3.91 million hectares, which accounted for 21.5% of the total planting area and produced 30.5% of the total yield in China [6,7].

To ensure maximum production from greenhouse vegetables, surplus fertilization is a common practice, which results in low fertilizer use efficiency [8,9]. Nitrogen (N) is the most important nutrient required by vegetables to maximize production; thus, it is essential to determine the appropriate fertilization rate. The rate of N uptake from different N sources by vegetables was observed to be less than 400 kg N ha^{-1} [10]. The average soil N application rate for greenhouse vegetables was 1732 kg ha^{-1} year^{-1} in Beijing, which is nearly four times the rate at which vegetables can absorb N [7]. The average N fertilization rate for greenhouse vegetables reached 4088 kg ha^{-1} year^{-1}, with an NUE less than 10%, in a vegetable production area in Shouguang, Northern China [11,12]. For vegetable production in Northern China, Ju et al. [13] reported annual average inputs of N from chemical fertilizer, manure, and irrigation water, of 1358, 1881, and 402 kg ha^{-1}, respectively, totaling 3641 kg ha^{-1}; this is nearly nine times more N than vegetables can absorb [13]. Excessive N input causes N to accumulate in the soil [14]. Whether from natural or anthropogenic, this leads to N leaching and volatilization losses that are the major sources of non-point source pollution [15,16]. Furthermore, it also is the major cause of severe nitrate leaching and increases the risk of groundwater pollution [17,18].

To address these problems, fertilizer reduction technology (FRT) has been investigated in the United States and other developed countries since the 1980s [19]. Reducing N fertilization to 200 kg ha^{-1} resulted in satisfactory production and good vegetable quality [20]. A reduction of 40% N fertilization in greenhouse vegetable production reduces N leaching loss by 39.6% without affecting the yield [21]. Reducing N application from 360 kg ha^{-1} to 240 kg ha^{-1} can enhance yield (8.8%) and N agronomic efficiency (51.3%) in a greenhouse cucumber experiment [21,22]. Fertilizer N reductions of 20% and 50% can reduce total N (TN) leaching by 18.3% and 43.0%, respectively in the cucumber–cabbage season [22]. Therefore, FRT is an effective technique to not only improve economic benefit but also ensure sustainable vegetable development with improved N use efficiency (NUE) and less N leaching.

The Ningxia Plain is a vital vegetable production area located upstream of the Yellow River and is developing rapidly in recent years. Higher demands for vegetables have forced local farmers to apply a higher dose of chemical fertilizers in this area [23]. The adaptation of FRT is difficult in Ningxia Plain as well as other rural regions in China due to fear of less production, ineffective soil testing services, ignorance of environmental conditions, and lack of expertise [24]. Some recent cases have proved that training/seminars are effective tools in guiding farmers to reduce N fertilizer input for agricultural production [25–27]. Nevertheless, after training, they may restart the previous practice of excessive N fertilizer application, believing that the continuous reduction of N fertilizer application in the long term may reduce the TN supply in the soil and have a negative impact on crop yield [28]. This gives rise to the need for this study to identify the long-term influences of FRT vegetable yield, environmental pollution, and profitability of vegetable production systems.

Keeping an eye on the need of this study, we conducted a six-year in situ study aiming to measure the effects of conventional, reduced fertilization, and organic fertilization. The major objectives of this study are to (1) measure the influence of reduced fertilization on vegetable yield and NUE, (2) identify the key characteristics of reduced fertilization with respect to N leaching, and (3) estimate the N fertilizer economic benefit.

2. Materials and Methods

2.1. Site Description

The greenhouse experiment was conducted at Yellow River Irrigation Region of Ningxia Plain at the NXL01 land block (38.4475 N, 106.3575 E), in Ningxia province, China, at an altitude of 1000 m. This experiment was launched over six years (December 2008 to November 2013). The mean annual rainfall and air temperatures were 233 mm and 9 °C, respectively. The tomato–cucumber rotation vegetable system was used: tomatoes were grown and harvested first, then fallow period, followed by cucumbers, with one rotation per year. The soil is classified as Sandy Loam Soil (USDA system) in the study area. The physicochemical properties of the soil were as: clay 14%, silt 30%, sand 56%, bulk density 1.37 g cm^{-3}, moisture content 10.3%, pH 8.27, 30.10 g kg^{-1} organic matter content, 2.42 g kg^{-1} total N, 2.14 g kg^{-1} total P, 302.40 mg kg^{-1} Olsen-P, and 390.00 mg kg^{-1} extractable K.

2.2. Experimental Design

Four treatments were applied during the tomato and cucumber growing seasons following the local experts' recommendations: control (CK: 0 kg N ha^{-1}); 100% chemical N reduction fertilization, using a single application of organic fertilizer in the form of manure (M: 362 kg N ha^{-1}); reduced conventional fertilization (RCF: chemical N 992 kg ha^{-1} + organic kg N 362 ha^{-1}), in which the chemical fertilizer N contents were reduced by, 38.7% and 28.8% on average, for tomatoes and cucumbers, respectively, relative to the CF treatment; conventional fertilization (CF: chemical N 1515 kg ha^{-1} + organic N 362 kg ha^{-1}). Phosphate (triple superphosphate, 46% P_2O_5) and manure fertilizers (chicken manure, with average N, P_2O_5, and K_2O contents of 1.00%, 1.32%, and 1.88%, respectively) were applied as base fertilizers before each tomato and cucumber season. Urea (46% N) and potassium sulfate (50% K_2O) fertilizers were split into base and topdressing applications and were hand broadcasted. The vegetables were transplanted to seedbeds (600 cm long × 130 cm wide) with plant spacing of 40 cm and row spacing of 75 cm for tomatoes, and plant spacing of 30 cm and row spacing of 70 cm for cucumbers. There were three replicates for each treatment and were arranged according to randomized complete block design (RCBD). The greenhouse was covered with plastic sheet. The light of Ningxia greenhouse is sunlight without any artificial light. The indoor temperature was controlled at 14–18 °C at night and 25–30 °C during the day by sunlight, covering quilt and uncovering shed film. Weeding and pesticide application were according to the local conventional practices. Yellow River water and groundwater were used for irrigation.

2.3. Evaluation of N Use Efficiency

Tomato and cucumber samples were collected, and yields were measured for each harvest. The Kjeldahl method was used to analyze N content, as described by Yang et al. [15]. The following indicators were calculated, using data collected over the entire study, to evaluate NUE: apparent recovery efficiency of applied N (RE_N, %), agronomic effectiveness of applied N (AE_N, kg kg^{-1}), physiological efficiency of applied N (PE_N, kg kg^{-1}), and partial factor productivity of applied N (PFP_N, kg kg^{-1}) [29]:

$$RE_N = \frac{TU_N - T_{CK}}{F_N}$$

$$AE_N = \frac{Y_N - Y_{CK}}{F_N}$$

$$PE_N = \frac{Y_N - Y_{CK}}{TU_N - T_{CK}}$$

$$PFP_N = \frac{Y_N}{F_N}$$

where TU_N is TN uptake, T_{CK} is TN uptake without N application, F_N is applied fertilizer N, Y_N is annual tomato/cucumber (tomato followed by cucumber) yield, and Y_{CK} is the tomato/cucumber yield without N application (all expressed in kg ha^{-1}). The relationships between these parameters and annual N application rate were examined using exponential, linear, logarithmic, and power functions.

2.4. Leachate Collection and Measurement of Nitrate Losses Due to Leaching

Using a leachate collection device as described by Zhao [23], leachate was collected 3 d after each irrigation. The leachate collection device and containers were cleaned before use. The samples were stored at −20 °C in a refrigerator, and the TN content of the leachate was analyzed using the alkaline potassium persulfate digestion UV spectrophotometric method [23]. Approximate N input and leaching were calculated as follows:

$$NL_{CC}\left(\text{kg ha}^{-1}\text{ year}^{-1}\right) = NL_{CT} - NL_{CO}$$

$$NLIR_{TN}\ (\%) = \frac{NI_{TN}}{NL_{TN}} \times 100\%$$

$$NLIR_{ON}\ (\%) = \frac{NI_{ON}}{NL_{ON}} \times 100\%$$

$$NLIR_{CN}\ (\%) = \frac{NI_{CN}}{NL_{CN}} \times 100\%$$

where NL_{CC} is N leaching caused by input of chemical fertilizer N, NL_{CT} is N leaching caused by TN input, NL_{CO} is N leaching caused by organic N input, $NLIR_{TN}$ is the TN leaching-to-input ratio, $NLIR_{ON}$ is the organic N leaching-to-input ratio, $NLIR_{CN}$ is the chemical N leaching-to-input ratio, NI_{TN} is TN input, NI_{ON} is organic N input, NI_{CN} is chemical N input, NL_{TN} is TN leached, NL_{ON} is organic N leached, and NL_{CN} is the amount of chemical fertilizer N leached.

2.5. Economic Analysis

The N fertilizer economic benefit (*NEB*, in USD t^{-1} N) and the input–output ratio, were used to calculate the economic benefits of N reduction.

$$NEB = \frac{B_N - B_{CK}}{TF_I}$$

$$Input - to - output\ ratio = \frac{T_I}{T_E}$$

where B_N benefits from N *input* (USD ha^{-1} year^{-1}), B_{CK} is benefited without N input, TF_I is total fertilizer N input (t ha^{-1} year^{-1}), T_I is total income (USD ha^{-1} year^{-1}), and T_E is total expenditure (USD ha^{-1} year^{-1}).

2.6. Statistical Analysis

Descriptive data analysis was conducted, and graphs were created, using Microsoft Excel 2013 (Microsoft Corp., Redmond, WA, USA). The results are expressed as means (with standard error, SE) of the three replicates. We estimated fruit yield, NUE, N leaching, and NEB. One-way ANOVA with Duncan multiple comparison test was used to assess differences among the treatments. All statistical analyses were performed using SPSS 19.0 (IBM Corporation, Armonk, NY, USA). p-values < 0.05 were considered statistically significant.

3. Results

3.1. Tomato and Cucumber Yield

The individual yield of tomato and cucumber, as well as the combined yield of both crops (Figure 1a–c), differed significantly ($p < 0.01$) between years, treatments, and with the interaction of year and treatment. The average annual yield for the treatments relative to the CF is shown in Figure 1d–f.

For the tomato rotations, there was no significant difference between the four treatments in the first year (2008). The yield of the control treatment was significantly lower in 2009 by 16.7% relative to the CF treatment, and the yield gap increased annually from 2009 to 2012. The yield of the M treatment did not decline until 2010, when it declined by 5.1%, relative to the CF treatment. There was no significant difference in yield between the CF and RCF treatments.

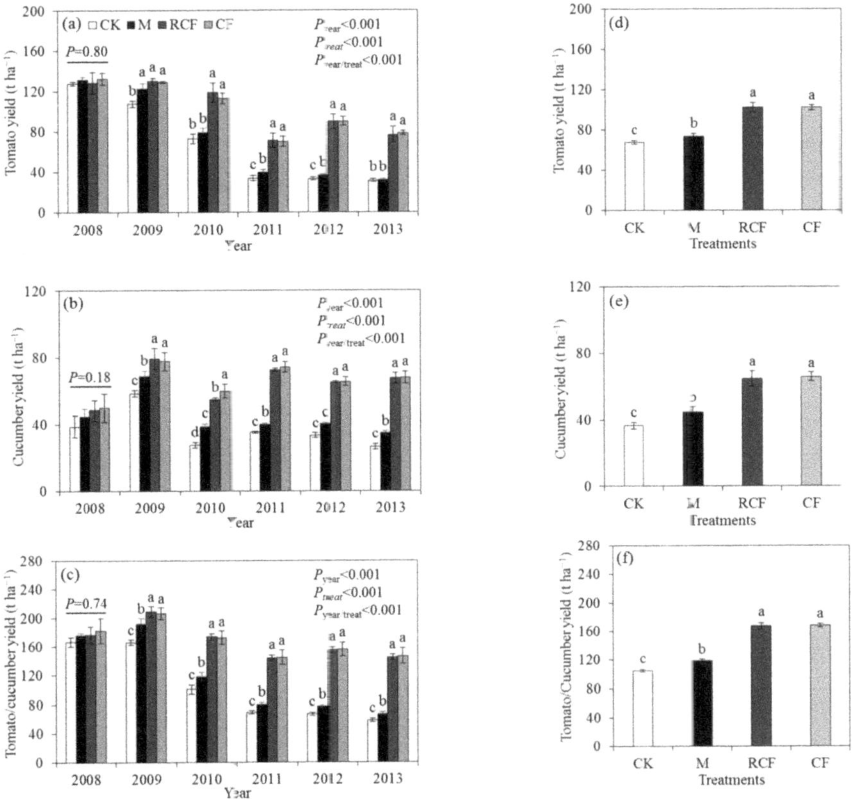

Figure 1. Fruit yield of (**a**) tomato, (**b**) cucumber, and (**c**) tomato/cucumber rotation. Average annual yield (**d–f**) in the control (CK), manure (M), and reduced conventional fertilization (RCF), relative to conventional fertilization (CF) for 2008–2013. The error bars indicate the standard deviations. CK: 0 kg ha^{-1} N; M: 362 kg ha^{-1} organic N; RCF: 992 chemical N + 362 kg ha^{-1} organic N; CF: 1515 kg ha^{-1} chemical N + 362 kg ha^{-1} organic N. The lowercase letters indicate the significant difference among the means each year, while each *p* value given upside right indicate the two-way ANOVA with year and treatment interaction.

Similarly, for cucumber rotations, there was no significant difference between treatments in the first year (2008). The yields of the CK and M treatments were significantly lower in 2009 (by 24.6% and 11.5%, respectively) relative to the CF treatment. There were no significant differences in yields between the CF and RCF treatments throughout the experimental period.

For the first year of tomato/cucumber rotation, there was no significant difference in yield between the four groups. Significant differences occurred from the second year (2009) to the end of the experiment. Relative to the CF treatment, the yield was reduced in the CK and M treatments and the yield gap increased annually. On average, the fruit yield of the M treatment was significantly higher (by 8.2%) than CK but was significantly lower than CF (by 28.2%) and RCF (by 28.4%) treatments. Moreover, there was no significant difference between the CF and RCF treatments.

3.2. Nitrogen Use Efficiency

RE_N, AE_N, PE_N, and PFP_N (Figure 2a–d) differed significantly ($p < 0.01$) between years and treatments. The RE_N, AE_N, and PFP_N of M showed a decreasing trend over the years. The RE_N, AE_N, and PE_N showed an upward trend to a relatively stable level year by year. By the end of this experiment, RE_N was 11.0, 9.0, and 5.7%, AE_N was 66.7, 43.2, and 18.5 kg kg^{-1}, and PE_N was 60.7, 43.2, and 31.6 kg kg^{-1} in high to low order of RCF, CF and M, respectively. PFP was highest for M (17.9 kg kg^{-1}), followed by RCF (11.1 kg kg^{-1}) and CF (7.2 kg kg^{-1}).

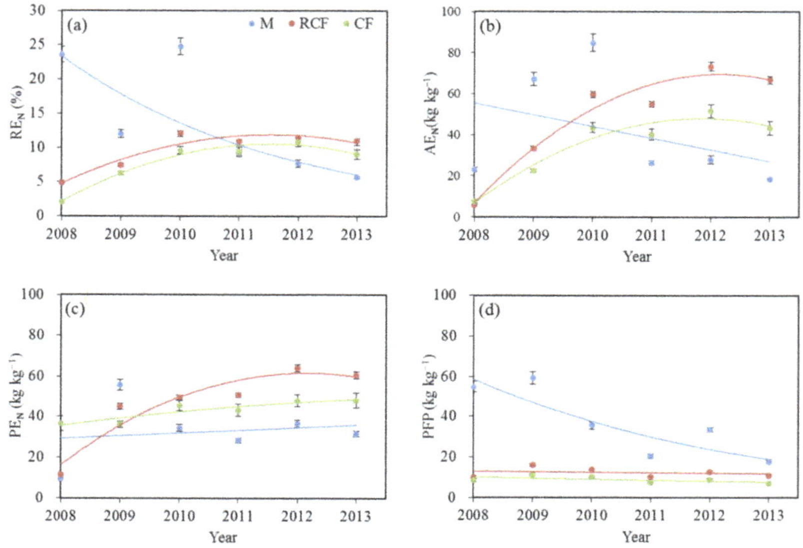

Figure 2. N use efficiency for 2008–2013. (**a**) Recovery efficiency (RE_N); (**b**) agronomic effectiveness. (AE_N); (**c**) physiological efficiency (PE_N); (**d**) partial factor productivity (PFP_N). Error bars reflect the standard deviations. Blue: M, 362 kg ha^{-1} organic N. Red: RCF, 992 kg ha^{-1} chemical N + 362 kg ha^{-1} organic N; Green: CF, 1515 kg ha^{-1} chemical N + 362 kg ha^{-1} organic N. The lines indicate the annual general tendency.

3.3. N Leaching

The N leaching data are shown in Figure 3. For all rotations, the amount of leached N is the sum of the leaching results measured during all four irrigation cycles throughout the tomato production period. Because of a shortage of water from the Yellow River in 2008, no irrigation was applied during the fallow period in that year. From 2009 to 2012, TN leaching showed an annual cycle for the CF and RCF treatments: it was high during the tomato stage, lowest in the fallow stage, and highest in the cucumber stage, each year. The average annual TN leaching rates from high to low were CF (170.7 kg N ha^{-1}), RCF (130.2 kg N ha^{-1}, 23.7% lower), M (92.0 kg N ha^{-1}, 46.1% lower), and CK (69.0 kg N ha^{-1}, 59.6% lower), respectively. For the CK, CF, RCF, and M treatments, the average annual leaching rates for the tomato stages were 28.5, 58.3, 45.4, and 35.7 kg N ha^{-1}, accounting for 41.2%, 34.1%, 34.8%, and 38.8%, respectively, of the applied N; for the fallow period, they were 15.8, 26.9, 21.9, and 19.4 kg N ha^{-1}, accounting for 22.9%, 15.7%, 16.8%, and 21.1%, respectively; for the cucumber stages, they were 24.8, 85.6, 63.0, and 39.6 kg N ha^{-1}, accounting for 35.8%, 50.1%, 48.3%, and 40.1%, respectively.

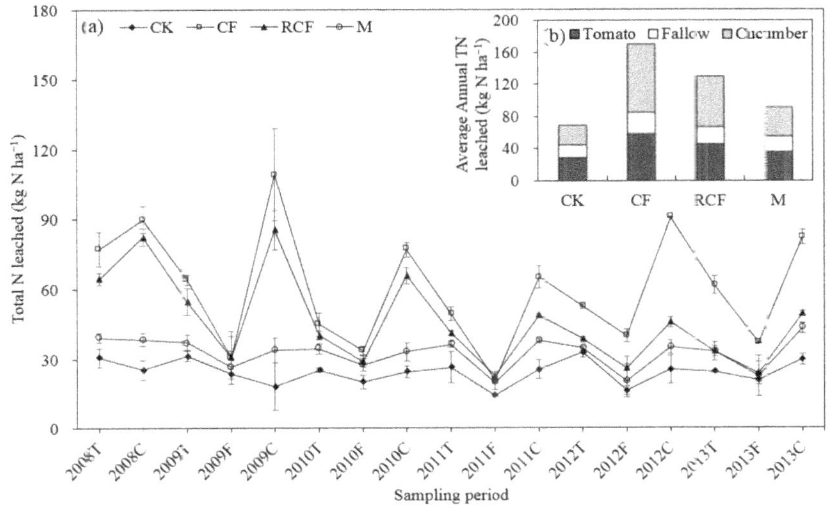

Figure 3. N leaching for 2008–2013. (**a**) Relationship between the sampling period and total nitrogen (TN) leaching. (**b**) TN leaching by treatment and vegetable rotation. T: tomato; C: cucumber; F: fallow. CK: 0 kg ha^{-1} N; CF: 1515 kg ha^{-1} chemical fertilizer N + 362 kg ha^{-1} organic N; RCF: 992 kg ha^{-1} chemical fertilizer N – 362 kg ha^{-1} organic N; M: 362 kg ha^{-1} organic N.

3.4. Economic Analysis

Total expenditure, total income, and TN input are shown in Table 1. The costs for field management were USD 98, 210, 210, and 112, for the CK, CF, RCF, and M treatments, respectively, based on USD 14 d^{-1} for labor times of 7, 15, 15, and 8 d, respectively. Each year, 45,000 tomato and 48,000 cucumber seedlings were planted per hectare, at a cost of USD 122.4 and 121.2 per thousand, respectively.

Table 1. Values used to calculate the average annual economic benefit (USD t^{-1}) based on six rotations (one rotation per year from 2008 to 2013) of greenhouse-cultivated tomato and cucumber plants.

Items	Details	Price (USD t^{-1})	Groups: CK	CF	RCF	M
Fertilizer	Urea	277	0	906	593	0
	$Ca(H_2PO_4)_2$	403	0	754	544	0
	K_2SO_4	337	0	809	792	0
	Chicken manure	88	0	1953	1953	1953
Seed	Tomato		5506	5506	5506	5506
	Cucumber		5817	5817	5817	5817
Pesticides and herbicides	Tomato		126	126	126	126
	Cucumber		84	84	84	84
Field management	Labor		98	210	210	112
Total expenditure			11,631	16,165	15,625	13,584
Fruit income	Tomato	281	10,309	18,511	18,174	11,713
	Cucumber	225	15,253	23,006	23,062	17,416
Total income			25,562	41,517	41,236	29,129
Net benefit (USD ha^{-1} year^{-1})			13,931	25,352	25,611	15,545
Total-N input (kg ha^{-1} year^{-1})			0	1877	1354	362
Benefit from N input (USD ha^{-1} year^{-1})			0	11,421	11,680	1614
NEB (USD t^{-1} N) [a]			0	6085	8626	4458
Input–output ratio			1:2.20	1:2.57	1:2.64	1:2.14

[a] NEB: N fertilizer economic benefit.

The net benefit, NEB, and the input–output ratios (Table 1) were used as the main factors in the economic analysis of N fertilization. The highest net benefit was RCF, which was 1, 64.8, and 83.8% higher than CF, M, and CK, respectively. The RCF group produced the highest NEB, which is 41.8% higher than CF and 93.5% higher than M. The input–output ratio was highest for RCF, followed by CF, CK, and M.

4. Discussion

4.1. Fruit Yield and N Use Efficiency

Nitrogen fertilizer is typically used to improve crop yield [30,31]. However, excessive N input does not increase yields [31–33]. In our six-year greenhouse experiment, we found that reducing a certain amount of N content did not significantly reduce fruit yield, relative to the conventional fertilization used by local farmers. This indicates that the local conventional fertilization practice provides excessive fertilization. Reducing chemical fertilizer application from 1027 kg ha^{-1} N to 692 kg ha^{-1} N by 38.7% (335 kg ha^{-1} N) for tomato and from 850 kg ha^{-1} N to 662 kg ha^{-1} N by 28.8% (188 kg ha^{-1} N) for cucumber did not negatively affect fruit yield.

In the first year, the yield of the control treatment was not significantly different from that of the other treatments. This reflects previous excessive fertilization, leading to N accumulation in the soil, which supported growth during the first tomato season. However, the yield of the control treatment was lower in 2009, indicating that there was insufficient accumulated N in the soil to support tomato growth after N absorption by tomatoes and cucumbers in 2008. Based on this, crop yields could be severely reduced when no fertilizer is applied.

Organic fertilizer supplementation can improve soil biological quality and function, thereby further improving crop yield [34,35]. This view has been generally accepted by local farmers in the region. When we applied organic fertilizer only, with an average annual N input of 362 kg ha^{-1}, tomato yield was the same as that of the reduced chemical fertilizer treatment for the first two years. The significant yield reduction occurred in the third year. We speculate that a single application of organic N can delay the reduction in vegetable yield, even when no chemical fertilizer N is used; however, the organic fertilizer level in this research was insufficient to support production in the third year.

The average fruit yield was 13.8% higher in the organic (M) treatment (at 119.2 t ha^{-1}) than in the control (CK) treatment (at 104.7 t ha^{-1}). However, only organic N application was substantially lower than those of the conventional and reduced fertilizer treatments (168.4 and 167.4 kg ha^{-1}, respectively). These results indicate that reducing 100% of chemical fertilizer reduces yield. As expected, 27.8% reduced fertilization (from 1877 to 1354 kg ha^{-1}) produced similar yields to conventional fertilization. This is consistent with previous findings that yields can be maintained under appropriate chemical fertilizer reduction [36–38], this could be associated with the slow or insufficient N supply from the organic source.

Excessive fertilization not only contributes little in terms of increasing yield but also increases N accumulation in soil [39]. Due to the high background nitrogen accumulation in the soil, long-term observation is needed to obtain relatively accurate results in the comparison of different nitrogen application levels. As shown in Figure 2, the RE_N and AE_N levels in CF and RCF exceeded M in the fourth year. And the PE_N level in CF and RCF exceeded M in the third year. Then, the situation was kept for the following years. Therefore, we believe that the evaluation of NUE with the results of the sixth year is reliable.

The relationship between CF and RCF for NUE in this research indicates that a moderate reduction in chemical fertilizer N application could improve NUE in an organic–inorganic fertilization situation. The main reason for this might be the balanced N supply from both sources, quicker at early stages from an inorganic source, and slow release throughout the cropping time by an organic source. Our observations are consistent with previous findings that moderate chemical fertilizer reductions can improve N use efficiency in areas that have been over-fertilized [40]. The NUE is typically negatively correlated with the N fertilizer application rate [41]. The PFP in our research have the same situation. Contrary to expectations, in spite of the N application being lower in M than that of CF, the RE_N, AE_N, and PE_N were also lower than that of CF in the end. It can be inferred that compared with the application of organic fertilizer alone, the combined application of organic and inorganic fertilizer can effectively improve RE_N, AE_N, and PE_N.

4.2. N Leaching

It has been reported that N leaching rates differ between years (2008–2014) for crops field [15]. Compared with short-term batch tests, long-term positioning tests are more reliable for comparing differences in N leaching among treatments [7,15]. Soil N leaching is related to excessive irrigation, heavy rainfall, over-fertilization, and poor tillage management [42]. In our six-year greenhouse positioning experiment, by controlling for irrigation and tillage mode, differences in leaching were explained mostly by differences in fertilizer input. Reducing chemical fertilizer N application by 34.5% from conventional levels reduced TN leaching significantly by 39.8%, and when using only organic manure fertilizer, reduced it by 77.4% (Figure 3b). This indicates that N leaching is directly related to the amount of chemical fertilizer used; therefore, reducing chemical fertilizer N input is an effective way to reduce N leaching, which is consistent with the previous studies [33,43].

During the fallow period, no fertilization was applied, and irrigation was reduced, causing TN leaching to be relatively low (15.7–22.9%). In the CK treatment, TN leaching was 5.4% higher for tomatoes than for cucumbers. In contrast, tomatoes had lower TN leaching in the CF, RCF, and M treatments by 16.0%, 13.5%, and 1.2%, respectively, compared with cucumbers. This indicates that relative to tomato cultivation, cucumbers have a higher risk of leaching caused by fertilization. Hence, reducing N application for cucumber cultivation has the potential to reduce N leaching.

Organic fertilizer application at 100 and 200 kg N ha^{-1} $year^{-1}$ N was shown to cause N leaching at 85.2 and 105.5 kg ha^{-1}, which were 18.3 and 38.6 kg ha^{-1} higher than control (66.9 kg ha^{-1}), respectively [44]. A greenhouse vegetable study found that applying poultry manure at less than 217.7 kg ha^{-2} N did not negatively affect groundwater; however, double N application will lead to an increase in leaching N [45]. In our study, the organic fertilizer N input was 362 kg ha^{-1} $year^{-1}$, which carries the risk of enhancing N loss. Nitrogen leaching was 92 kg ha^{-1} $year^{-1}$ in the M treatment and 69 kg ha^{-1} $year^{-1}$ in the control counterpart (Table 2); note that this difference of 23 kg ha^{-1} $year^{-1}$ was caused by organic N input. Then, the N leaching caused by the chemical N input part of CF and RCF were 78 and 39 kg ha^{-1} $year^{-1}$, respectively. Furthermore, the organic fertilizer N leaching-to-input ratio ($NLIR_{ON}$) was 6.4%, which was significantly higher (by 42.2%) than that of the conventional fertilizer treatment (4.5%), and 178.3% greater than that of the reduced fertilizer treatment (77.8%). The N leaching-to-input ratio ($NLIR_{CN}$) in CF and RCF was 3.5% and 2.3%, which were 45.3% and 64.0% lower than $NLIR_{ON}$, respectively. Therefore, we found that organic fertilization carries a higher risk of N leaching than chemical fertilization, which is consistent with some earlier findings [46,47], despite more research having widely shown that the application of organic N fertilizer would cause a decrease in N leaching than inorganic [48–50]. Most of the nitrogen in organic fertilizer exists in the form of macromolecular, which can only be absorbed by plants after a certain period of dissolution by a series of microbial-mediated steps [51,52]. Therefore, compared with inorganic nitrogen from chemical N, it is difficult to be directly used by crops. These macromolecular N and dissolved N from organic fertilizer would also be lost by leaching when irrigation happened [53]. Therefore, we speculate that slow nitrogen dissolution, limited uptake by plants, and solubility of organic N are the main reasons for the high N leaching loss of organic fertilizer. However, further research is still needed to confirm the details. Previous studies have found that a combination of chemical and organic fertilizer is a sustainable fertilization approach [54], which can promote crop productivity and N uptake, and reduces N losses [55,56]. Similarly, our findings support combined chemical and organic fertilization, while improving the activity of N-transforming microbes in the soil by organic manure, the deficiency of N leaching was balanced.

Table 2. Approximate fertilizer-N input and N leaching in the greenhouse vegetable cultivation system used in this study, from 2008 to 2013.

Items	CK	CF	RCF	M
N input (kg ha^{-1} year^{-1})				
Organic	0 ^{b}	362	362	362
Chemical	0	1515	992	0
Total	0	1877	1354	362
N leaching (kg ha^{-1} year^{-1})				
Total	69 ^{d}	170 ^{a}	131 ^{b}	92 ^{c}
NL_{TNC}	0	101 ^{a}	62 ^{b}	23 ^{c}
NL_{ON}	0	23	23	23
NL_{CN}	0	78 ^{a}	39 ^{b}	0.00
N leaching-input ratio (%)				
$NLIR_{TN}$	-	4.5 ^{b}	3.6 ^{c}	6.4 ^{a}
$NLIR_{ON}$	-	-	-	6.4 ^{a}
$NLIR_{CN}$	-	3.5	2.3	-

CK: control (0 kg ha^{-1} N); CF: conventional fertilization (1515 chemical fertilizer + 362 kg ha^{-1} organic N); RCF: reduced conventional fertilization (992 chemical fertilizer + 362 organic kg N ha^{-1}); M: manure fertilization (362 kg ha^{-1} organic N). Results are expressed as means. Lowercase letters after means identify groups that differ significantly between treatments ($p < 0.05$), by row. NL_{TN}: N leaching caused by TN input; NL_{ON}: N leaching caused by organic N input; NL_{CN}: N leaching caused by chemical N input; $NLIR_{TN}$: TN leaching-to-input ratio; $NLIR_{ON}$: organic N leaching-to-input ratio; $NLIR_{CN}$: chemical N leaching-to-input ratio.

4.3. Economic Benefit Analysis

Fertilizer N input is the main factor affecting crop yield and economic benefits [57]. We found that reduced chemical fertilization produced a similar economic net benefit to conventional fertilization, whereas 100% chemical fertilization reduction produced 38.7% less net benefit than conventional fertilization (Table 1). Although organic fertilization alone reduces production costs (the chemical fertilizer part), it also significantly reduces profitability. Therefore, moderate chemical fertilizer reduction was an effective way to balance production costs and maintain profitability. Reduced chemical fertilization also increased NEB by 41.8%, whereas organic fertilization alone reduced it by 26.7%. It can be inferred that the NEB of a single application of organic fertilizer was much lower than that of a mixed application of organic and chemical fertilizer. Further, it can be seen that organic fertilizer alone is inferior to chemical fertilizer for increasing economic growth. Although the advantages of organic fertilizer application are generally recognized, its relatively low N content and the large amounts required make its cost far higher than that of chemical fertilizer [58–60].

Ranking the treatments in terms of their input–output ratios, in ascending order, yielded the following order: M, CK, CF, and RCF. This result showed that only organic fertilizer will not bring economic benefits (M < CK), and an appropriate reduction in chemical fertilizer in areas where there has been excessive fertilization can effectively improve the economic benefit of N fertilization (RCF < CF). Similarly, for a rice–wheat rotation system, Wang et al. [61] found that a 50% reduction in chemical fertilizer N raised NEB by 320.8% and increased the input–output ratio from 1:3.0 to 1:4.0, relative to conventional tillage [62].

4.4. Selecting the Optimum N Application Rate

In conventional agricultural production in China, the yield and economic benefits related to the N application rate are the most important driving factors [63–65]. To reduce the high production costs and environmental pollution risks caused by excessive N input, sustainable agriculture aims to balance the N application rate, with both ecological and agronomic benefits [66–68]. Therefore, we aimed to determine the optimum range of N application rates to balance N leaching, vegetable yield, net benefit, and NEB in greenhouse vegetable farming. Organic fertilizer was applied at the same rate in the organic fertilization, conventional fertilization, and reduced conventional fertilization treatments. Figure 4 illustrates the effects of each treatment on yield, N leaching, net benefit, and the N fertilizer economic benefit.

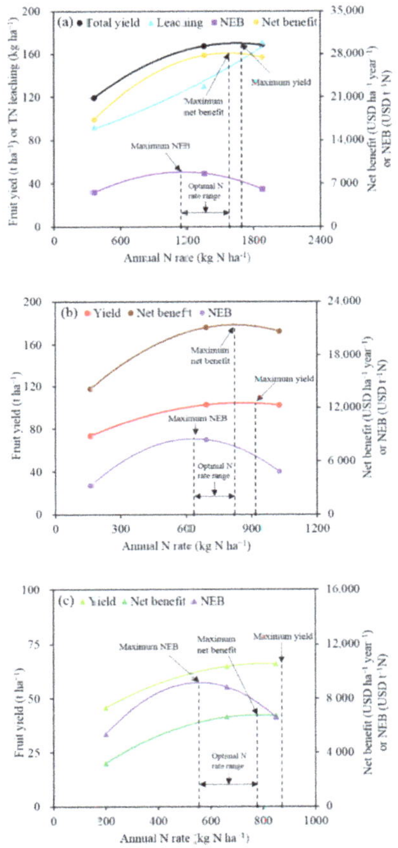

Figure 4. The optimum range of N application. (**a**) Regression models of tomato/cucumber yield, N leaching, N fertilizer economic benefit (NEB), and net benefit against the annual N application rate (2008–2013). Regression models of net benefit, N fertilizer economic benefit (NEB), tomato yield (**b**), and cucumber yield (**c**) against the annual N application rate (2008–2013).

Yang et al. [15] found that TN leaching at less than 170 kg ha^{-1} would cause no environmental harm. In our study, the highest average annual N leaching rate was 170.7 kg ha^{-1} in conventional fertilization treatment, at the N application rate of 1877 kg ha^{-1}, and N leaching was lower when the N application rate was reduced. Therefore, we consider a total average annual N input of less than 1877 kg ha^{-1} to be safe for the environment. In the tomato/cucumber rotation, the maximum yield was obtained at an annual N application rate of 1691 kg ha^{-1}, using 12.3% less chemical fertilizer N than conventional fertilization (Figure 4a). Generally, the N application rate that maximizes the crop's economic benefit is lower than that which maximizes yield [62,69]. Our findings were similar: the maximum net benefit occurred at an annual N application rate of 1583 kg ha^{-1} (108 kg ha^{-1} lower than the rate that maximized fruit yield), and using 19.4% less chemical fertilizer

N than conventional fertilization. The maximum NEB was obtained at the annual N rate of 1143 kg N ha^{-1}, using 48.4% less chemical fertilizer N than conventional fertilization. Maximum NEB reflects the most economical use of chemical N fertilizer. Further, using less N reduces N leaching and the risk of environmental pollution; hence, NEB may become a key indicator for determining chemical fertilizer N use in the future. We, therefore, recommend N application rates of 1143 and 1583 kg ha^{-1} for tomato and cucumber greenhouse farming, respectively, and reducing chemical fertilizer N application rates by 19.4% and 48.4%, respectively, relative to conventional fertilization. The maximum yield, net benefit, and NEB for tomatoes occurred at N application rates of 918, 821, and 634 kg ha^{-1}, respectively (Figure 4b); for cucumbers, they occurred at 871, 778, and 556 kg ha^{-1}, respectively (Figure 4c). In summary, our findings indicate that the optimal ranges of N application rates for greenhouse production of tomatoes and cucumbers are 634–821 kg ha^{-1} and 556–778 kg ha^{-1}, respectively, under current experimental conditions.

4.5. Practical Implementation of the Study

The practical application of this study will be an effective tool to control N pollution. Several studies, including the current one, discuss effectively the FRT reducing the crop yield. Some of the published material indicates that even a huge decrease in chemical fertilization can control the yield losses in combination to the environmental safety [21–23]. The integrative use of controlled chemical fertilization in combination with organic fertilization can significantly improve the yield attributes and cost effectiveness and control environmental pollution. It is further recommended to conduct future research experiments on combined fertilization of organic and inorganic N, trying various fertilization resources, site-specific and crop-specific studies, following the short-term and long-term studies.

5. Conclusions

Because it produces the highest yield and total income, conventional fertilization is highly preferred by farmers. In our study, conventional fertilization had the lowest N use efficiency and highest N leaching among the treatments, and lower economic benefit and NEB than reduced conventional fertilization. Reduced conventional fertilization produced the optimal N fertilization approach. Although organic fertilization provides benefits such as improving soil texture and soil microorganism diversity, the high N leaching ratio and low economic benefits caused by organic fertilization mean that it is not ideal. Therefore, it is feasible to use organic fertilizer with chemical fertilizer to produce an optimum fertilization scheme. Further studies are needed to explore the optimal reduction in chemical fertilizer use and optimal proportions of chemical and organic fertilizer.

Author Contributions: Conceptualization, Y.G., M.A.B., X.L., H.W., X.Z. and H.L.; data curation, Y.Z., J.L., F.L., Q.-U.-A.R. and A.R.; formal analysis, Y.G., J.L. and F.L.; funding acquisition, X.Z.; Investigation, Y.G. and Y.Z.; methodology, M.A.B., X.L., H.W., A.R. and H.L.; project administration, H.W. and X.Z.; Resources, X.L. and H.L.; software, M.A.B., Y.Z., J.L. and Q.-U.-A.R.; supervision, H.L.; validation, F.L.; writing—original draft, Y.G.; writing—review & editing, M.A.B., Y.Z., J.L., X.L., F.L., H.W., Q.-U.-A.R., A.R., X.Z. and H.L. All authors have read and agreed to the published version of the manuscript.

Funding: This study was supported by the National Natural Science Foundation of China [grant number: U20A20114].

Institutional Review Board Statement: Not applicable.

Informed Consent Statement: Not applicable.

Acknowledgments: We sincerely thank Jungai Li for her substantial help in the data analysis. We also thank Andre Brito and Luiz Silva for the help of data analysis and language polishing work.

Conflicts of Interest: The authors declare no conflict of interest.

References

1. Nordey, T.; Basset-mens, C.; de Bon, H.; Martin, T.; Déletré, E.; Simon, S.; Parrot, L.; Despretz, H.; Huat, J.; Biard, Y.; et al. Protected cultivation of vegetable crops in sub-Saharan Africa: Limits and prospects for smallholders: A review. *Agron. Sustain. Dev.* **2017**, *37*, 53. [CrossRef]
2. Chang, J.; Wu, X.; Liu, A.; Wang, Y.; Xu, B.; Yang, W.; Meyerson, L.; Gu, B.; Peng, C.; Ge, Y. Assessment of net ecosystem services of plastic greenhouse vegetable cultivation in China. *Ecol. Econ.* **2011**, *70*, 740–748. [CrossRef]
3. Xu, L.; Lu, A.; Wang, J.; Ma, Z.; Pan, L.; Feng, X.; Luan, Y. Accumulation status, sources and phytoavailability of metals in greenhouse vegetable production systems in Beijing, China. *Ecotoxicol. Environ. Saf.* **2015**, *122*, 214–220. [CrossRef] [PubMed]
4. Ministry of Agriculture China. *China Agriculture Statistical Report 2008*; China Agriculture Press: Beijing, China, 2009.
5. Hu, W.; Huang, B.; Tian, K.; Holm, P.E.; Zhang, Y. Heavy metals in intensive greenhouse vegetable production systems along Yellow Sea of China: Levels, transfer and health risk. *Chemosphere* **2017**, *167*, 82–90. [CrossRef]
6. Li, J.; Wan, X.; Liu, X.; Chen, Y.; Slaughter, L.C.; Weindorf, D.C.; Dong, Y. Changes in soil physical and chemical characteristics in intensively cultivated greenhouse vegetable fields in North China. *Soil Tillage Res.* **2019**, *195*, 104366. [CrossRef]
7. Li, J.; Liu, H.; Wang, H.; Luo, J.; Zhang, X.; Liu, Z.; Zhang, Y.; Zhai, L.; Lei, Q.; Ren, T.; et al. Managing irrigation and fertilization for the sustainable cultivation of greenhouse vegetables. *Agric. Water Manag.* **2018**, *210*, 354–363. [CrossRef]
8. Yang, X.; Lu, Y.; Ding, Y.; Yin, X.; Raza, S.; Tong, Y. Optimising nitrogen fertilisation: A key to improving nitrogen-use efficiency and minimising nitrate leaching losses in an intensive wheat/maize rotation (2008–2014). *Field Crop. Res.* **2017**, *206*, 1–10. [CrossRef]
9. Liang, H.; Hu, K.; Batchelor, W.; Qin, W.; Modelling, B.L.-E. Developing a water and nitrogen management model for greenhouse vegetable production in China: Sensitivity analysis and evaluation. *Ecol. Model.* **2018**, *367*, 24–33. [CrossRef]
10. Zhu, J.; Li, X.; Christie, P.; Agriculture, J.L. Environmental implications of low nitrogen use efficiency in excessively fertilized hot pepper (*Capsicum frutescens* L.) cropping systems. *Agric. Ecosyst. Environ.* **2005**, *111*, 70–80. [CrossRef]
11. Yu, H.; Li, T.; Zhang, X.Z. Nutrient budget and soil nutrient status in greenhouse system. *Agric. Sci. China* **2010**, *9*, 871–879. [CrossRef]
12. Hui-min, J.; Jian-feng, Z.; Xiao-zong, S.; Zhao-hui, L.I.U.; Li-hua, J. Responses of Agronomic Benefit and Soil Quality to Better Management of Nitrogen Fertilizer Application in Greenhouse Vegetable Land ∗ 1. *Pedosph. Int. J.* **2012**, *22*, 650–660. [CrossRef]
13. Ju, X.; Kou, C.; Zhang, F.; Christie, P. Nitrogen balance and groundwater nitrate contamination: Comparison among three intensive cropping systems on the North China Plain. *Environ. Pollut.* **2006**, *143*, 117–125. [CrossRef] [PubMed]
14. Wang, Y.; Li, K.; Tanaka, T.S.T.; Yang, D.; Inamura, T. Soil nitrate accumulation and leaching to groundwater during the entire vegetable phase following conversion from paddy rice. *Nutr. Cycl. Agroecosyst.* **2016**, *106*, 325–334. [CrossRef]
15. Lu, J.; Hu, T.; Zhang, B.; Wang, L.; Yang, S.; Fan, J.; Yan, S.; Zhang, F. Nitrogen fertilizer management effects on soil nitrate leaching, grain yield and economic benefit of summer maize in Northwest China. *Agric. Water Manag.* **2021**, *247*, 106739. [CrossRef]
16. Min, J.; Shi, W. Nitrogen discharge pathways in vegetable production as non-point sources of pollution and measures to control it. *Sci. Total Environ.* **2018**, *613–614*, 123–130. [CrossRef]
17. Kaushal, S.S.; Gro, P.M.; Band, L.E.; Elliott, E.M.; Shields, C.A.; Kendall, C. Tracking Nonpoint Source Nitrogen Pollution in Human-Impacted Watersheds. *Environ. Sci. Technol.* **2011**, *45*, 8225–8232. [CrossRef]
18. Perego, A.; Basile, A.; Bonfante, A.; de Mascellis, R.; Terribile, F.; Brenna, S.; Acutis, M. Nitrate leaching under maize cropping systems in Po Valley (Italy). *Agric. Ecosyst. Environ.* **2012**, *147*, 57–65. [CrossRef]
19. Lambert, D.K.; Lambert, D.K. Risk Considerations in the Reduction of Nitrogen Fertilizer Use in Agricultural Production. *West. J. Agric. Econ.* **1990**, *15*, 234–244.
20. Vieira, I.S.; Vasconcelos, E.P.; Monteiro, A. Nitrate accumulation, yield and leaf quality of turnip greens in response to nitrogen fertilisation. *Nutr. Cycl. Agroecosyst.* **1998**, *51*, 249–258. [CrossRef]
21. Min, J.; Zhang, H.; Shi, W. Optimizing nitrogen input to reduce nitrate leaching loss in greenhouse vegetable production. *Agric. Water Manag.* **2012**, *111*, 53–59. [CrossRef]
22. Wang, D.; Guo, L.; Zheng, L.; Zhang, Y.; Yang, R.; Li, M.; Ma, F.; Zhang, X.; Li, Y. Effects of nitrogen fertilizer and water management practices on nitrogen leaching from a typical open field used for vegetable planting in northern China. *Agric. Water Manag.* **2019**, *213*, 913–921. [CrossRef]
23. Zhao, Y.; Luo, J.H.; Chen, X.Q.; Zhang, X.J.; Zhang, W.L. Greenhouse tomato-cucumber yield and soil N leaching as affected by reducing N rate and adding manure: A case study in the Yellow River Irrigation Region China. *Nutr. Cycl. Agroecosyst.* **2012**, *94*, 221–235. [CrossRef]
24. Lu, Y.; Chadwick, D.R.; Norse, D.; Powlson, D.S.; Shi, W. Sustainable intensification of China's agriculture: The key role of nutrient management and climate change mitigation and adaptation. *Agric. Ecosyst. Environ.* **2005**, *209*, 1–4. [CrossRef]
25. Peng, S.; Buresh, R.J.; Huang, J.; Yang, J.; Zou, Y.; Zhong, X.; Wang, G.; Zhang, F. Strategies for overcoming low agronomic nitrogen use efficiency in irrigated rice systems in China. *Field Crop. Res.* **2006**, *96*, 37–47. [CrossRef]
26. Huang, J.; Xiang, C.; Jia, X.; Water, R.H.-J. Impacts of training on farmers' nitrogen use in maize production in Shandong, China. *J. Soil Water Conserv.* **2012**, *67*, 321–327. [CrossRef]
27. Jia, X.-p.; Huang, J.K.; Xiang, C.; Hou, L.-k.; Zhang, F.-s.; Chen, X.-p.; Cui, Z.-l.; Bergmann, H. Farmer's Adoption of Improved Nitrogen Management Strategies in Maize Production in China: An Experimental Knowledge Training. *J. Integr. Agric.* **2013**, *12*, 364–373. [CrossRef]

28. Huang, J.; Huang, Z.; Jia, X.; Hu, R.; Xiang, C. Long-term reduction of nitrogen fertilizer use through knowledge training in rice production in China. *Agric. Syst.* **2015**, *135*, 105–111. [CrossRef]
29. Qiao, J.; Yang, L.; Yan, T.; Xue, F.; Zhao, D. Nitrogen fertilizer reduction in rice production for two consecutive years in the Taihu Lake area. *Agric. Ecosyst. Environ.* **2012**, *146*, 103–112. [CrossRef]
30. Erisman, J.; Sutton, M.; Galloway, J.; Klimont, Z.; Winiwarter, W. How a century of ammonia synthesis changed the world. *Nat. Geosci.* **2008**, *1*, 636–639. [CrossRef]
31. Du, Y.D.; Gu, X.B.; Wang, J.W.; Niu, W.Q. Yield and gas exchange of greenhouse tomato at different nitrogen levels under aerated irrigation. *Sci. Total Environ.* **2019**, *668*, 1156–1164. [CrossRef]
32. Badr, M.A.; Abou-Hussein, S.D.; El-Tohamy, W.A. Tomato yield, nitrogen uptake and water use efficiency as affected by planting geometry and level of nitrogen in an arid region. *Agric. Water Manag.* **2016**, *169*, 90–97. [CrossRef]
33. Wang, H.; Li, J.; Cheng, M.; Zhang, F.; Wang, X.; Fan, J.; Wu, L.; Fang, D.; Zou, H.; Xiang, Y. Optimal drip fertigation management improves yield, quality, water and nitrogen use efficiency of greenhouse cucumber. *Sci. Hortic.* **2019**, *243*, 357–366. [CrossRef]
34. Sánchez-Monedero, M.A.; Mondini, C.; Cayuela, M.L.; Roig, A.; Contin, M.; De Nobili, M. Fluorescein diacetate hydrolysis, respiration and microbial biomass in freshly amended soils. *Biol. Fertil. Soils* **2008**, *44*, 885–890. [CrossRef]
35. Tao, R.; Liang, Y.; Wakelin, S.A.; Chu, G. Supplementing chemical fertilizer with an organic component increases soil biological function and quality. *Appl. Soil Ecol.* **2015**, *96*, 42–51. [CrossRef]
36. Gai, X.; Liu, H.; Zhai, L.; Tan, G.; Liu, J.; Ren, T.; Wang, H. Vegetable yields and soil biochemical properties as influenced by fertilization in Southern China. *Appl. Soil Ecol.* **2016**, *107*, 170–181. [CrossRef]
37. Yang, X.Y.; Sun, B.H.; Zhang, S.L. Trends of Yield and Soil Fertility in a Long-Term Wheat-Maize System. *J. Integr. Agric.* **2014**, *13*, 402–414. [CrossRef]
38. Zhang, W.; Xu, M.; Wang, X.; Huang, Q.; Nie, J.; Li, Z.; Li, S.; Hwang, S.W.; Lee, K.B. Effects of organic amendments on soil carbon sequestration in paddy fields of subtropical China. *J. Soils Sediments* **2012**, *12*, 457–470. [CrossRef]
39. Zhou, J.; Gu, B.; Schlesinger, W.H.; Ju, X. Significant accumulation of nitrate in Chinese semi-humid croplands. *Sci. Rep.* **2016**, *6*, 25088. [CrossRef]
40. Zhang, H.; Yu, C.; Kong, X.; Hou, D.; Gu, J.; Liu, L.; Wang, Z.; Yang, J. Progressive integrative crop managements increase grain yield, nitrogen use e ffi ciency and irrigation water productivity in rice. *Field Crop. Res.* **2018**, *215*, 1–11. [CrossRef]
41. Zhang, X.; Davidson, E.; Mauzerall, D.; Searchinger, T.D.; Dumas, P.; Shen, Y. Managing nitrogen for sustainable development. *Nature* **2015**, *528*, 51–59. [CrossRef]
42. Zhang, Y.; Wang, H.; Lei, Q.; Luo, J.; Lindsey, S.; Zhang, J.; Zhai, L.; Wu, S.; Zhang, J.; Liu, X.; et al. Optimizing the nitrogen application rate for maize and wheat based on yield and environment on the Northern China Plain. *Sci. Total Environ.* **2018**, *618*, 1173–1183. [CrossRef] [PubMed]
43. Carneiro, J.; Coutinho, J.; Trindade, H. Nitrate leaching from a maize× oats double-cropping forage system fertilized with organic residues under Mediterranean conditions. *Agric. Ecosyst. Environ.* **2012**, *160*, 29–39. [CrossRef]
44. Manfred, K.; Muller, J.; Isselstein, J. Nitrogen management in organic farming: Comparison of crop rotation residual effects on yields, N leaching and soil conditions. *Nutr. Cycl. Agroecosyst.* **2010**, *87*, 21–31. [CrossRef]
45. Alvarez González, C.E.; Gil, E.; Fernández-Falcón, M.; Hernández, M.M. Water leachates of nitrate nitrogen and cations from poultry manure added to an Alfisol Udalf soil. *Water Air Soil Pollut.* **2009**, *202*, 273–288. [CrossRef]
46. Chambers, B.J.; Smith, K.A.; Pain, B.F. Strategies to encourage better use of nitrogen in animal manures. *Soil Use Manag.* **2000**, *16*, 157–166. [CrossRef]
47. Shepherd, M.; Newell-Price, P. Manure management practices applied to a seven-course rotation on a sandy soil: Effects on nitrate leaching. *Soil Use Manag.* **2013**, *29*, 210–219. [CrossRef]
48. Daudén, A.; Quílez, D. Pig slurry versus mineral fertilization on corn yield and nitrate leaching in a Mediterranean irrigated environment. *Eur. J. Agron.* **2004**, *21*, 7–19. [CrossRef]
49. Zavattaro, L.; Monaco, S.; Sacco, D.; Grignani, C. Options to reduce N loss from maize in intensive cropping systems in Northern Italy. *Agric. Ecosyst. Environ.* **2012**, *147*, 24–35. [CrossRef]
50. Zhou, M.; Zhu, B.; Brüggemann, N.; Bergmann, J.; Wang, Y.; Butterbach-Bahl, K. N_2O and CH_4 emissions, and NO3- leaching on a crop-yield basis from a subtropical rain-fed wheat-maize rotation in response to different types of nitrogen fertilizer. *Ecosystems* **2014**, *17*, 286–301. [CrossRef]
51. Fangueiro, D.; Coutinho, J.; Borges, L.; Vasconcelos, E. Recovery efficiency of nitrogen from liquid and solid fractions of pig slurry obtained using different separation technologies. *J. Plant Nutr. Soil Sci.* **2015**, *178*, 229–236. [CrossRef]
52. Prendergast-Miller, M.; de Menezes, A.B.; Farrell, M.; Macdonald, L.M.; Richardson, A.E.; Bissett, A.; Toscas, P.; Baker, G.; Wark, T.; Thrall, P.H. Soil nitrogen pools and turnover in native woodland and managed pasture soils. *Soil Biol. Biochem.* **2015**, *85*, 63–71. [CrossRef]
53. Macdonald, B.; Ringrose-Voase, A.; Nadelko, A.J.; Farrell, M.; Tuomi, S.; Nachimuthu, G. Dissolved organic nitrogen contributes significantly to leaching from furrow-irrigated cotton–wheat–maize rotations. *Soil Res.* **2016**, *55*, 70–77. [CrossRef]
54. Zhuang, M.; Kee, S.; Zhang, J.; Li, H.; Shan, N.; Yuan, Y.; Wang, L. Effect of full substituting compound fertilizer with different organic manure on reactive nitrogen losses and crop productivity in intensive vegetable production system of China. *J. Environ. Manag.* **2019**, *243*, 381–384. [CrossRef] [PubMed]

55. Zhou, P.; Sheng, H.; Li, Y.; Tong, C.; Ge, T.; Wu, J. Lower C sequestration and N use ef fi ciency by straw incorporation than manure amendment on paddy soils. *Agric. Ecosyst. Environ.* **2016**, *219*, 93–100. [CrossRef]
56. Xia, L.; Lam, S.K.; Yan, X.; Chen, D. How Does Recycling of Livestock Manure in Agroecosystems Affect Crop Productivity, Reactive Nitrogen Losses, and Soil Carbon Balance? *Environ. Sci. Technol.* **2017**, *51*, 7450–7457. [CrossRef]
57. Zhang, M.; Yao, Y.; Tian, Y.; Ceng, K.; Zhao, M.; Zhao, M.; Yin, B. Increasing yield and N use efficiency with organic fertilizer in Chinese intensive rice cropping systems. *Field Crop. Res.* **2018**, *227*, 102–109. [CrossRef]
58. Zhang, M.; Li, B.; Xiong, Z.Q. Effects of organic fertilizer on net global warming potential under an intensively managed vegetable field in southeastern China: A three-year field study. *Atmos. Environ.* **2016**, *145*, 92–103. [CrossRef]
59. Wang, H.; Xu, J.; Liu, X.; Zhang, D.; Li, L.; Li, W.; Sheng, L. Effects of long-term application of organic fertilizer on improving organic matter content and retarding acidity in red soil from China. *Soil Tillage Res.* **2019**, *195*, 104382. [CrossRef]
60. Tang, Q.; Ti, C.; Xia, L.; Xia, Y.; Wei, Z.; Yan, X. Ecosystem services of partial organic substitution for chemical fertilizer in a peri-urban zone in China. *J. Clean. Prod.* **2019**, *224*, 779–788. [CrossRef]
61. Wang, S.; Yang, L.; Su, M.; Ma, X.; Sun, Y.; Yang, M.; Zhao, P.; Shen, J.; Zhang, F.; Goulding, K.; et al. Increasing the agricultural, environmental and economic benefits of farming based on suitable crop rotations and optimum fertilizer applications. *Field Crop. Res.* **2019**, *240*, 78–85. [CrossRef]
62. Zhao, J.; Liu, Z.; Gao, F.; Wang, Y.; Lai, H.; Pan, X.; Yang, D.; Li, X. A 2-year study on the effects of tillage and straw management on the soil quality and peanut yield in a wheat–peanut rotation system. *J. Soils Sediments* **2021**, *21*, 1698–1712. [CrossRef]
63. Song, H.; Guo, J.; Ren, T.; Chen, Q.; Li, B.; Wang, J. Increase of soil ph in a solar greenhouse vegetable production system. *Soil Sci. Soc. Am. J.* **2012**, *76*, 2074–2082. [CrossRef]
64. Xia, Y.; Yan, X. Ecologically optimal nitrogen application rates for rice cropping in the Taihu Lake region of China. *Sustain. Sci.* **2012**, *7*, 33–44. [CrossRef]
65. Fan, Z.; Lin, S.; Zhang, X.; Jiang, Z.; Yang, K.; Jian, D.; Chen, Y.; Li, J.; Chen, Q.; Wang, J. Conventional flooding irrigation causes an overuse of nitrogen fertilizer and low nitrogen use efficiency in intensively used solar greenhouse vegetable production. *Agric. Water Manag.* **2014**, *144*, 11–19. [CrossRef]
66. Norse, D.; Ju, X. Environmental costs of China's food security. *Agric. Ecosyst. Environ.* **2015**, *209*, 5–14. [CrossRef]
67. Tian, Y.; Wang, Q.; Zhang, W.; Gao, L. Reducing environmental risk of excessively fertilized soils and improving cucumber growth by Caragana microphylla-straw compost application in long-term. *Sci. Total Environ.* **2016**, *544*, 251–261. [CrossRef]
68. Liu, Y.; Han, M.; Zhou, X.; Li, W.; Du, C.; Zhang, Y.; Zhang, Y.; Sun, Z.; Wang, Z. Optimizing nitrogen fertilizer application under reduced irrigation strategies for winter wheat of the north China plain. *Irrig. Sci.* **2022**, *40*, 255–265. [CrossRef]
69. Xia, L.; Ti, C.; Li, B.; Xia, Y.; Yan, X. Greenhouse gas emissions and reactive nitrogen releases during the life-cycles of staple food production in China and their mitigation potential. *Sci. Total Environ.* **2016**, *556*, 116–125. [CrossRef]

sustainability MDPI

Review

Aquaculture—Production System and Waste Management for Agriculture Fertilization—A Review

Roberto G. Chiquito-Contreras [1], Luis Hernandez-Adame [2,3], Gerardo Alvarado-Castillo [1], María de J. Martínez-Hernández [1], Gabriela Sánchez-Viveros [1], César J. Chiquito-Contreras [1] and Luis G. Hernandez-Montiel [2,*]

1 Facultad de Ciencias Agrícolas, Universidad Veracruzana, Xalapa 91090, Veracruz, Mexico; rchiquito@uv.mx (R.G.C.-C.); gealvarado@uv.mx (G.A.-C.); jesmartinez@uv.mx (M.d.J.M.-H.); gabsanchez@uv.mx (G.S.-V.); cchiquito@uv.mx (C.J.C.-C.)

2 Nanotechnology & Microbial Biocontrol Group, Centro de Investigaciones Biológicas del Noroeste, La Paz 23096, Baja California Sur, Mexico; ladame@cibnor.mx

3 CONACYT-Centro de Investigaciones Biológicas del Noroeste, La Paz 23096, Baja California Sur, Mexico

* Correspondence: lhernandez@cibnor.mx

Citation: Chiquito-Contreras, R.G.; Hernandez-Adame, L.; Alvarado-Castillo, G.; Martínez-Hernández, M.d.J.; Sánchez-Viveros, G.; Chiquito-Contreras, C.J.; Hernandez-Montiel, L.G. Aquaculture—Production System and Waste Management for Agriculture Fertilization—A Review. *Sustainability* **2022**, *14*, 7257. https://doi.org/10.3390/su14127257

Academic Editors: Muhammad Sultan, Yuguang Zhou, Walter Den and Uzair Sajjad

Received: 3 May 2022
Accepted: 10 June 2022
Published: 14 June 2022

Abstract: Aquaculture is the fastest growing animal food production sector worldwide and is becoming the main source of aquatic animal foodstuff for human consumption. However, the aquaculture sector has been strongly criticized for its environmental impacts. It can cause discharge and accumulation of residual nutrients in the areas surrounding the production farms. This is because, of the total nutrients supplied to production ponds, only 30% are converted into product, while the rest is usually discharged into the environment to maintain water quality in aquaculture culture systems, thereby altering the physic-chemical characteristics of the receiving water. In contrast, this same accumulation of nutrients is gaining importance within the agricultural sector, as it has been reported that the main nutrients required by plants for their development are found in this aquaculture waste. The purpose of this review article is to indicate the different aquaculture production systems, the waste they generate, as well as the negative effects of their discharge into the environment. Biofiltration and bioremediation processes are mentioned as alternatives for aquaculture waste management. Furthermore, the state of the art in the treatment and utilization of aquaculture waste as a mineral source for agricultural nutrition through biodigestion and biomineralization processes is described. Finally, aquaponics is referred to as a biological production approach that, through efficient use of water and recycling of accumulated organic nutrients in aquaculture systems, can contribute to addressing the goals of sustainable aquaculture development.

Keywords: environment; eutrophication; particulate fraction; effluent; treatment

1. Introduction

Aquaculture is an activity aimed at the cultivation of aquatic animals such as freshwater or marine fish, molluscs, crustaceans, and emerges as a strategy to replace traditional fishing, reducing the pressure exerted on natural freshwater or marine populations [1]. Aquaculture has experienced the highest average annual growth in the last 10 years, with a projected contribution of 52% of fishery products for human consumption by 2025 [2]. Every year, this sector generated around 171 million tons with an approximate value of 36,000 million US dollars, of which 47% came from the aquaculture sector, with Asia being the largest representative at 89% [3].

The aquaculture sector generates jobs and food products; however, it causes environmental problems due to the discharge of organic matter (OM) and nutrient-rich waste that pollutes the water [4]. The nutrients supplied to the farmed animals are not fully consumed, with only 30% being utilized by fish, molluscs or crustaceans, the rest settling and accumulating as a particulate fraction (commonly referred to as "sediment" or "sludge") composed mainly of OM, nitrogen (N), and phosphorus (P) [5]. Previous research reports an annual discharge of 27.0 kg/ha of N and 9.0 kg/ha of P in Norwegian fish farms [6]. Another report an annual discharge of 84.0 kg/ha of N, 21.0 kg/ha of P and 2400 kg/ha of OM [7]. Whereas, in fish farms in Japan, the rate of OM accumulation is between 3.9 and 11.7 mg/day [8].

Generally, these nutrients are removed to maintain water quality in aquaculture systems by discharging them into the environment altering the physico-chemical characteristics of the receiving water, decreasing dissolved oxygen (DO) concentration, but increasing; the total suspended solids (TSS), the biological oxygen demand (BOD), and chemical oxygen demand (COD). Furthermore, it decreases benthic fauna [9]. In contrast, this same accumulation of nutrients is gaining importance, as it has been reported that it contains the main nutrients required by plants, with a high potential for their treatment and reuse as sources for agricultural fertilisation [10].

Therefore, this article indicates the different aquaculture production systems, the waste they generate, as well as the negative effects produced by their discharge into the environment. Biofiltration and bioremediation processes are mentioned as alternatives for the management of aquaculture waste. Furthermore, the state of the art in the treatment and use of the aquaculture particulate fraction as mineral sources for agricultural fertilisation by means of biodigestion and biomineralization processes is described. Finally, aquaponics is eluded as a biological production approach which, through the efficient use of water and the recycling of organic nutrients accumulated in aquaculture systems, can help to address the objectives of sustainable aquaculture development.

2. Aquaculture Production Systems

Worldwide, the aquaculture is classified according to the degree of production intensity (balanced feed, sowing density, artificial aeration, among others), such as extensive, semi–intensive, and intensive. The extensive systems use basic management levels, since they do not make use of ingredients for the production of this type of systems, the organisms grow up on their own and productivity is limited by natural water conditions, stocking density is low, which generate a production not greater than 500 kg/ha, the semi-intensive systems make use of additional ingredients, such as feed with high protein content, therapeutic products to prevent disease, as well as chemical and organic fertilizers to increase natural productivity of the system, support stocking densities from 10–30 fish/m^3, and reach production oscillating from 1000 to 2500 kg/ha, whereas in intensive aquaculture, greater yields are achieved than what the capacity of the natural environment allows, by means of techniques, such as balanced feed, artificial aeration, as well as mechanical and biological filtration supporting densities from 60–120 fish/m^3 and reaching productions from 10,000 to 80,000 kg/ha; the aquaculture systems are classified such as flow, pond, recirculating, weir and net cages and floating and bottom farming (Table 1) [2,11].

Table 1. Aquaculture systems used in the production of marine and freshwater organisms and waste production.

Aquaculture System	Characteristics	Species Production	Waste Production	Reference
Flow	This system has rectangular canals with an outlet drop at the end of the structure allowing elevating O concentration and releasing CO_2. The flow or canal system use run-off waters coming from rivers or springs.	– Siluriformes – *Solea solea* – *Oncorhynchus mykiss*	The water is not retained the sufficient time for significant OM biological decomposition processes to develop, thus continuous waste produced is discharged to the environment.	[12]

Table 1. *Cont.*

Aquaculture System	Characteristics	Species Production	Waste Production	Reference
Pond	This system is made up of artificial structures covered with high-density plastic to retain water for long periods of time, water quality is controlled by natural, chemical, and biological processes that occur in ponds. A constant water source is necessary to guarantee sufficient capacity to achieve a daily recharge of at least 10% of total pond volume to allow eliminating NH_4^+ and OM excess	– *Cyprinus carpio* – *Cherax quadricarinatus* – Dendrobranchiata – *Oreochromis niloticus* – Caridea	Around 80 to 90% of dry matter and C, as well as 70 to 30% of N and P end up as waste. From 1 to 100 kg/ha of daily feed rate, approximately 350 mg/m^2/day is excreted by fish as waste.	[13–15]
Recirculating aquaculture system (RAS)	This system consists of intensive fish production that uses water treatments to facilitate recycling. RAS generally include: (1) Settlers and micro-screens for collecting sediment and suspended particles, (2) Nitrifying biofilters and (3) Gas exchange devices to eliminate dissolved CO_2 and add the O.	– *Maccullochella peelii* – *Lates calcarifer* – *Oreochromis niloticus* – *Solae senegalensis* – *Coregonus lavaretus*	RAS consume a small quantity of water (only 5% per day to compensate for the loss caused by evaporation, solid elimination, and plant absorption) and generate pollutants of small volume but with a high nutrient concentrate.	[16–18]
Open-net pen or net cage	This system basically represents "fencing" a portion of water. Net cages are systems that retain farmed species in a confined area, excluding unwanted animals from the surrounding water body, this system depends on the water course where this type of system is located, in which the number of pollutants dumped in the environment cannot be controlled.	– *Salmo salar* – *Cyclopterus lumpus* – *Oplegnathus punctatus* – *Lates calcarifer*	Sites with bad circulation imply low DO concentration conditions, and the accumulation of metabolic waste promotes algal growth and many other benthic organisms that adhere and colonize around the cage, reducing water movement through the cage severely and deteriorating water quality.	[19]
Floating and bottom	This system uses similar principles to those of open–net pen or cage-net systems, which is why they also depend on water movement as well as its natural quality to supply the necessary nutrients and conditions for the development of farming bivalves.	This system is those destined for bivalve mollusk production (oysters, mussels, clams, and scallops)	Likewise, they cannot also control the number of pollutants dumped in the environment	[20]

3. Aquaculture Waste

Waste produced by aquaculture is classified into four forms: gases (H_2S), liquids (effluents), semisolids, and solids (particulate fraction), of which the last two are known as sediments or sludge [21]. Solid waste or sludge is further divided into two categories: suspended solids and settleable solids [22].

3.1. Gas Emission from Aquaculture

Within the aquaculture systems, sulfur (S) is a residual chemical element that originated from metabolic waste produced by farmed organisms; its form is mainly as a sulfate ion since, in aerobic sediment conditions in suspension, S decomposes as sulfide (S^{2-}) and oxidizes to sulfate (SO_4^{2-}). However, in aquaculture, as feed applied to culture systems increases, the accumulation of organic detritus is promoted, causing severe anoxia conditions (lack of oxygen) in sediments, this situation makes anaerobic bacteria use the oxygen molecules present in sulfate ions, increasing H_2S production, any concentration of H_2S interrupts the respiration of the aquatic animals, causing them stress and making them susceptible to diseases [23].

3.2. Aquaculture Effluents

In most aquaculture systems, food supply is the main cause of water pollution and deterioration. Only 30% of the provided nutrients turn into a product, whereas the rest must be removed and generally dumped in the environment in the form of effluents (fluids loaded with solid, liquid, or gaseous waste) [24]. Aquaculture effluents include organic compounds, such as proteins, lipids, carbohydrates, vitamins, and minerals, while inorganic waste products accumulate mainly as NH_4^+, NO_2^-, NO_3^-, bicarbonates, and phosphates, of which N and P are the main components from effluents that cause environmental pollution [25]. The rate of pollutants released to the environment is directly ruled in function of the amount of feed consumed and digestibility. Generally, pond and recirculation systems produce a smaller number of effluents to be discharged but with much higher OM and nutrient concentrations, while flow, net cage, or open-net pen, floating and bottom farming systems emit greater flow but with a lower concentration of these pollutants [26].

3.3. Aquaculture Particulate Fraction

Waste conformed by N, P, and dissolved organic carbon compounds negatively affects the environment [27], these particles are mainly formed from unconsumed food, waste produced by fish, and the residual part where unassimilated forms accumulate the greatest content of incoming nutrients to the aquaculture systems. Additional treatments are thus necessary for the good use of minerals [28].

Within aquaculture production systems, up to 70% of the feed supplied may end up as a particulate fraction at a daily average of 0.4–12.3% [29]. This matter usually contains approximately 7–32% N, as well as 30–84% of P provided for the development of the cultured organisms. Furthermore, the aquaculture particulate fraction is divided into two categories; suspended solids and settleable solids [30]. Suspended solids are fine particles ranging from 30 to 100 micrometers (μm), so they do not settle and remain suspended in the water of aquaculture systems, making them very difficult to collect [22]. In contrast, settleable solids are larger particles (100 > μm), which form sediment in a short period of time, making them easier to collect and remove from culture systems [31].

4. Aquaculture Waste Effect on the Environment

One of the main effects on the environment caused by aquaculture is the eutrophication of the surrounding areas of fish farms; this is because only 30% of supplied N is used in fish farms. This is because the rest is discharged as effluent with each water recharge in this system. Nutrient levels in the receiving bodies are thus elevated above normal and start an ideal environment for anoxic sediments and changes in benthic blooms in the communities in the areas where these residuals are dumped [32].

Change is generated by suspended solids, which reduce light penetration through water, inhibiting the photosynthesis process of phytoplankton and marine grass, and thus generating an increase in mortality of these organisms [33]. Subsequently, bacterial degradation of dead plants consumes oxygen in water, affecting aquatic species farming negatively. In extreme circumstances, profiles of aquatic organisms may transform into species tolerant to sediments, which affects the aquatic food chain on its root. Furthermore, when the particulate fraction settles in the bottom, it tends to biologically degrade due to its OM content and, in consequence, transforms the bottom of ponds or cultivation areas to anaerobic conditions [34].

Alterations may provoke significant changes in the community composition of benthic organisms. For example, a report found that water quality and sediments were negatively affected by effluents dumped at 50 and, 150 m while studying the impact of shrimp effluents dumped on white clams (Dosinia ponderosa) at distances of 50, 150, and 300 m from the discharge area. Physiological and stress conditions of clams in the affected areas deteriorated from the discharge area; glucose, lactate, cholesterol, and aminotransferase alanine were altered, and thermal shock protein transcriptions were expressed in these clams [35]. Another investigation evaluated the environmental impact caused in part by yellowtail (Seriola quinqueradiata) farming on sediments and water quality during low and high feeding times. They observed that the OM charge in sediments was significantly higher than the control site (100 m in distance), covering an impact area of 10 m surrounding the fish farm, accumulating a high level of enriched organic sediments. It subsequently increases in high volatile sulfur acid in superficial sediments, as well as elevated NH_4^+ and phosphate (PO_4^{3-}) concentrations [8].

5. Aquaculture Waste Treatment

In order to mitigate the impacts of waste in the environment, and at the same time take advantage of the high degree of biodegradable organic substances and nutrients they contain, the main treatment methods currently used are: biofiltration by means of artificial systems made up of substrates and plants with the capacity to absorb and reduce the content of nutrients, OM and toxic substances in wastewater [36]. Another method is bioremediation by means of microbiological agents attached to a surface through a matrix of extracellular polymeric substances with the ability to remove, attenuate or transform pollutants in water [37]. This is alongside the use of deposit feeders such as polychaetes [38] and sea cucumbers [39,40] due to their ability to assimilate particulate organic residues, as they accelerate the depletion of organic matter pools through bioturbation, thus improving sediment quality [41].

Moreover, it should be noted that the particulate fraction is the most harmful type of waste produced by aquaculture systems. Therefore, if it is not removed from the ponds, it can degrade, significantly increasing the concentration of TSS, causing a detriment to water quality. In addition, the aquaculture particulate fraction is the residual part where most of the nutrients entering the aquaculture systems accumulate in a non-assimilable form. It is thus necessary to carry out additional treatments for the correct use of these minerals [42]. Biodigestion and biomineralization are 2 of the most practiced strategies for aquaculture particulate fraction treatment, where treatment results are expressed in percentage reduction in pollutants such as COD, PO_4^{3-}, NH_4^+, NO_2^-, NO_3^-, TN, and TP, as well as in quantity of recovered macro/micronutrients of agricultural interest [43–45].

5.1. Biofiltration of Aquaculture Waste

A microbial oxidative process transforms toxic metabolites such as NH_3^+ or NO_2^- into chemical forms less toxic (ammonium or nitrate) to culture organisms through the intervention of nitrifying bacteria [46]. Biofiltration of aquaculture waste consists of substrate and plant systems used for filtration, reduction, and removal of suspended solids [47] macro and micronutrients [48] as well as heavy metals [49]. Where the removal of these components depends on a complex interaction of physical, chemical, biological processes (sedimentation, adsorption, coprecipitation, cation exchange, photodegradation, phytoaccumulation, biodegradation, and microbial activity) and mainly on the type of plant used, as well as its absorption rate [50] in each retention time [51]. In recent years, the use of artificial systems associated with halophytes [52] and macrophytes has been highlighted [53–55].

5.2. Bioremediation of Aquaculture Waste

The bioremediation is defined as the elimination, attenuation or transformation of pollutants present in aquaculture waste, through the application of biological processes carried out by autotrophic and heterotrophic communities. cyanobacteria, bacteria (purple, sulphate reducing and non-reducing) and diatoms among another taxonomic groups, agglutinated in a "biofilm" or "microbial mat" [56]. Understood as any group of organisms in which cells stick together and adhere to a surface by excreting a matrix of extracellular polymeric substances, these communities act simultaneously and synergistically on each of the organic and inorganic pollutants present in the water [57]. In recent years, the use of beneficial biological agents such as bacteria [58,59], biopolymers [10,60,61], microalgae [51,62–70] and macroalgae [71,72] have been used in bioremediation

5.3. Biodigestion of Aquaculture Waste

Biodigestion is a simple and efficient process; it is commonly used to stabilize municipal and industrial organic waste. However, in recent years this approach has gained importance as a form of aquaculture waste treatment. This process requires low energy cost, and results in high methane (CH_4) recovery, CO_2 and H_2S used in biogas production, as well as achieving a reduction in the mass and volume of aquaculture particulate fractions. In anaerobic digestion, nutrients such as NH_4^+ and P are released from the nitrogenous OM, which offers the feasibility of recovering these minerals [73].

Previous studies in the treatment of aquaculture particulate fractions have used anaerobic digesters such as batch-fed sequencing digesters [74] and fully stirred tank reactors [75]. They have been shown to facilitate the release of intracellular material, increasing its biodegradability, thereby improving the biogas production obtained, with a shorter retention time for complete digestion. Evaluating the anaerobic digestion of particulate fractions of a rainbow trout culture by batch reactors, found that, at 10 days, anaerobic digestion solubilized 23.5% of the total Kjeldahl N as total ammonia nitrogen (TAN) and 53.0% of the total P as orthophosphates, and the biochemical methane potential was 318 g CH_4 g TVS_0, representing 65% digestion [74,76]. Otherwise, the addition of carbohydrates at a C/N ratio 1 to 15 as a pre-treatment for anaerobic digestion of brackish aquaculture particulate fractions in an anaerobic sequencing reactor (ASBR) increased gas production and COD removal efficiency by 80% compared to untreated residuals. In addition, the concentrations of soluble oxygen and PO_4^- increased, generating an average gas production of 0.08 g COD/L per day [77]. By gradually increasing the organic load (OLR) in aerobic digestion of particulate fractions from brackish media, observed a 45% improvement in the yield of methane produced was observed [78]. Another report, by applying four different pre-treatments (chemical, mechanical, thermal, and biological), as elements of improvement for anaerobic digestion of Nile tilapia waste, observed an increase in TAN release, as well as an improvement in NO_2^- and STT removal of 90 and 20%, respectively [76]. The applied biodigestion processes in Atlantic salmon (*Salmo salar*) aquaculture waste and a resulting solution as nitrogen fertilization source on barley (*Hordeum vulgare*) cultivation, expressed aquaculture sludge reduction average of 20%, as well as relative agronomic efficiency from 50–80% in compared with the traditional mineral fertilizers [77]. Studying the effect of anaerobic digestion on particulate fractions of Nile tilapia culture as fertilizer sources in lettuce (*Lactuca sativa*) culture, plants grown in the system supplied with the anaerobic solution expressed significantly higher yields than the hydroponic control. This result was attributed to the presence of NH_4^+, OM, rhizobacteria, fungi and humic acids predominantly in the anaerobic residues. They play an important role in nutrient uptake and are utilized by agricultural crops [79]. Another report, studying anaerobic digestion in particulate fractions of lesser weever (*Echiichthys vipera*), observed that CH_4 production increased in relation to the amount of particulate fraction used, achieving an 8% increase in yield by increasing the maximum methane potential and maximum methane production rate from 66.8 mL CH_4/g VS_{fed} a 70.9 mL CH_4/g VS_{fed} y de 4.40 mL CH_4/g $VS_{fed\text{-}d}$ a 5.59 mL CH_4/g $VS_{fed\text{-}d}$, respectively [42].

5.4. Biomineralization of Aquaculture Waste

This is a strategy used for aquaculture particulate fraction treatment that consists of any reaction series such as hydrolysis, acidogenic, and methanogenic. These have as an objective to recover macro/micronutrients of the particulate organic fractions by means of aerobic and anaerobic bioreactors (organic matter containers) that lead to the formation of assimilable mineral elements for plants. By using organic carbon contained in residual OM per microorganism in aerobic and anaerobic environments, transformation of organic phosphorus into phosphates occurs whose accumulation in aquaculture systems may reach similar levels to hydroponic solutions [80].

By valuating nutrient mobilization under aerobic and anaerobic conditions for aquaculture particulate fractions, we found that treatment resulted in a 3.2-fold increase in reactive soluble P, while anaerobic treatment was unaffected. Both aerobic and anaerobic treatment resulted in an increase in K^+ concentrations from 1 to 28.1 and 36.8 mg/L, respectively. It is concluded that conditions support the mobilization of P and K^+ with lower losses of NO_3^-, improving the delivery of these nutrients for plant production, thus reducing the emission of nutrients by the aquaculture particulate fractions. In contrast, anaerobic conditions revealed a complete loss of NO_3^-, posing the risk of unwanted by-products and more complicated to manage under commercial conditions [45]. In determining the organic reduction (COD and TSS) and nutrient recycling performance of a Nile tilapia culture, observed that, the system was able to remove at least 50% of TSS and COD, as well as obtaining consistent mineralization in the range of 10–60% for all of the macro and micronutrients [81] Moreover, aerobic reactor yield in aquaculture particulate fractions and mineralization of macro/microelements as a nutrient supplement for commercial hydroponics and demonstrated that acid conditions (pH below 6) could increase nutrient mineralization and mobilization significantly, such as P, K and, Ca. However, the opposite effect was observed with respect to waste particulate reduction. A better elimination yield was obtained in the high pH reactor [82].

6. Aquaponics

Currently, the aquaculture sector has been searching for alternatives in development and technology transfer with a vision directed to treatment and maximum use of resources (food, water, soil, and energy) to achieve sustainability of this productive activity [83]. The use of aquaponics recirculating systems has been identified as a biological productive approach, which, through the efficient use of water and recycling of organic nutrients accumulated in the aquaculture systems, may help to deal with the objectives of sustainable aquaculture development [84–86] (Figure 1).

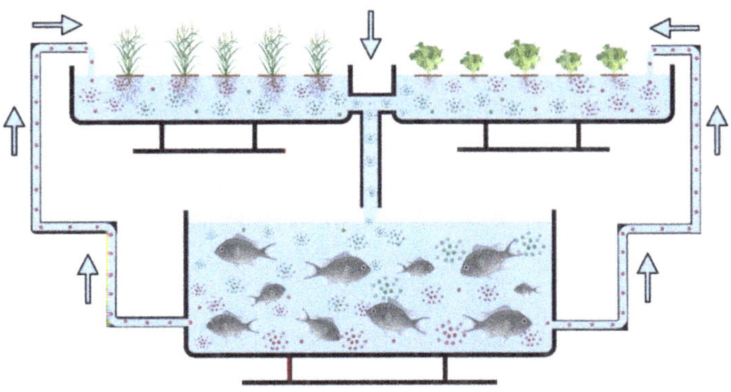

Figure 1. Aquaponic system for food production: Plants and fishes.

In particular, these systems should be helpful for arid regions with non-cultivable soil [87] with greatly brackish waters not suitable for irrigation [88], as well as for marginal land and urban areas [89]. The efficiency of these systems have been demonstrated to achieve an efficiency of 99% in water recycling, reaching demand of use lower than 100 L/kg of harvested fish [90].

In aquaponics, the metabolic waste produced by an aquatic organism is converted to NO_2^- through nitrifying bacteria (*Nitrosomonas* and *Nitrobacter*), which are used as a fertilizer source for plant production [44]. For example, used an aquaponics system for Nile tilapia and tomato (*Solanum lycopersicum*) production, tilapia production was similar to that of the conventional RAS systems while tomato production was similar to that obtained by conventional hydroponics [91]. Furthermore, with a conventional RAS system, 3.4 t of fish can be produced per year, as well as sufficient residual nutrients to harvest 35 t of tomato per year [92]. Another report compared lettuce production between cultivated in conventional hydroponic and aquaponic solutions of Nile tilapia revealed that the aquaponics solution increased plant growth by 39% [93]. A study evaluated the nutritional quality obtained in basil (*Ocimum basilicum*) hydroponic and aquaponic production associated with crayfish (*Procambarus* spp.), where chlorophyll and nutrient content in leaves did not show significant differences between the productive systems [94].

Comparing the quality and production of aquaponics and hydroponics tomato fruit, the different cultivation systems reached similar production yields. Furthermore, the parameters, such as lycopene and ß–carotene were similar in both systems [95]. A study reported the potential of aquaponics systems in the reduction in aquaculture particulate fractions and their use was as a fertilizer source in tomato (*S. lycopersicum*) cultivation [96], these authors found that the system assessed expressed a weekly collection capacity of dry OM, of 2.7–3.0 kg, as well as a production yield of 36% higher than hydroponics. Another report contrasted lettuce cultivation yield by fertilizing with only one traditional hydroponics and aquaculture solution made from the waste of common carp farming. They observed that on average, final fresh and dry weights were 7.9 and 33.2%, respectively, higher than in the fertilized culture with the aquaponics solution [97].

Assessed growth rates of juniper (*Anethum graveolens*), eruca (*Eruca sativa*), coriander (*Coriandrum sativum*), and parsley (*Petroselinum crispum*) between hydroponics and aquaponics systems associated with herbivore carp (*Ctenopharyngodon idella*) farming and found that throughout the three seasons the aquaponics method had similar productions to those of the hydroponics method [98]. A study reported the use aquaponics solutions as an alternative to those of hydroponics in lettuce production and found that leaf mineral content did not show significant differences between both treatments. Furthermore, in the fertilized system with aquaponics solution, water savings of 62.8% were obtained, as well as a reduction in fertilization demand of 72% [99]. Another study reported the nutrient recovery starting from particulate fractions in an aquaponics system of crucian carp (*Carassius auratus*), observing a recovery capacity of macronutrients of up to 46% and 18% for micronutrients [30]. Comparing aquaponics with hydroponics in the distribution of N and P, as well as their use efficiency in cherry tomato, basil, and lettuce crops, observed that, in aquaponics between 59–70% of the total N input was lost and between 30–41% was assimilated as biomass, while in hydroponics a loss of 76–87% was estimated, and only 14–24% was assimilated. Of the total P input, in aquaponics 38–54% was lost and 46–62% was assimilated as biomass, while in hydroponics 79–89% was lost, and only 11–21% was assimilated. It is concluded that hydroponics is less efficient in nutrient use by expressing a 2 times higher N loss through off-gassing and up to 3 times higher P loss through inorganic P compared to aquaponics [100]. Evaluating the yield of common chicory (*Cichorium intybus*), grown in aquaponics, in soil fertilized with particulate fractions from Nile tilapia farming, as well as with chemical fertilization, observed that, the aquaponic system expressed higher yields during the first harvest cycle, during the second harvest cycle, the parameters of number of leaves, fresh matter and dry matter showed higher values for the plants fertilized with the aquaculture particulate fractions than those treated with chemical fertilization. These results suggest a cumulative effect of nutrients in the soil after successive applications of aquaponic particulate fractions, therefore, they can be a viable option to fertilize vegetables in the soil and obtain similar and possibly higher yields than those of traditional mineral fertilization [101]. Therefore, metabolic waste generated in aquaculture aquaponics practices is not seen as a pollutant but rather as a strategic sector to make fertilizing sources for culture nutrition, avoiding the damages caused by eutrophication in the environment generated by aquaculture [18,43,102].

7. Conclusions

Achieving sustainable development in the aquaculture sector means allowing a certain permissible level of pollutants in water, but without reaching values that deteriorate the waters of the culture systems, thus avoiding their discharge into the environment. For this reason, the aquaculture sector is currently looking for alternatives in the development and transfer of technologies, with a vision directed towards the treatment and use of these wastes. In this sense, several treatment technologies have been presented to mitigate the problem mentioned above in this work. As a first approach, biofiltration using artificial systems associated with halophytic and macrophyte plants was studied as one of the most efficient and straightforward methods to implement in aquaculture systems to reduce and eliminate suspended solids, macro and micronutrients, as well as heavy metals. Furthermore, bioremediation was discussed through beneficial biological agents such as biopolymers, bacteria, microalgae, and macroalgae for the transformation of pollutants through the application of biological processes. Besides, anaerobic biodigestion protocols were set in perspective as systems representing a low energy cost in reducing the mass and volume of aquaculture waste and produce a high recovery of CH_4, CO_2 and H_2S used in biogas production. With regards to biomineralization through aerobic and anaerobic bioreactors, this leads to the formation of plant-assimilable mineral elements by utilizing the organic carbon and phosphorus contained in the aquaculture waste. Finally, acuaponics practices have been addressed as an alternative to create a more sustainable aquaculture industry, in which the flora not only acts as a treatment system, but also provides a valuable source of food and energy. Therefore, in aquaponics, metabolic wastes generated in aquaculture practices are not seen as a pollutant but rather as a strategic sector for the manufacture of fertilizer sources for crop nutrition, thus avoiding eutrophication damage to the environment generated by aquaculture.

Author Contributions: Conceptualization, R.G.C.-C. and L.G.H.-M.; methodology, R.G.C.-C., L.G.H.-M., G.A.-C., M.d.J.M.-H. and L.H.-A.; investigation, R.G.C.-C., L.G.H.-M., G.A.-C., M.d.J.M.-H., L.H.-A. and G.S.-V.; resources, G.A.-C., L.H.-A., G.S.-V. and C.J.C.-C.; writing—original draft preparation, R.G.C.-C. and L.G.H.-M.; writing—review and editing, R.G.C.-C., L.G.H.-M., G.A.-C. and L.H.-A.; visualization, G.A.-C., M.d.J.M.-H. and C.J.C.-C.; supervision, R.G.C.-C. and L.G.H.-M.; project administration L.G.H.-M.; funding acquisition, R.G.C.-C., G.S.-V. and C.J.C.-C. All authors have read and agreed to the published version of the manuscript.

Funding: This research received no external funding.

Institutional Review Board Statement: Not applicable.

Informed Consent Statement: Not applicable.

Data Availability Statement: Not applicable.

Conflicts of Interest: The authors declare no conflict of interest.

References

1. El-Saadony, M.T.; Alagawany, M.; Patra, A.K.; Kar, I.; Tiwari, R.; Dawood, M.A.O.; Dhama, K.; Abdel-Latif, H.M. The functionality of probiotics in aquaculture: An overview. *Fish Shellfish Immunol.* **2021**, *117*, 36–52. [CrossRef] [PubMed]
2. El Estado Mundial de la Pesca y la Acuicultura (SOFIA). Cumplir los Objetivos de Desarrollo Sostenible. Available online: www.fao.org/3/i9540es/i9540es.pdf (accessed on 1 March 2022).
3. Ahmed, N.; Thompson, S.; Glaser, M. Global aquaculture productivity, environmental sustainability, and climate change adaptability. *Environ. Manag.* **2019**, *63*, 159–172. [CrossRef] [PubMed]
4. Bergheim, A.; Schumann, M.; Brinker, A. Water pollution from fish farms. *Encycl. Water Sci. Technol. Soc.* **2019**, *1*, 1–10. [CrossRef]
5. Bao, W.; Zhu, S.; Jin, G.; Ye, Z. Generation, characterization, perniciousness, removal and reutilization of solids in aquaculture water: A review from the whole process perspective. *Rev. Aquac.* **2018**, *11*, 1342–1366. [CrossRef]
6. Hamilton, H.A.; Brod, E.; Hanserud, O.S.; Gracey, E.O.; Vestrum, M.I.; Bøen, A.; Brattebø, H. Investigating cross-sectoral synergies through integrated aquaculture, fisheries, and agriculture phosphorus assessments: A case study of Norway. *J. Ind. Ecol.* **2016**, *20*, 867–881. [CrossRef]
7. Gál, D.; Pekár, F.; Kerepeczki, É. A survey on the environmental impact of pond aquaculture in Hungary. *Aquacult. Int.* **2016**, *24*, 1543–1554. [CrossRef]
8. Srithongouthai, S.; Tada, K. Impacts of organic waste from a yellowtail cage farm on surface sediment and bottom water in Shido Bay (the Seto Inland Sea, Japan). *Aquaculture* **2017**, *471*, 140–145. [CrossRef]

9. Páez-Osuna, F.; Álvarez-Borrego, S.; Ruiz-Fernández, A.C.; García-Hernández, J.; Jara-Marini, M.E.; Bergés-Tiznado, M.E.; Ruelas-Inzunza, J.R. Environmental status of the Gulf of California: A pollution review. *Earth Sci. Rev.* **2017**, *166*, 181–205. [CrossRef]
10. Ezziddine, M.; Liltved, H.; Homme, J.M. A method for reclaiming nutrients from aquacultural waste for use in soilless growth systems. *Water Sci. Technol.* **2020**, *81*, 81–90. [CrossRef]
11. Stabili, L.; Di Salvo, M.; Alifano, P.; Talà, A. An integrative, multiparametric approach for the comprehensive assessment of microbial quality and pollution in aquaculture systems. *Microb. Ecol.* **2021**, *83*, 271–283. [CrossRef]
12. Fornshell, G.; Hinshaw, J.; Tidwell, J.H. Flow-through raceways. In *Aquaculture Production Systems*, 1st ed.; Tidwell, J.H., Ed.; Wiley-Blackwell: Oxford, UK, 2012; Volume 1, pp. 173–189.
13. Liu, X.G.; Wang, J.; Wu, Z.F.; Cheng, G.F.; Gu, Z.J. Anaerobic ammonium oxidation bacteria in a freshwater recirculating pond aquaculture system. *Int. J. Environ. Res. Public Health* **2021**, *18*, 4941. [CrossRef] [PubMed]
14. Boyd, C.E. Hydrology and pond construction of freshwater catfish. In *Channel Catfish Culture*, 1st ed.; Tucker, C.S., Ed.; Elsevier: Amsterdam, The Netherlands, 1985; Volume 1, pp. 107–134.
15. Cremer, M.; Chappell, J.; Zhang, J.; Zhou, E.H. New intensive pond aquaculture technology demonstrated in China. In *Global Aquaculture Advocate*; Global Aquaculture Alliance: Portsmouth, NH, USA, 2014; pp. 60–62.
16. Drønen, K.; Roalkvam, I.; Dahle, H.; Olsen, A.B.; Nilsen, H.; Wergeland, H. Microbiome dataset from a marine recirculating aquaculture system (RAS) for salmon post-smolt production in Norway. *Data Brief* **2021**, *40*, 107767. [CrossRef] [PubMed]
17. Espinal, C.A.; Matulić, D. Recirculating aquaculture technologies. In *Aquaponics Food Production Systems: Combined Aquaculture and Hydroponic Production Technologies for the Future*, 1st ed.; Goddek, S., Joyce, A., Kotzen, B., Burnell, G.M., Eds.; Springer: Cham, Switzerland, 2019; Volume 1, pp. 35–76. [CrossRef]
18. Edwards, P. Aquaculture environment interactions: Past, present and likely future trends. *Aquaculture* **2015**, *447*, 2–14. [CrossRef]
19. Masser, M.P. Cage culture in freshwater and protected marine areas. In *Aquaculture Production Systems*, 1st ed.; Tidwell, J.H., Ed.; Wiley-Blackwell: Oxford, UK, 2012; Volume 1, pp. 119–130.
20. Helm, M.M.; Bourne, N.; Lovatelli, A. Cultivo de bivalvos en criadero: Un manual práctico. In *FAO Documento Técnico de Pesca*, 1st ed.; Helm, M.M., Bourne, N., Lovatelli, A., Eds.; FAO: Rome, Italy, 2006; Volume 471, pp. 31–82. (In Spanish)
21. Dauda, A.B.; Ajadi, A.; Tola-Fabunmi, A.S.; Akinwole, A.O. Waste production in aquaculture: Sources, components and managements in different culture systems. *Aquac. Fish.* **2019**, *4*, 81–88. [CrossRef]
22. Schumann, M.; Brinker, A. Understanding and managing suspended solids in intensive salmonid aquaculture: A review. *Rev. Aquac.* **2020**, *12*, 2109–2139. [CrossRef]
23. Boyd, C.E. Hydrogen Sulfide Toxic, But Manageable. In *Global Aquaculture Advocate*; Global Aquaculture Alliance: Portsmouth, NH, USA, 2014; pp. 34–36.
24. Boyd, C.E. Aquaculture effluent management at the farm level. *Aquaculture* **2003**, *226*, 101–112. [CrossRef]
25. Herath, S.S.; Satoh, S. Environmental impact of phosphorus and nitrogen from aquaculture. In *Feed and Feeding Practices in Aquaculture*; Allen, D., Ed.; Woodhead Publishing Series in Food Science, Technology and Nutrition; Woodhead Publishing: Sawston, UK, 2015; Volume 1, pp. 369–386. [CrossRef]
26. Ngo, H.H.; Guo, W.; Vo, T.T.; Nghiem, L.D.; Hai, F.I. Aerobic treatment of effluents from the aquaculture industry. In *Current Developments in Biotechnology and Bioengineering: Biological Treatment of Industrial Effluents*; Larroche, C., Sanroman, M., Guocheng, D., Pandey, A., Eds.; Elsevier: Amsterdam, The Netherlands, 2017; Volume 1, pp. 35–77. [CrossRef]
27. Madariaga, S.T.; Marín, S.L. Sanitary and environmental conditions of aquaculture sludge. *Aquac. Res.* **2017**, *48*, 1744–1750. [CrossRef]
28. Kokou, F.; Fountoulaki, E. Aquaculture waste production associated with antinutrient presence in common fish feed plant ingredients. *Aquaculture* **2018**, *495*, 295–310. [CrossRef]
29. Chen, S.; Coffin, D.E.; Malone, R.F. Sludge production and management for recirculating aquacultural systems. *J. World Aquacult. Soc.* **1997**, *28*, 303–315. [CrossRef]
30. Zhang, H.; Gao, Y.; Shi, H.; Lee, C.T.; Hashim, H.; Zhang, Z.; Wu, W.M.; Li, C. Recovery of nutrients from fish sludge in an aquaponic system using biological aerated filters with ceramsite plus lignocellulosic material media. *J. Clean. Prod.* **2020**, *258*, 120886. [CrossRef]
31. Fernandes, P.; Pedersen, L.F.; Pedersen, P.B. Microscreen effects on water quality in replicated recirculating aquaculture systems. *Aquac. Eng.* **2015**, *65*, 17–26. [CrossRef]
32. Bannister, R.J.; Johnsen, I.A.; Hansen, P.K.; Kutti, T.; Asplin, L. Near-and far-field dispersal modelling of organic waste from Atlantic salmon aquaculture in fjord systems. *ICES J. Mar. Sci.* **2016**, *73*, 2408–2419. [CrossRef]
33. King, O.C.; Smith, R.A.; Warne, M.S.J.; van de Merwe, J.P.; Connolly, R.M.; Brown, C.J. Combined impacts of photosystem II-inhibiting herbicides and light availability on seagrass and marine microalgae. *Mar. Ecol. Prog. Ser.* **2021**, *668*, 215–230. [CrossRef]
34. Sugiura, S.H. Phosphorus, aquaculture, and the environment. *Rev. Fish. Sci. Aquac.* **2018**, *26*, 515–521. [CrossRef]
35. Martínez-Porchas, M.; Scheuren-Acevedo, S.M.; Martínez-Córdova, L.R.; Gollas-Galvan, T.; Barraza-Guardado, R.H.; Enríquez-Ocaña, F.; Cortés-Jacinto, E.; Porchas-Cornejo, M.A. Physiological and sanitary condition of the white clam *Dosinia ponderosa* collected from a coastal area impacted by shrimp farm effluent. *Aquacult. Int.* **2016**, *24*, 243–256. [CrossRef]

36. Sikder, M.N.A.; Min, W.W.; Ziyad, A.O.; Kumar, P.P.; Kumar, R.D. Sustainable treatment of aquaculture effluents in future—A review. *Int. Res. J. Adv. Eng. Sci.* **2016**, *1*, 190–193.
37. Minnikova, T.; Kolesnikov, S.; Minkina, T.; Mandzhieva, S. Assessment of ecological condition of haplic chernozem calcic contaminated with petroleum hydrocarbons during application of bioremediation agents of various natures. *Land* **2021**, *10*, 169. [CrossRef]
38. Gómez, S.; Hurtado, C.F.; Orellana, J.; Valenzuela-Olea, G.; Turner, A. *Abarenicola pusilla* (Quatrefages, 1866): A novel species for fish waste bioremediation from marine recirculating aquaculture systems. *Aquac. Res.* **2018**, *49*, 1363–1367. [CrossRef]
39. Chen, L.; Wang, X.Y.; Liu, R.Z.; Wang, G.Y. Culturable microorganisms associated with sea cucumbers and microbial natural products. *Mar. Drugs* **2021**, *19*, 461. [CrossRef]
40. Robinson, G.; Caldwell, G.S.; Jones, C.L.W.; Stead, S.M. The effect of resource quality on the growth of *Holothuria scabra* during aquaculture waste bioremediation. *Aquaculture* **2019**, *499*, 101–108. [CrossRef]
41. Robinson, G.; Caldwell, G.S.; Wade, M.J.; Free, A.; Jones, C.L.W.; Stead, S.M. Profiling bacterial communities associated with sediment-based aquaculture bioremediation systems under contrasting redox regimes. *Sci. Rep.* **2016**, *6*, 38850. [CrossRef] [PubMed]
42. Wu, Y.; Song, K. Process performance of anaerobic co-digestion of waste activated sludge and aquaculture sludge. *Aquac. Eng.* **2020**, *90*, 1–6. [CrossRef]
43. Delaide, B.; Monsees, H.; Gross, A.; Goddek, S. Aerobic and anaerobic treatments for aquaponic sludge reduction and mineralisation. In *Aquaponics Food Production Systems*; Goddek, S., Joyce, A., Kotzen, B., Burnell, G.M., Eds.; Springer: Cham, Switzerland, 2019; Volume 1, pp. 247–266. [CrossRef]
44. Wongkiew, S.; Hu, Z.; Chandran, K.; Woo-Lee, J.; Khanal, S.K. Nitrogen transformations in aquaponic systems: A review. *Aquac. Eng.* **2017**, *76*, 9–19. [CrossRef]
45. Monsees, H.; Keitel, J.; Paul, M.; Kloas, W.; Wuertz, S. Potential of aquacultural sludge treatment for aquaponics: Evaluation of nutrient mobilization under aerobic and anaerobic conditions. *Aquacult. Environ. Interact.* **2017**, *9*, 9–18. [CrossRef]
46. Ni, Z.; Wu, X.; Li, L.; Lv, Z.; Zhang, Z.; Hao, A.; Iseri, Y.; Kuba, T.; Zhang, X.; Wu, W.M.; et al. Pollution control and in situ bioremediation for lake aquaculture using an ecological dam. *J. Clean. Prod.* **2018**, *172*, 2256–2265. [CrossRef]
47. Godoy-Olmos, S.; Jauralde, I.; Monge-Ortiz, R.; Milián-Sorribes, M.C.; Jover-Cerdá, M.; Tomás-Vidal, A.; Martínez-Llorens, S. Influence of diet and feeding strategy on the performance of nitrifying trickling filter, oxygen consumption and ammonia excretion of gilthead sea bream (*Sparus aurata*) raised in recirculating aquaculture systems. *Aquac. Int.* **2022**, *30*, 581–606. [CrossRef]
48. Resende, L.; Flores, J.; Moreira, C.; Pacheco, D.; Baeta, A.; Garcia, A.C.; Rocha, A.C.S. Effective and low-maintenance IMTA system as effluent treatment unit for promoting sustainability in coastal aquaculture. *Appl. Sci.* **2022**, *12*, 398. [CrossRef]
49. Gorito, A.M.; Ribeiro, A.R.; Gomes, C.R.; Almeida, C.M.R.; Silva, A.M.T. Constructed wetland microcosms for the removal of organic micropollutants from freshwater aquaculture effluents. *Sci. Total Environ.* **2018**, *644*, 1171–1180. [CrossRef]
50. Ouyang, X.; Guo, F. Paradigms of mangroves in treatment of anthropogenic wastewater pollution. *Sci. Total Environ.* **2016**, *544*, 971–979. [CrossRef]
51. Daneshvar, E.; Antikainen, L.; Koutra, E.; Kornaros, M.; Bhatnagar, A. Investigation on the feasibility of Chlorella vulgaris cultivation in a mixture of pulp and aquaculture effluents: Treatment of wastewater and lipid extraction. *Bioresour. Technol.* **2018**, *255*, 104–110. [CrossRef]
52. Beyer, C.P.; Gómez, S.; Lara, G.; Monsalve, J.P.; Orellana, J.; Hurtado, C.F. *Sarcocornia neei*: A novel halophyte species for bioremediation of marine aquaculture wastewater and production diversification in integrated systems. *Aquaculture* **2021**, *543*, 736971. [CrossRef]
53. Wijewardene, L.; Wu, N.; Fohrer, N.; Riis, T. Epiphytic biofilms in freshwater and interactions with macrophytes: Current understanding and future directions. *Aquat. Bot.* **2022**, *176*, 103467. [CrossRef]
54. Hua, Z.L.; Li, X.Q.; Zhang, J.Y.; Gu, L. Removal potential of multiple perfluoroalkyl acids (PFAAs) by submerged macrophytes in aquatic environments: Tolerance of *Vallisneria natans* and PFAA removal in submerged macrophyte-microbiota systems. *J. Hazard. Mater.* **2022**, *424*, 127695. [CrossRef] [PubMed]
55. Sun, J.; Doeser, A.; Cao, Y.; Lv, X.; Li, W.; Liu, F. Regional macrophyte diversity is shaped by accumulative effects across waterbody types in southern China. *Aquat. Bot.* **2022**, *176*, 103468. [CrossRef]
56. Barnharst, T.; Rajendran, A.; Hu, B. Bioremediation of synthetic intensive aquaculture wastewater by a novel feed-grade composite biofilm. *Int. Biodeterior. Biodegrad.* **2018**, *126*, 131–142. [CrossRef]
57. Musyoka, S.N. Concept of microbial bioremediation in aquaculture wastes—Review. *Int. J. Adv. Sci. Tech. Res.* **2016**, *5*, 1–10.
58. Dong, D.; Sun, H.; Qi, Z.; Liu, X. Improving microbial bioremediation efficiency of intensive aquacultural wastewater based on bacterial pollutant metabolism kinetics analysis. *Chemosphere* **2021**, *265*, 129151. [CrossRef]
59. Irhayyim, T.; Beliczky, G.; Bercsényi, M. Nutrient bioremediation efficiency of bacterial biofilms and plant based biofilters in a recirculating common carp (*Cyprinus carpio* L.) culture system. *Iran. J. Fish. Sci.* **2021**, *20*, 828–845.
60. Zadinelo, I.V.; dos Santos, L.D.; Cagol, L.; de Muniz, G.I.B.; Ellendersen, L.D.S.N.; Alves, H.J.; Bombardelli, R.A. Adsorption of aquaculture pollutants using a sustainable biopolymer. *Environ. Sci. Pollut. Res.* **2018**, *25*, 4361–4370. [CrossRef]
61. Bernardi, F.; Zadinelo, I.V.; Alves, H.J.; Meurer, F.; dos Santos, L.D. Chitins and chitosans for the removal of total ammonia of aquaculture effluents. *Aquaculture* **2018**, *483*, 203–212. [CrossRef]

62. Calderini, M.L.; Stevčić, Č.; Taipale, S.; Pulkkinen, K. Filtration of Nordic recirculating aquaculture system wastewater: Effects on microalgal growth, nutrient removal, and nutritional value. *Algal Res.* **2021**, *60*, 102486. [CrossRef]
63. Kafil, M.; Berninger, F.; Koutra, E.; Kornaros, M. Utilization of the microalga *Scenedesmus quadricauda* for hexavalent chromium bioremediation and biodiesel production. *Bioresour. Technol.* **2022**, *346*, 126665. [CrossRef] [PubMed]
64. Viegas, C.; Gouveia, L.; Gonçalves, M. Aquaculture wastewater treatment through microalgal. Biomass potential applications on animal feed, agriculture, and energy. *J. Environ. Manag.* **2021**, *286*, 112–187. [CrossRef] [PubMed]
65. Al-Jabri, H.; Das, P.; Khan, S.; Thaher, M.; AbdulQuadir, M. Treatment of wastewaters by microalgae and the potential applications of the produced biomass—A review. *Water* **2021**, *13*, 27. [CrossRef]
66. Kalra, R.; Gaur, S.; Goel, M. Microalgae bioremediation: A perspective towards wastewater treatment along with industrial carotenoids production. *J. Water Process Eng.* **2021**, *40*, 101794. [CrossRef]
67. Andreotti, V.; Chindris, A.; Brundu, G.; Vallainc, D.; Francavilla, M.; García, J. Bioremediation of aquaculture wastewater from *Mugil cephalus* (Linnaeus, 1758) with different microalgae species. *Chem. Ecol.* **2017**, *33*, 750–761. [CrossRef]
68. Tossavainen, M.; Lahti, K.; Edelmann, M.; Eskola, R.; Lampi, A.M.; Piironen, V.; Korvonen, P.; Ojala, A.; Romantschuk, M. Integrated utilization of microalgae cultured in aquaculture wastewater: Wastewater treatment and production of valuable fatty acids and tocopherols. *J. Appl. Phycol.* **2019**, *31*, 1753–1763. [CrossRef]
69. Arumugam, K.; Ahmad, M.F.; Yaacob, N.S.; Ikram, W.M.; Maniyam, M.N.; Abdullah, H.; Katayama, T.; Komatsu, K.; Kuwahara, V.S. Enhancement of targeted microalgae species growth using aquaculture sludge extracts. *Heliyon* **2020**, *6*, 45–56. [CrossRef]
70. Priyadarshani, I.; Sahu, D.; Rath, B. Microalgal bioremediation: Current practices and perspectives. *J. Biochem. Tech.* **2012**, *3*, 299–304.
71. Amosu, A.O.; Robertson-Andersson, D.V.; Kean, E.; Maneveldt, G.W.; Cyster, L. Biofiltering and uptake of dissolved nutrients by *Ulva armoricana* (Chlorophyta) in a land-based aquaculture system. *Int. J. Agric. Biol.* **2016**, *18*, 298–304. [CrossRef]
72. Wang, C.; Jiang, C.; Gao, T.; Peng, X.; Ma, S.; Sun, Q.; Xia, B.; Xie, X.; Bai, Z.; Xu, S.; et al. Improvement of fish production and water quality in a recirculating aquaculture pond enhanced with bacteria-microalgae association. *Aquaculture* **2022**, *547*, 737420. [CrossRef]
73. Jasmin, M.Y.; Syukri, F.; Kamarudin, M.S.; Karim, M. Potential of bioremediation in treating aquaculture sludge: Review article. *Aquaculture* **2019**, *519*, 734–905. [CrossRef]
74. Suhr, K.I.; Letelier-Gordo, C.O.; Lund, I. Anaerobic digestion of solid waste in RAS: Effect of reactor type on the biochemical acidogenic potential (BAP) and assessment of the biochemical methane potential (BMP) by a batch assay. *Aquac. Eng.* **2015**, *65*, 65–71. [CrossRef]
75. Luo, G.Z.; Ma, N.; Li, P.; Tan, H.X.; Liu, W. Enhancement of anaerobic digestion to treat saline sludge from recirculating aquaculture systems. *Sci. World J.* **2015**, *2015*, 479101. [CrossRef]
76. González-Hermoso, J.P.; Segovia, M. Effect of four different pretreatments in anaerobic digestion and nutrient removal of effluents from a recirculating aquaculture system. *Lat. Am. J. Aquat. Res.* **2017**, *45*, 276–292. [CrossRef]
77. Brod, E.; Oppen, J.; Kristoffersen, A.Ø.; Haraldsen, T.K.; Krogstad, T. Drying or anaerobic digestion of fish sludge: Nitrogen fertilisation effects and logistics. *Ambio* **2017**, *46*, 852–864. [CrossRef]
78. Zhang, X.; Tao, Y.; Hu, J.; Liu, G.; Spanjers, H.; van Lier, J.B. Biomethanation and microbial community changes in a digester treating sludge from a brackish aquaculture recirculation system. *Bioresour. Technol.* **2016**, *214*, 338–347. [CrossRef]
79. Goddek, S.; Schmautz, Z.; Scott, B.; Delaide, B.; Keesman, K.J.; Wuertz, S.; Junge, R. The effect of anaerobic and aerobic fish sludge supernatant on hydroponic lettuce. *Agronomy* **2016**, *6*, 37. [CrossRef]
80. Khiari, Z.; Kaluthota, S.; Savidov, N. Aerobic bioconversion of aquaculture solid waste into liquid fertilizer: Effects of bioprocess parameters on kinetics of nitrogen mineralization. *Aquaculture* **2019**, *500*, 492–499. [CrossRef]
81. Delaide, B.; Goddek, S.; Keesman, K.; Jijakli, M.H. A methodology to quantify aerobic and anaerobic sludge digestion performances for nutrient recycling in aquaponics. *Biotechnol. Agron. Soc. Environ.* **2018**, *22*, 106–112. [CrossRef]
82. Goddek, S.; Delaide, B.P.L.; Joyce, A.; Wuertz, S.; Jijakli, M.H.; Gross, A.; Eding, E.H.; Bläser, I.; Reuter, M.; Keizer, L.C.P.; et al. Nutrient mineralization and organic matter reduction performance of RAS-based sludge in sequential UASB-EGSB reactors. *Aquac. Eng.* **2018**, *83*, 10–19. [CrossRef]
83. Greenfeld, A.; Becker, N.; Bornman, J.F.; Spatari, S.; Angel, D.L. Monetizing environmental impact of integrated aquaponic farming compared to separate systems. *Sci. Total Environ.* **2021**, *792*, 59–84. [CrossRef] [PubMed]
84. Goddek, S.; Espinal, C.A.; Delaide, B.; Jijakli, M.H.; Schmautz, Z.; Wuertz, S.; Keesman, K.J. Navigating towards decoupled aquaponic systems: A system dynamics design approach. *Water* **2016**, *8*, 303. [CrossRef]
85. Lennard, W.; Goddek, S. Aquaponics: The basics. In *Aquaponics Food Production Systems: Combined Aquaculture and Hydroponic Production Technologies for the Future*; Goddek, S., Joyce, A., Kotzen, B., Burnell, G.M., Eds.; Springer Nature: Cham, Switzerland, 2019; Volume 1, pp. 113–144. [CrossRef]
86. Abusin, S.A.A.; Mandikiana, B.W. Towards sustainable food production systems in Qatar: Assessment of the viability of aquaponics. *Glob. Food Sec.* **2020**, *25*, 1–7. [CrossRef]
87. Goddek, S.; Körner, O. A fully integrated simulation model of multi-loop aquaponics: A case study for system sizing in different environments. *Agric. Syst.* **2019**, *171*, 143–154. [CrossRef]
88. Goddek, S.; Keesman, K.J. The necessity of desalination technology for designing and sizing multi-loop aquaponics systems. *Desalination* **2018**, *428*, 76–85. [CrossRef]

89. Van-Gorcum, B.; Goddek, S.; Keesman, K.J. Gaining market insights for aquaponically produced vegetables in Kenya. *Aquac. Inter.* **2019**, *27*, 1–7. [CrossRef]
90. Love, D.C.; Uhl, M.S.; Genello, L. Energy and water use of a small-scale raft aquaponics system in Baltimore, Maryland, United States. *Aquacult. Eng.* **2015**, *68*, 19–27. [CrossRef]
91. Kloas, W.; Groß, R.; Baganz, D.; Graupner, J.; Monsees, H.; Schmidt, U.; Staaks, G.; Suhl, J.; Tschirner, M.; Wittstock, B.; et al. A new concept for aquaponic systems to improve sustainability, increase productivity, and reduce environmental impacts. *Aquac. Environ. Interact.* **2015**, *7*, 179–192. [CrossRef]
92. Yogev, U.; Barnes, A.; Gross, A. Nutrients and energy balance analysis for a conceptual model of three loops off grid, acuaponics. *Water* **2016**, *8*, 589. [CrossRef]
93. Delaide, B.; Goddek, S.; Gott, J.; Soyeurt, H.; Jijakli, M.H. Lettuce (*Lactuca sativa* L. var. Sucrine) growth performance in complemented aquaponic solution outperforms hydroponics. *Water* **2016**, *8*, 467. [CrossRef]
94. Saha, S.; Monroe, A.; Day, M.R. Growth, yield, plant quality and nutrition of basil (*Ocimum basilicum* L.) under soilless agricultural systems. *Ann. Agric. Sci.* **2016**, *61*, 181–186. [CrossRef]
95. Suhl, J.; Dannehl, D.; Kloas, W.; Baganz, D.; Jobs, S.; Schiebe, G.; Schmidt, U. Advanced Aquaponics: Evaluation of intensive tomato production in aquaponics vs. conventional hydroponics. *Agric. Water Manag.* **2016**, *178*, 335–344. [CrossRef]
96. Monsees, H.; Kloas, W.; Wuertz, S. Decoupled systems on trial: Eliminating bottlenecks to improve aquaponic processes. *PLoS ONE* **2017**, *12*, e0183056. [CrossRef]
97. Goddek, S.; Vermeulen, T. Comparison of *Lactuca sativa* growth performance in conventional and RAS-based hydroponic systems. *Aquac. Int.* **2018**, *26*, 1–10. [CrossRef]
98. Lennard, W.; Ward, J. A comparison of plant growth rates between an NFT hydroponic system and an NFT aquaponic System. *Horticulturae* **2019**, *5*, 27. [CrossRef]
99. Monsees, H.; Suhl, J.; Paul, M.; Kloas, W.; Dannehl, D.; Würtz, S. Lettuce (*Lactuca sativa*, variety Salanova) production in decoupled aquaponic systems: Same yield and similar quality as in conventional hydroponic systems but drastically reduced greenhouse gas emissions by saving inorganic fertilizer. *PLoS ONE* **2019**, *14*, e0218368. [CrossRef]
100. Yang, T.; Kim, H.J. Comparisons of nitrogen and phosphorus mass balance for tomato-, basil-, and lettuce-based aquaponic and hydroponic systems. *J. Clean. Prod.* **2020**, *274*, 122–135. [CrossRef]
101. Lenz, G.L.; Loss, A.; Lourenzi, C.R.; Lopes, D.L.D.A.; Siebeneichler, L.D.M.; Brunetto, G. Common chicory production in aquaponics and in soil fertilized with aquaponic sludge. *Sci. Hortic.* **2021**, *281*, 109–146. [CrossRef]
102. Barrett, L.T.; Theuerkauf, S.J.; Rose, J.M.; Alleway, H.K.; Bricker, S.B.; Parker, M.; Petrolia, D.R.; Jones, R.C. Sustainable growth of non-fed aquaculture can generate valuable ecosystem benefits. *Ecosyst. Serv.* **2022**, *53*, 101395. [CrossRef]

sustainability

MDPI

Article

Spatial Prediction of Total Nitrogen in Soil Surface Layer Based on Machine Learning

Zunfang Liu [1], Haochuan Lei [1,*], Lei Lei [1] and Haiyan Sheng [2]

[1] Department of Geological Engineering, Qinghai University, Xining 810016, China
[2] College of Agriculture and Animal Husbandry, Qinghai University, Xining 810016, China
* Correspondence: lhcqhdx@163.com

Abstract: In order to satisfy the basic requirements of sustainable agricultural development, it is important to understand the spatial distribution characteristics of soil total nitrogen (TN) content to better guide accurate fertilization to increase grain yield. To this end, this paper constructs three inversion models of partial least squares regression (PLSR), back propagation neural network (BPNN) and support vector machines (SVM) with remote sensing data to predict the TN content in Datong County, Xining City, Qinghai Province, China. The results showed that the average TN content was 1.864 g/kg, and the coefficient of variation (CV) was 30.596%. The prediction accuracy of the SVM model (R^2 = 0.676, RMSE = 0.296) among the three inversion models was higher than that of the BPNN model (R^2 = 0.560, RMSE = 0.305) and the PLSR model (R^2 = 0.374, RMSE = 0.334). The model with the highest accuracy predicted the spatial distribution of TN, and TN content showed a spatial distribution trend which was high in the northwest and low in the southeast, and gradually decreased from north to south. This study provides reference basis and support for soil fertility evaluations and sustainable agricultural development.

Keywords: soil total nitrogen; BP neural network; support vector machines; spatial distribution; remote sensing

Citation: Liu, Z.; Lei, H.; Lei, L.; Sheng, H. Spatial Prediction of Total Nitrogen in Soil Surface Layer Based on Machine Learning. *Sustainability* **2022**, *14*, 11998. https://doi.org/10.3390/su141911998

Academic Editors: Muhammad Sultan, Yuguang Zhou, Walter Den and Uzair Sajjad

Received: 3 August 2022
Accepted: 20 September 2022
Published: 22 September 2022

Publisher's Note: MDPI stays neutral with regard to jurisdictional claims in published maps and institutional affiliations.

1. Introduction

Soil is a significant natural resource in human production and life and is an important carrier of the human living environment. As the main material basis of land resources, the inherently non-renewable nature of soil determines the limited carrying capacity [1]. One of the critical factors reflecting the quality of farmland is the soil nutrient content, and the normal growth and development of plants and soil nutrients are closely related [2]. In the era of rapid development of digital agriculture, accurate, fast and dynamic acquisition of soil information on demand is the guarantee of modern precision agriculture and understanding the spatial distribution characteristics of nutrients is a basic requirement for sustainable agriculture, which plays an important role in sustainable agro-ecological development [2–4] Total nitrogen (TN) is a key indicator of soil nitrogen availability and one of the measures of soil fertility. Nitrogen is a nutrient element that is required in high amounts during vegetation growth [3,5,6]. With the progress of modern agriculture, it has been an important problem for researchers to obtain the required soil information in a limited time to assess the land fertility in a timely manner and to guide the scientific fertilization of agricultural production.

Traditional nutrient content testing is expensive, time-consuming, and basically obtained in laboratory analysis, the nutrient content obtained at the point scale is not conducive to good development of sustainable agriculture, and remote-sensing technology is an important technique for obtaining spatial characteristics of soil nutrients [3,7]. Combining remote sensing images to predict the spatial distribution of soil nutrients meets the current requirements for sustainable agricultural development. Presently, most of the models used for spatial prediction of TN are divided into two categories. The first type is linear models, which are constructed by simulating the linear relationship between the reflectance of remote sensing image bands and the TN content, and thus inverse models, including partial least squares regression (PLSR) [8–10], multiple linear regression (MLR) [11,12], and other models. However, due to the multiple and complex relationships between the reflectance of multispectral image bands and soil nutrient content, the constructed linear models are not sufficient to reflect the spatial distribution of nutrients well and are lacking in prediction accuracy. In this case, the second type of models for nutrient prediction, that is, machine learning methods, is needed. Machine learning techniques have the best performance in soil nutrient spatial inversion prediction [13], which can make up for the deficiencies of linear models and solve the complex nonlinear relationship between band reflectance and nutrient content and can well reflect the characteristics of nutrient spatial distribution in the study area. The models commonly used in soil nutrient spatial distribution prediction are neural networks (NN) [4,14], support vector machines (SVM) [15,16], random forest (RF) [3,17]. Models such as BPNN and SVM can solve the nonlinear problem well, which makes the prediction accuracy of the model higher and the nutrient spatial distribution information more accurate. Nonlinear models, mainly BPNN and SVM, have been widely used in soil moisture [18,19], soil organic matter [20], soil heavy metal [21], and soil quality [22].

The existing studies on TN basically extract sensitive bands through correlation analysis and select the best inversion model based on the deviation between measured and predicted values [23], which contain both linear and nonlinear models, both of which can be used for spatial prediction of soil nutrients in a region. Prediction accuracy of the BPNN model in inversion of TN content in black soil is 6.5% higher than that of PLSR [24]. Xiao et al. [25] used MODIS data to construct a stochastic forest model to invert the spatial distribution of TN content in Shandong province with a model coefficient of determination of 0.570. Lin et al. [26] estimated soil total nitrogen based on the synthetic color learning machine (SCLM) method. Machine learning models have useful application prospects in the study of predicting the spatial distribution of TN, but machine learning models have different results for different nutrient species and study areas. It is especially important to study the spatial prediction model of soil total nitrogen suitable for this study area, which will be of great help in predicting the spatial distribution of soil total nitrogen in this study area.

The study of spatial characteristics of soil nutrients is an urgent problem in agriculture and ecology, and the construction of a model suitable for predicting the spatial distribution of TN in this study area is equally important. Therefore, in this study, three linear and nonlinear prediction models, PLSR, BPNN and SVM, were constructed based on Landsat 5 TM multispectral remote sensing images, the original reflectance and the mathematically transformed reflectance to analyze the spatial distribution of soil TN in Datong County, Xining City, Qinghai Province, China. The prediction accuracy of the three models was also compared, and the optimal model was selected to predict the spatial distribution of TN, which provides data support and basis for soil quality evaluation, grain yield estimation and sustainable agriculture in Datong County.

2. Materials and Methods

2.1. Study Area

Datong County, Xining City, Qinghai Province is located in the eastern part of Qinghai Province, Hehuang Valley (100°51′–101°56′ E, 36°43′–37°23′ N), the southern foot of Qilian Mountains, Huangshui River upstream of the Beichuan River basin, and is the transition zone of Qinghai–Tibet Plateau and Loess Plateau. Its elevation ranges from 2178 to 4444 m; the topography of the northwest is high and the southeast low. The study area belongs to the plateau continental climate, the maximum and minimum temperature are 35.6 °C and −26.1 °C, the annual average temperature is 4.9 °C, with the most precipitation being in August. The region has a typical highland continental climate with dry spring and humid summer.

2.2. Soil Sampling and Analysis

Soil samples for the study were collected in September 2012 at a sampling depth of 0–20 cm, approximately 2 kg per sample, for a total of 73 soil samples, which were sampled while the latitude and longitude data of the sampling points were recorded by handheld Global Positioning System (GPS) (Figure 1). The samples were naturally dried and finely ground in the laboratory and employed in chemical analysis of soil composition. Soil total nitrogen mass fraction was given in the semi-micro Kjeldahl method. The maximum values of TN in the study area were 0.639 g/kg and 3.375 g/kg, the mean value was 1.864 g/kg, the skewnesswas 0.326, the kurtosis was −0.106, and the study data was consistent with a normal distribution The coefficient of variation (CV) was 30.596%, with a moderate degree of dispersion (Table 1). In this study, soil samples were randomly divided into modeling and validation samples in the ratio of 7:3.

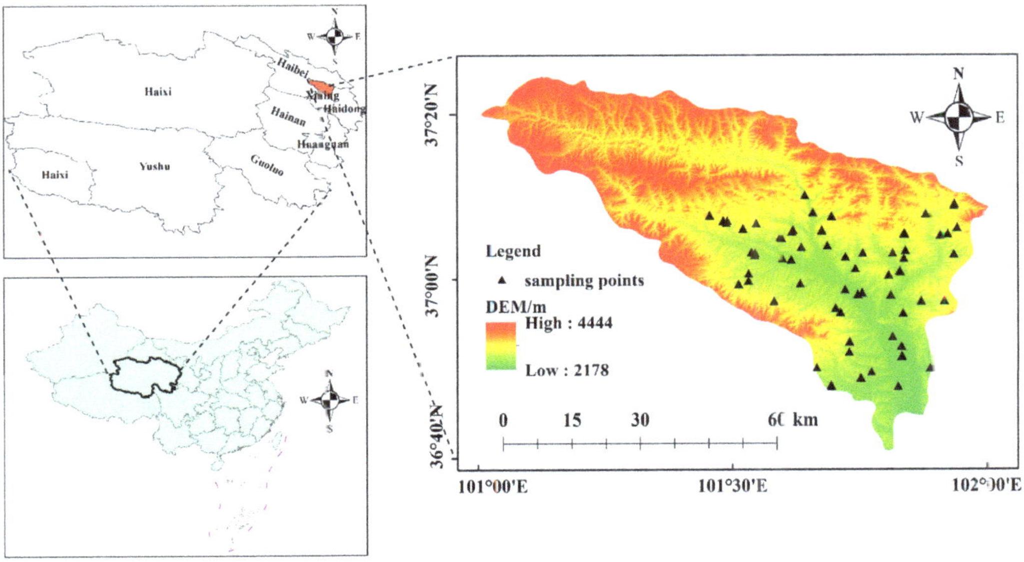

Figure 1. Study area and sampling points.

Table 1. Description statistics of TN content.

Variable	Sample Numbers	Max/(g/kg)	Min/(g/kg)	Mean/(g/kg)	CV/%	Skewness	Kurtosis
TN	73	3.375	0.639	1.864	30.596	0.326	−0.106

2.3. Image Acquisition and Processing

The Landsat 5 TM remote sensing image data used in this paper were obtained quickly from the Google Earth Engine (GEE) remote-sensing cloud platform (https://earthengine.google.com/ (accessed on 21 April 2021)). The image observation parameters and waveband characteristics are shown in Table 2. In order to ensure the accuracy of the prediction of TN content spatial distribution, the imaging time of remote sensing data selected in this paper is consistent with the collection time of soil samples. Moreover, the month when crops have been harvested is chosen for when there is no considerable amount of vegetation and crop shading on the soil surface and it is more convenient to obtain soil spectral information. The correlation analysis of the nutrient data and the image bands obtained from the field sampling and assay was utilized to obtain the best inversion band for the soil total nitrogen content.

Table 2. Landsat5 TM observation parameters and band characteristics.

Band	Name	Wavelength/μm	Spatial Resolution/m	Band Characteristics
B1	Blue	0.45–0.52	30	High penetration ability for water bodies, which is conducive to detecting water depth.
B2	Green	0.52–0.60	30	Distinguish between tree species and vegetation.
B3	Red	0.63–0.69	30	Chlorophyll is the main absorption band and responds to the health status of vegetation.
B4	Near Infrared	0.77–0.90	30	Crop yield estimation, identification of green vegetation.
B5	Shortwave infrared 1	1.55–1.75	30	Moisture absorption zone, commonly used for soil moisture surveys.
B6	Thermal infrared	10.40–12.50	120	Record the surface thermal radiation capacity.
B7	Shortwave infrared 2	2.08–2.35	30	Strong moisture absorption zones, distinguishing the main rock types.

Geometric correction, radiometric correction, image mosaic and cropping are the principal processes of remote sensing image preprocessing. After image pre-processing, the Digital Number (DN) of the original image is converted into the true Surface Reflectance (SR). The Google Earth Engine (GEE) platform can provide the TOA image set and the SR image set, which can save a lot of time in processing data. Image preprocessing stage calls Landsat 5 SR images (LANDSAT/LT05/C02/T1_L2) from the GEE platform using the function ee.ImageCollection, through which only the called images are reprocessed for de-clouding, mosaicking and cropping. Based on the findings of previous studies, it is difficult to extract the characteristic bands only from the original reflectance, but the full nitrogen characteristic bands can be extracted effectively by processing the original reflectance (1/R, log(1/R), 1/log R) [27,28].

2.4. Methods

The technical roadmap of this study is shown in Figure 2. Firstly, remote sensing images were obtained from GEE, combined with actual soil sampling data after a series of preprocessing, and mathematically transformed the image band reflectance. Then, correlation analysis was performed to extract the feature bands to construct three prediction models, PLSR, BPNN and SVM. Finally, the model prediction accuracy was evaluated by the coefficient of determination (R^2) and root mean square error (RMSE); the model with the highest accuracy was selected for total soil nitrogen spatial prediction.

2.4.1. Partial Least Squares Regression

The PLSR algorithm, first proposed by Wold et al. [29]. The algorithm covers common multiple regression analysis, principal component analysis and correlation analysis, while retaining the advantages of the three regression analyses, and is an optimized algorithm for the previous linear regression. In this paper, the original reflectance or the mathematically transformed reflectance of a TM image in multispectral band was used as the basis, and each principal component was obtained by the Karhunen–Loeve (KL) transformation. Then, with the total nitrogen content of soil in the study area as the dependent variable and the real reflectance value of the surface of 51 sampling points or the mathematically transformed reflectance as the independent variable, the PLSR model was established using SPSS software, and the regression equation is shown in Equation (1).

$$Y = a_1X_1 + a_2X_2 + \ldots + a_mX_m. \quad (1)$$

where Y represents the TN content, X_m represents the independent variable in the regression equation, that is, the characteristic band, and a represents the coefficient of each characteristic band, which is determined by the regression results.

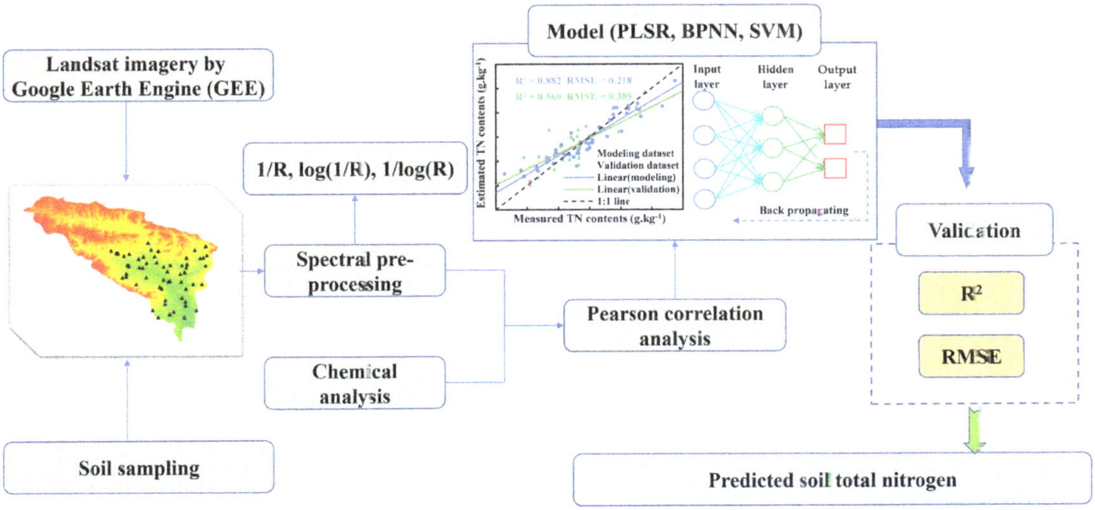

Figure 2. Methodology flow chart.

2.4.2. Neural Networks

A neural network is a system that simulates the structure and function of the neuronal network in human brain by engineering technology and consists of a large number of simple nonlinear processing units [30]. The artificial neural network can process fuzzy, nonlinear and noisy information through the interaction of neuron groups in the network, and is especially suitable for processing nonlinear problems, which is widely used in pattern recognition, image processing and automatic control. The neural network model consists of an input layer, a hidden layer, and an output layer, of which there are one input layer and one output layer, and multiple hidden layers. The BPNN model used in this paper is one of the more commonly used neural network models.

The neural network model constructed in this paper contains 3 layers, and the hidden layer has 1 layer. The number of neurons in the input layer is the number of feature bands extracted by Pearson correlation analysis. The number of neurons in the output layer is 1, which is the TN content, and the number of nodes in the input layer is the number of feature spectral bands, which is obtained by correlation analysis, and the number of nodes in the hidden layer is 100. The learning rate of the model is placed at 0.001, the maximum number of 8000 iterations, and the transfer function of the BP neural network model uses the sigmoid function. Each parameter in the model needs to be determined by comparing the operation results after repeated trials.

2.4.3. Support Vector Machines

Support vector machines (SVM) is supervised machine learning method proposed based on statistical theory, and the algorithm can solve nonlinear problems well. The key technique of the algorithm is the selection of the kernel function, and there are four kernel functions available: polynomial, sigmoid, linear and radial basis function (RBF) [31]. The kernel functions can be selected reasonably according to the different research problems in the practical problem solving. The radial basis function RBF is the most commonly used kernel function in soil mapping studies [32], in which two parameters penalty (cost) and kernel width (sigma) are included [5]. The Python language is used to determine the optimal parameters in order to obtain a better model.

2.4.4. Pearson Correlation Coefficient

Correlation analysis is a statistical analysis that determines the linear relationship between two variables [33]. In this paper, the Pearson correlation coefficient is applied in order to determine the correlation between TN and band reflectance, in order to extract the TN characteristic band.

2.4.5. Model Validation

In this paper, the R^2 in the standard regression evaluation and the RMSE in the error index evaluation to evaluate the stability and prediction accuracy of the model. The value of R^2 is [0, 1]; when the fit of the model is higher, the closer the measured value is to the predicted value, the closer R^2 is to 1. Conversely, when the value of RMSE is smaller, the deviation of the predicted value from the measured value is smaller, the prediction accuracy is higher, and the inversion error of TN content is smaller.

3. Results and Analysis

3.1. Correlation Analysis

A Pearson correlation analysis showed that there was a correlation between TN and band reflectance (Table 3). The original reflectance of the band showed a negative correlation with the soil total N content, and the highest correlation between the reflectance of the B2 band and TN ($r = -0.534$). The correlation between reflectance and TN after spectral transformation treatment was improved, and the correlation between reflectance and TN in the visible band was higher than that in other bands. In particular, the highest correlation between reflectance and TN was found after inverse treatment ($r = 0.584$). The highest correlation between the transformed reflectance and TN was 9.4% higher than the original reflectance. It is clear from the above analysis that the most sensitive band of TN is the visible band.

Table 3. Pearson correlation and *p*-Value.

Spectral Transformation	Indicators	B1	B2	B3	B4	B5	B7
REF	r	−0.483 **	−0.534 **	−0.517 **	−0.231	−0.310 *	−0.378 **
	p	0.000	0.000	0.000	0.106	0.028	0.007
1/R	r	0.566 **	0.584 **	0.514 **	0.185	0.284 *	0.349 *
	p	0.000	0.000	0.000	0.199	0.046	0.013
log(1/R)	r	0.527 **	0.562 **	0.518 **	0.209	0.300 *	0.366 **
	p	0.000	0.000	0.000	0.145	0.034	0.009
1/log(R)	r	0.499 **	0.541 **	0.517 **	0.243	0.310 *	0.375 **
	p	0.000	0.000	0.000	0.088	0.029	0.007

Notes: ** = $p \leq 0.01$; * = $p \leq 0.05$.

3.2. Partial Least Squares Regression

Characteristic bands identified by Pearson correlation analysis were used for model construction. The highest correlation visible band was used to construct the PLSR model. The regression coefficients of determination (R^2) and root mean square errors (RMSE) of the best principal components, modeling datasets and validation dataset, are shown in Table 4. The constructed models all had R^2 greater than 0.5. The model constructed with the inverse of reflectance (1/R) had the highest accuracy ($R^2 = 0.604$, RMSE = 0.285). The accuracy of the model constructed with the reflectance after mathematical transformation was higher than that of the model constructed with the original reflectance, indicating that appropriate mathematical transformation can improve the prediction accuracy of the model. The model can perform a rough inversion of the spatial distribution of TN in Datong County.

Table 4. Comparison of evaluation results of TN PLSR model.

Spectral Transformation	Principal Component Number	Modeling Dataset		Validation Dataset	
		R^2	RMSE	R^2	RMSE
REF	3	0.564	0.299	0.356	0.315
1/R	3	0.604	0.285	0.374	0.334
log(1/R)	4	0.598	0.287	0.357	0.331
1/log(R)	4	0.557	0.334	0.255	0.430

3.3. BP Neural Network

A BPNN TN spatial inversion model was constructed with four-band reflectance as the input variables of the model, and the inversion accuracy results were evaluated (Table 5). The results show that the inversion accuracy of the model constructed with the inverse of reflectance as the independent variable was the highest (R^2 = 0.882, RMSE = 0.218). The model constructed with the original reflectance was the next highest (R^2 = 0.806, RMSE = 0.273). The R^2 of the models constructed with the four reflectance was greater than 0.7. Therefore, the model constructed in this paper can quantitatively invert the TN content of soil in the Datong County and provide favorable conditions for precise nutrient mapping of agricultural farmland.

Table 5. Accuracy evaluation results of BPNN model.

Spectral Transformation	Modeling Dataset		Validation Dataset	
	R^2	RMSE	R^2	RMSE
REF	0.806	0.273	0.522	0.313
1/R	0.882	0.218	0.560	0.305
log(1/R)	0.771	0.291	0.425	0.349
1/log(R)	0.787	0.282	0.466	0.331

Meanwhile, a scatter plot of the model inversion results was plotted in this study (Figure 3). From the figure, it can be observed that the sampling points of the BPNN model built with the inverse of reflectance as the input variable are distributed around the 1:1 line and are more aggregated. Several other transformations from models also have TN spatial prediction capability, but the prediction accuracy is not very satisfactory, and the scatter plot is more scattered. Moreover, these models are suitable for the prediction of TN content in the range of 1.5 to 3.0 g/kg, beyond which the prediction accuracy is low. In general, the PLSR model has weaker prediction ability than the BPNN model, and the inversion results of the constructed BPNN model have less error and higher accuracy. The R^2 of the validation dataset was less than that of the modeling dataset, which was probably due to the uneven distribution of TN at the sampling points and the small number of soil sampling points used for validation. In summary, the predictive ability of the 1/R model is high. Compared to the PLSR model, the nonlinear model has a greater advantage in terms of predictive power and stability, which is consistent with the findings of many scholars [3,34].

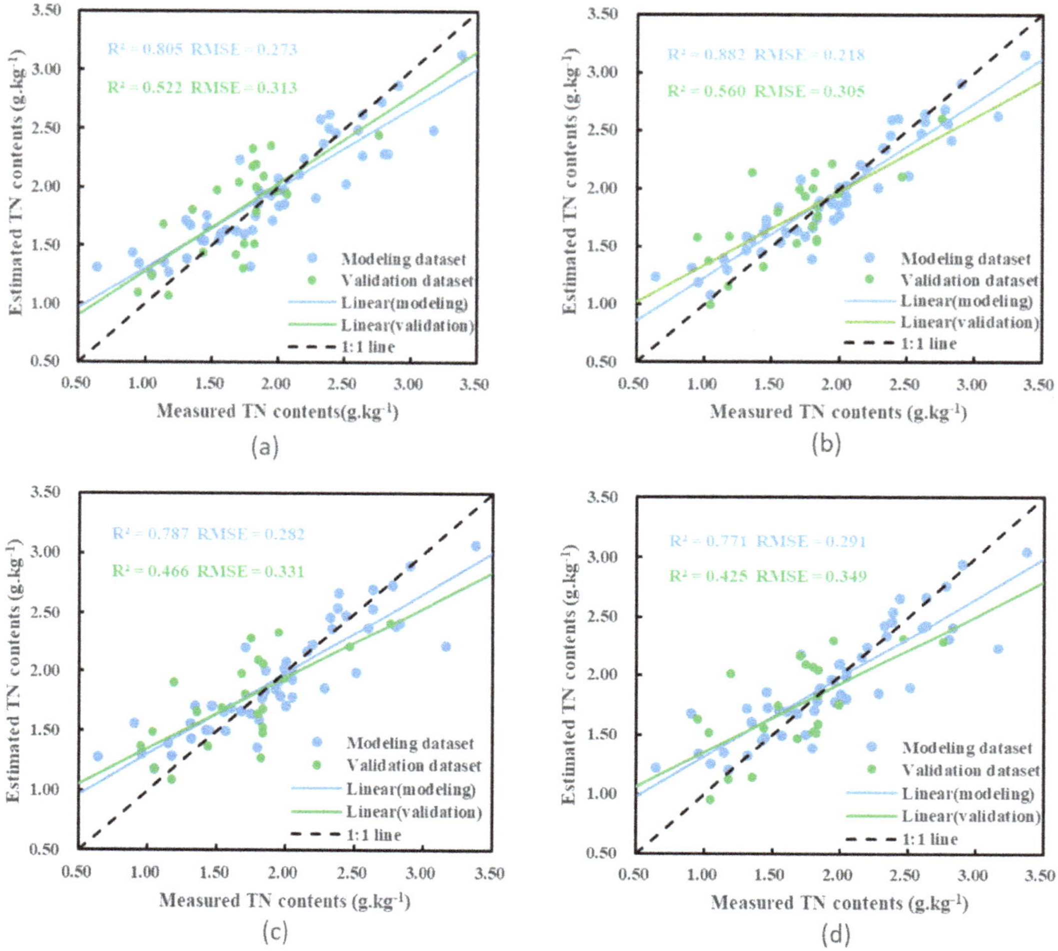

Figure 3. Inversion scatter plot of BPNN model (**a**): REF; (**b**): 1/R; (**c**): 1/log(R); (**d**): log(1/R).

3.4. Support Vector Machines

The prediction accuracy of the TN inversion model constructed with the SVM algorithm is higher than that of the BPNN model, as seen from the accuracy of the modeling samples and the validation samples (Table 6). The inversion accuracy of the model constructed after logarithmic and then inverse processing of reflectance is the highest (R^2 = 0.879, RMSE = 0.233), which is different from that of the BPNN model. In this instance, logR (R^2 = 0.879, RMSE = 0.233) > 1/R (R^2 = 0.837, RMSE = 0.288) > log1/R (R^2 = 0.824, RMSE = 0.296) > REF (R^2 = 0.763, RMSE = 0.336). The R2 of the validation samples of this model are all greater than 0.6, while the BPNN model is all below 0.6, indicating that for modeling, the BPNN model has the advantage and the SVM model is superior in predictive inversion.

Table 6. The prediction results of TN using SVM model.

Spectral Transformation	Modeling Dataset		Validation Dataset	
	R^2	RMSE	R^2	RMSE
REF	0.763	0.336	0.626	0.385
1/R	0.837	0.288	0.635	0.302
log(1/R)	0.824	0.296	0.645	0.312
1/log(R)	0.879	0.233	0.676	0.296

As can be seen from the scatter plot (Figure 4), the model constructed by the original reflectance has a higher prediction than the measured value when the TN content is lower than 2.0 g/kg, and a lower prediction when it is greater than 2.0 g/kg; all other models have similar prediction errors. However, in general, the model prediction and the actual value are basically around the 1:1 line, with less deviation, better tightness of prediction points and fewer outliers.

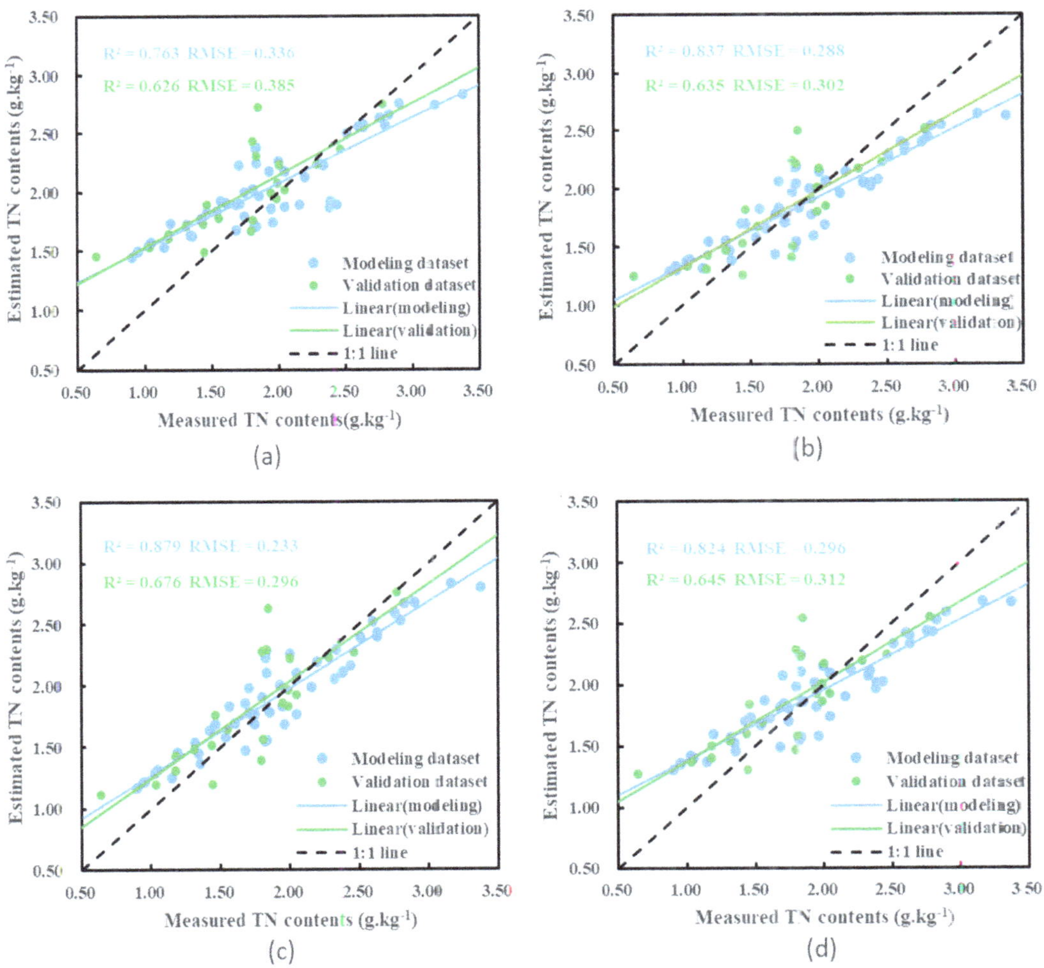

Figure 4. SVM model constructed with four reflectance (**a**): REF; (**b**): 1/R; (**c**): 1/log(R); (**d**): log(1/R).

3.5. Predicting the Spatial Distribution of TN

When the three models of PLSR, BPNN and SVM are compared together, SVM is more advantageous and has higher prediction accuracy than PLSR and BPNN models and can reflect the spatial distribution of TN content in the study area well. Figure 5 shows the spatial distribution of TN prediction by the three models for Datong County. From the prediction map, it can be seen that the spatial distribution of TN predicted by the PLSR model differs greatly from the other two models, and the prediction result is high, with the maximum value of 3.724 g/kg and the minimum value of 1.456 g/kg; the maximum value of TN spatial distribution predicted by the BPNN model is 3.163 g/kg and 0.124 g/kg, respectively; the predicted content of SVM model ranges from 0.028–2.955 g/kg. By comparison, SVM model predicts the most accurate TN content in Datong County, which can provide a good understanding of the spatial distribution of TN; the three models show the same spatial prediction trend of TN, with the overall trend of TN being high in the northwest and low in the southeast.

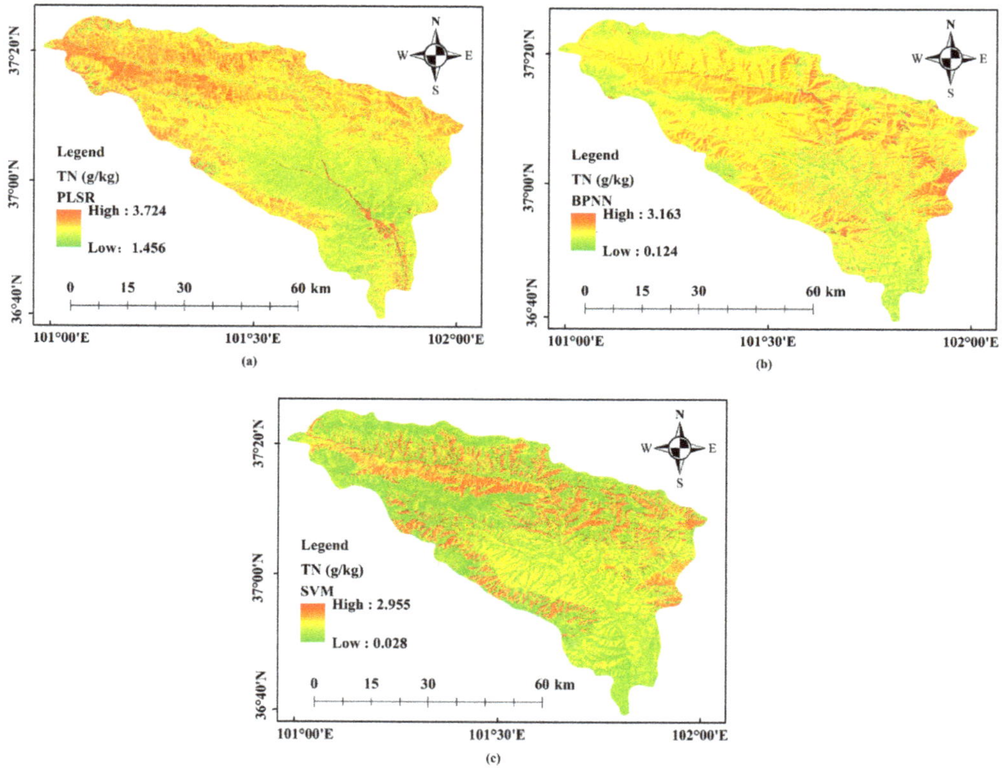

Figure 5. Predictive maps of TN using (**a**) PLSR; (**b**) BPNN; (**c**) SVM model.

4. Discussion

Soil nutrients are an important factor in measuring soil fertility, and traditional farm management and agricultural systems have led to polarization of soil nutrients in farmland, i.e., excess nutrients in fertile soils reduce the utilization of chemical fertilizers, and nutrient-poor soil vegetation does not receive adequate nutrients. Soil nutrient monitoring has evolved from qualitative to quantitative studies [35], and the determination of TN inversion models plays an important role in understanding their spatial distribution characteristics, guiding the implementation of precision agriculture, and promoting the development of agricultural production [2,36]. Monitoring of total nitrogen content has further improved and updated its inversion means from traditional laboratory analysis to large-scale spatial prediction based on regional scale. The most important of them is the application of remote-sensing technology, which has the advantages of accuracy, speed and economy [35] and can predict well the spatial distribution of total soil nitrogen in the study area. The establishment of models for prediction has gradually developed from linear to nonlinear models, for example, Liu et al. [28] established a PLSR model to predict the soil total nitrogen content in Shaanxi Province. Li et al. [37] established a stepwise linear regression model to invert the soil total nitrogen content based on vegetation information. Xu et al. [38] used Landsat remote sensing images to construct the regression kriging (RK) to predict the soil total nitrogen content of a farm. Wang et al. [39] mapped the distribution of TN content in northeastern Liaoning Province, China, based on a random forest model with three different kernel functions and a multiple linear stepwise regression (MLSR) model. Ye et al. [40] predicted soil total nitrogen content based on a radial basis function neural network (RBFNN). Further development of the prediction model can be seen.

The spatial distribution of TN is largely influenced by anthropogenic and natural factors, and there is spatial heterogeneity [16,40]. Therefore, the spatial prediction of TN cannot only use image band reflectance as the model independent variable, but needs to consider more environmental and anthropogenic factors, such as topographic factors [36], climatic factors [41], soil types [6], tillage practices [4,42], and crop types [43]. Nonlinear models play an important role in regions subject to anthropogenic disturbances [40]. Sun et al. [44] used a random forest model to analyze the relative importance of factors affecting soil nutrients. The results showed that temperature, precipitation and elevation were significant influencing factors for soil nutrient prediction. Different soil types [45] and different topographic trends will directly affect the TN spatial distribution. Dong et al. [17] showed that predicting soil nutrients based on environmental and anthropogenic factors is more important than the common interpolation technique to obtain better prediction accuracy. Zhang et al. [43] found that the TN spatial pattern is related to elevation. Peri et al. [41] found that TN decreased with increasing soil desertification, indicating that TN content was closely related to the degree of vegetation denseness. The growth and development of vegetation cannot be separated from the influence of climatic conditions, and climate will ultimately affect the spatial distribution of TN content. In summary, spatial distribution of TN content is influenced by many factors and there is a complex relationship between them. Future research will focus on the spatial prediction of TN under the synergistic conditions of environmental and anthropogenic factors, and understand the relationship between each influencing factor, and extract TN sensitive bands and factors with high correlation with the help of higher resolution remote sensing data to accurately predict soil TN content.

5. Conclusions

Based on the field sampling data of Datong County, this paper uses statistical software to correlate the TN of sampling points with the reflectance of the corresponding TM image bands, as well as to construct PLSR, BPNN and SVM models to invert the spatial distribution of TN in the study area, respectively. The study demonstrates that the average TN content in Datong County is 1.864 g/kg with a coefficient of variation of 30.596%, which is a medium degree of variation. The correlation was significantly improved (r = 0. 584, $p < 0.01$) and the prediction accuracy of the model were improved after preprocessing the mathematical transformation of the raw spectral reflectance. Comparing the prediction accuracy of the three models, PLSR, BPNN and SVM, the prediction accuracy from highest to lowest was SVM ($R^2 = 0.676$, RMSE = 0.296), BPNN ($R^2 = 0.560$, RMSE = 0.305) and PLSR ($R^2 = 0.374$, RMSE = 0.334), it can be seen that the prediction accuracy of the nonlinear model is better than the linear model, and the prediction accuracy of the SVM model among the nonlinear models is higher than that of the BPNN model, which can accurately predict the spatial distribution of soil TN in Datong County. The prediction of TN spatial distribution by the SVM model shows an overall spatial distribution trend of high in the north and low in the south, and slightly lower in the middle of the study area. Soil TN is an important indicator of soil fertility and predicting soil TN spatial distribution is beneficial to soil quality evaluation and effective implementation of precision agriculture. Prediction the spatial distribution of TN based on SVM algorithm is an effective technical tool for sustainable agriculture at the county scale, which can effectively map the distribution of soil TN. It provides a basis and technical support for soil precision fertilization and sustainable agricultural management.

Author Contributions: Conceived and designed the research, Z.L. and H.L.; data curation, Z.L. and L.L.; writing—original draft preparation, Z.L.; visualization, Z.L.; supervision, H.S.; funding acquisition, H.L. while the other authors supported writing the paper. All authors have read and agreed to the published version of the manuscript.

Funding: This research was funded by the National Natural Science Foundation of China (U20A20115).

Institutional Review Board Statement: Not applicable.

Informed Consent Statement: Not applicable.

Data Availability Statement: The source of remote sensing images used in this paper is the same as Google Earth Engine (https://earthengine.google.com/ (accessed on 21 April 2021)); additional data may be obtained with the consent of the corresponding author.

Acknowledgments: We acknowledge the anonymous reviewers for their valuable comments on our study.

Conflicts of Interest: The authors declare no conflict of interest.

References

1. Chen, B.; Jiang, Q.; Wang, K. Application and progress in estimating soil organic matter content based on remote sensing. *J. Shandong Agr. Univ. Nat. Sci. Ed.* **2011**, *42*, 317–321.
2. Wang, R.; Zou, R.; Liu, J.; Liu, L.; Hu, Y. Spatial distribution of soil nutrients in farmland in a hilly region of the pearl river delta in China based on geostatistics and the inverse distance weighting method. *Agriculture* **2021**, *11*, 50. [CrossRef]
3. Song, Y.; Zhao, X.; Su, H.; Li, B.; Hu, Y.; Cui, X. Predicting Spatial Variations in Soil Nutrients with Hyperspectral Remote Sensing at Regional Scale. *Sensors* **2018**, *18*, 3086. [CrossRef]
4. Alemu, L.; Mesfin, B. Performance of mid infrared spectroscopy to predict nutrients for agricultural soils in selected areas of Ethiopia. *Heliyon* **2022**, *8*, e09050.
5. Zhou, T.; Geng, Y.; Chen, J.; Pan, J.; Haase, D.; Lausch, A. High-resolution digital mapping of soil organic carbon and soil total nitrogen using DEM derivatives, Sentinel-1 and Sentinel-2 data based on machine learning algorithms. *Sci. Total Environ.* **2020**, *729*, 138244. [CrossRef]
6. Mobasheri, M.; Amani, M.; Ranjbaran, M.; Mahdavi, S.; Zabihi, H.R. Introducing an index in determination of doil total nitrogen content in an agricultural soil using laboratory spectrometry. *Commun. Soil Sci. Plan.* **2020**, *51*, 288–296. [CrossRef]
7. Gulhane, V.; Rode, S.; Pande, C. Wavelet for predicting soil nutrients using remotely sensed satellite images. *Int. J. Comput. Appl.* **2017**, *174*, 35–38.
8. Yu, S.; Bu, H.; Dong, W.; Jiang, Z.; Zhang, L.; Xia, Y. Construction and evaluation of prediction model of main soil nutrients based on spectral information. *Appl. Sci.* **2022**, *12*, 6298. [CrossRef]
9. Xu, L.; Xie, D.; Wei, C.; Li, B. Prediction of total nitrogen and total phosphorus concentrations in purple soil using hyperspectral data. *Spectrosc. Spect. Anal.* **2013**, *33*, 723–727.
10. Wang, S.; Shi, P.; Zhang, H.; Wang, X. Retrieval of soil total nitrogen content in reclaimed farmland of mining area based on hyperspectral imaging. *Chin. J. Ecol.* **2019**, *38*, 294–301.
11. Komolafe, A.A.; Olorunfemi, I.E.; Oloruntoba, C.; Akinluyi, F.O. Spatial prediction of soil nutrients from soil, topography and environmental attributes in the northern part of Ekiti State, Nigeria. *Remote Sens. Appl. Soc. Environ.* **2021**, *21*, 100450. [CrossRef]
12. Miran, N.; RasouliSadaghiani, M.H.; Feiziasl, V.; Sepehr, E.; Rahmati, M.; Mirzaee, S. Predicting soil nutrient contents using Landsat OLI satellite images in rain-fed agricultural lands, northwest of Iran. *Environ. Monit. Assess.* **2021**, *193*, 607. [CrossRef]
13. Swapna, B.; Manivannan, S.; Kamalahasan, M. Prognostic of soil nutrients and soil fertility index using machine learning classifier techniques. *Int. J. Collab.* **2022**, *18*, 3. [CrossRef]
14. Li, Y.; Zhao, Z.; Wei, S.; Sun, D.; Yang, Q.; Ding, X. Prediction of regional forest soil nutrients based on Gaofen-1 remote sensing data. *Forests* **2021**, *12*, 1430. [CrossRef]
15. Xu, Y.; Li, B.; Shen, X.; Li, K.; Cao, X.; Cui, G.; Yao, Z. Digital soil mapping of soil total nitrogen based on Landsat 8, Sentinel 2, and WorldView-2 images in smallholder farms in Yellow River Basin, China. *Environ. Monit. Assess.* **2022**, *194*, 282. [CrossRef]

16. Dharumarajan, S.; Lalitha, M.; Niranjana, K.; Hegde, R. Evaluation of digital soil mapping approach for predicting soil fertility parameters—A case study from Karnataka Plateau, India. *Arab. J. Geosci.* **2022**, *15*, 386. [CrossRef]
17. Dong, W.; Wu, T.; Luo, J.; Sun, Y.; Xia, L. Land parcel-based digital soil mapping of soil nutrient properties in an alluvial-diluvia plain agricultural area in China. *Geoderma* **2019**, *340*, 234–248. [CrossRef]
18. Wang, X.; Lü, H.; Crow, W.T.; Zhu, Y.; Wang, Q.; Su, J.; Zheng, J.; Gou, Q. Assessment of SMOS and SMAP soil moisture products against new estimates combining physical model, a statistical model, and in-situ observations: A case study over the Huai River Basin, China. *J. Hydrol.* **2021**, *598*, 126468. [CrossRef]
19. Zhu, Q.; Wang, Y.; Luo, Y. Improvement of multi-layer soil moisture prediction using support vector machines and ensemble Kalman filter coupled with remote sensing soil moisture datasets over an agriculture dominant basin in China. *Hydrol. Process.* **2021**, *35*, 14154. [CrossRef]
20. De Santana, F.B.; Otani, S.K.; De-Souza, A.M.; Poppi, R.J. Comparison of PLS and SVM models for soil organic matter and particle size using vis-NIR spectral libraries. *Geoderma Reg.* **2021**, *27*, e00436. [CrossRef]
21. Zhang, H.; Yin, S.; Chen, Y.; Shao, S.; Wu, J.; Fan, M.; Chen, F.; Gao, C. Machine learning-based source identification and spatial prediction of heavy metals in soil in a rapid urbanization area, eastern China. *J. Clean. Prod.* **2020**, *273*, 122858. [CrossRef]
22. Niu, Y.; Ye, S. Data Prediction Based on Support Vector Machine (SVM)—Taking Soil Quality Improvement Test Soil Organic Matter as an Example. *IOP Conf. Ser. Earth Environ. Sci.* **2019**, *295*, 012021. [CrossRef]
23. Qiu, H. Hyperspectral Remote Sensing Inversion of Organic Matter, Available Nitrogen, Phosphorus and Potassium Contents in Cropland Soil. M.D. Thesis, Fujian Agriculture and Forestry University, Fuzhou, China, 2017.
24. Yang, Y.; Zhao, J.; Qin, K.; Zhao, N.; Yang, C.; Zhang, D.; Cui, X. Prediction of black soil nutrient content based on airborne hyperspectral remote sensing. *Trans. Chin. Soc. Agric. Eng.* **2019**, *35*, 94–101.
25. Xiao, W.; Chen, W.; He, T.; Ruan, L.; Guo, J. Multi-Temporal mapping of soil total nitrogen using Google Earth Engine across the Shandong province of China. *Sustainability* **2020**, *12*, 10274. [CrossRef]
26. Lin, L.; Gao, Z.; Liu, X. Estimation of soil total nitrogen using the synthetic color learning machine (SCLM) method and hyperspectral data. *Geoderma* **2020**, *380*, 114664. [CrossRef]
27. Zhang, S.; Lu, X.; Nie, G.; Li, Y.; Shao, Y.; Tian, Y.; Fan, L.; Zhang, Y. Estimation of soil organic matter in coastal wetlands by SVM and BP based on hyperspectral remote sensing. *Spectrosc. Spect. Anal.* **2020**, *40*, 556–561.
28. Liu, J.; Dong, Z.; Chen, X. Study on hyperspectral estimation model of total nitrogen content in soil of Shaanxi province. *IOP Conf. Ser. Earth Environ. Sci.* **2018**, *108*, 042025. [CrossRef]
29. Wold, S.; Sjöström, M.; Eriksson, L. PLS-regression: A basic tool of chemometrics. *Chemom. Intell. Lab. Syst.* **2001**, *58*, 109–130. [CrossRef]
30. Irmak, A.; Jones, J.W.; Batchelor, W.D.; Irmak, S.; Boote, K.J.; Paz, J.O. Artificial neural network model as a data analysis tool in precision farming. *Trans. ASABE* **2006**, *49*, 2027–2037. [CrossRef]
31. Pradhan, B. A comparative study on the predictive ability of the decision tree, support vector machine and neuro-fuzzy models in landslide susceptibility mapping using GIS. *Comput. Geosci.* **2012**, *51*, 350–365. [CrossRef]
32. Taghizadeh-Mehrjardi, R.; Neupane, R.; Sood, K.; Kumar, S. Artificial bee colony feature selection algorithm combined with machine learning algorithms to predict vertical and lateral distribution of soil organic matter in South Dakota, USA. *Carbon Manag.* **2017**, *8*, 277–291. [CrossRef]
33. Yang, Z.; Chen, X.; Jing, F.; Guo, B.; Lin, G. Spatial variability of nutrients and heavy metals in paddy field soils based on GIS and Geostatistics. *Chin. J. Appl. Ecol.* **2018**, *29*, 1893–1901.
34. Morellos, A.; Pantazi, X.E.; Moshou, D.; Alexandridis, T.; Whetton, R.; Tziotzios, G.; Wiebensohn, J.; Bill, R.; Mouazen, A.M. Machine learning based prediction of soil total nitrogen, organic carbon and moisture content by using VIS-NIR spectroscopy. *Biosyst. Eng.* **2016**, *152*, 104–116. [CrossRef]
35. Ji, W.; Liu, Y. Research on Quantitative Evaluation of Remote Sensing and Statistics Based on Wireless Sensors and Farmland Soil Nutrient Variability. *Comput. Intell. Neurosc.* **2022**, *2022*, 3646264. [CrossRef] [PubMed]
36. Wang, S.; Adhikari, K.; Wang, Q.; Jin, X.; Li, H. Role of environmental variables in the spatial distribution of soil carbon (C), nitrogen (N), and C:N ratio from the northeastern coastal agroecosystems in China. *Ecol. Indic.* **2018**, *84*, 263–272. [CrossRef]
37. Li, Y.; Pan, X.; Wang, C.; Liu, Y.; Zhao, Q. Monitoring changes of soil organic matter and total nitrogen in cultivated land in Guangxi by remote sensing. *Acta Ecol. Sin.* **2014**, *34*, 5283–5291.
38. Xu, Y.; Smith, S.E.; Grunwald, S.; Abd-Elrahman, A.; Wani, S.P.; Nair, V.D. Estimating soil total nitrogen in smallholder farm settings using remote sensing spectral indices and regression kriging. *Catena* **2018**, *163*, 111–122. [CrossRef]
39. Wang, S.; Jin, X.; Adhikari, K.; Li, W.; Yu, M.; Bian, Z.; Wang, Q. Mapping total soil nitrogen from a site in northeastern China. *Catena* **2018**, *166*, 134–146. [CrossRef]
40. Ye, Y.; Jiang, Y.; Kuang, L.; Han, Y.; Xu, Z.; Guo, X. Predicting spatial distribution of soil organic carbon and total nitrogen in a typical human impacted area. *Geocarto Int.* **2022**, *37*, 4465–4482. [CrossRef]
41. Peri, P.L.; Rosas, Y.M.; Ladd, B.; Toledo, S.; Lasagno, R.G.; Pastur, G.M. Modeling soil nitrogen content in South Patagonia across a climate gradient, vegetation type, and grazing. *Sustainability* **2019**, *11*, 2707. [CrossRef]
42. Ojoyi, M.M.; Mutanga, O.; Odindi, J.; Kahinda, J.M.M.; Abdelrahman, E.M. Implications of land use transitions on soil nitrogen in dynamic landscapes in Tanzania. *Land Use Policy* **2017**, *64*, 95–100. [CrossRef]

43. Zhang, Y.; Sui, B.; Shen, H.; Ouyang, L. Mapping stocks of soil total nitrogen using remote sensing data: A comparison of random forest models with different predictors. *Comput. Electron. Agr.* **2019**, *160*, 23–30. [CrossRef]
44. Sun, M.; Hou, E.; Wu, J.; Huang, J.; Huang, X.; Xu, X. Spatial patterns and drivers of soil chemical properties in typical hickory plantations. *Forests* **2022**, *13*, 457. [CrossRef]
45. Xiao, S.; He, Y.; Dong, T.; Nie, P. Spectral analysis and sensitive waveband determination based on nitrogen detection of different soil types using near infrared sensors. *Sensors* **2018**, *18*, 523. [CrossRef]

MDPI
St. Alban-Anlage 66
4052 Basel
Switzerland
Tel. +41 61 683 77 34
Fax +41 61 302 89 18
www.mdpi.com

Sustainability Editorial Office
E-mail: sustainability@mdpi.com
www.mdpi.com/journal/sustainability

www.ingramcontent.com/pod-product-compliance
Lightning Source LLC
Chambersburg PA
CBHW040801250926
51850CB00007BA/1050
9783036558905